New Multinational Enterprises From Korea and Taiwan

This volume represents the first substantive study of emerging multinationals (MNEs) from Asian economies, drawing on the unique experiences of South Korea and Taiwan. Combining an econometric investigation with detailed case studies of leading Korean and Taiwanese electronics companies, it aims to demystify the nature and theoretical implications of these dynamic economic regions.

The author argues that many of these 'new multinationals' do not possess the characteristics typically ascribed to MNEs, such as technological leadership, strong marketing positions or advanced managerial capabilities. This is highlighted by documentation of recent developments in outward investment from Korea and Taiwan at both micro and macro level. The implications of the recent Asian crisis for the internationalisation of Korean and Taiwanese firms are also explored.

The unique perspective and comprehensive nature of *New Multinational Enterprises from Korea and Taiwan* make it essential reading for those interested in the field of international business, economic development and Asian studies.

Roger van Hoesel is at present a Consultant at Buck Consultants International, Nijmegen, The Netherlands. The research for this book was done while he worked at Erasmus University Rotterdam. He has been published widely in journals and books focusing on economic developments in Korea and Taiwan, and is co-editor of *Multinational Enterprises from the Netherlands* (1999), also published by Routledge.

Routledge Studies in International Business and the World Economy

1 **States and Firms**
Multinational enterprises in institutional competition
R. Sally

2 **Multinational Restructuring, Internationalization and Small Economies**
The Swedish case
T. Andersson, T. Fredriksson and R. Svensson

3 **Foreign Direct Investment and Governments**
Catalysts for economic restructuring
J. H. Dunning and R. Narula

4 **Multinational Investment and Economic Structure**
Globalization and competitiveness
R. Narula

5 **Entrepreneurship in a Global Context**
S. Birley and I. Macmillan

6 **The Global Structure of Financial Markets**
An overview
D. K. Ghosh and E. Ortiz

7 **Alliance Capitalism and Global Business**
J. H. Dunning

8 **Multinational Enterprises from the Netherlands**
R. van Hoesel and R. Narula

9 **Competition, Growth Strategies and the Globalization of Services**
Real estate advisory services in Japan, Europe and the United States
T. LaPier

10 **European Integration and Foreign Direct Investment in the EU**
The case of the Korean consumer electronics industry
Sang Hyup Shin

11 **New Multinational Enterprises from Korea and Taiwan**
Beyond Export-led Growth
R. van Hoesel

12 **Competitive Industrial Development in the Age of Information**
The role of co-operation in the technology sector
R. J. Braudo and J. G. MacIntosh

13 **The Global Restructuring of the Steel Industry**
Innovations, Institutions and Industrial Change
A. P. D'Costa

14 **Privatisation and Liberalisation in European Telecommunications**
Comparing Britain, the Netherlands and France
W. Hulsink

New Multinational Enterprises From Korea and Taiwan

Beyond export-led growth

Roger van Hoesel

ROUTLEDGE

London and New York

First published in 1999
by Routledge
11 New Fetter Lane, London EC4P 4EE

Simultaneously published in the USA and Canada
by Routledge
29 West 35th Street, New York, NY 10001

Typeset in Baskerville by
J&L Composition Ltd, Filey, North Yorkshire
Printed and bound in Great Britain by
Biddles Ltd., Guildford and King's Lynn

British Library Cataloguing in Publication Data
A catalogue record for this book is available from the British Library

Library of Congress Cataloguing in Publication Data
Hoesel, Roger van, 1965–
New multinational enterprises from Korea and Taiwan : beyond export-led growth/Roger van Hoesel.
p. cm. —(Routledge studies in international business and the world economy; 11)
Includes index.
1. International business enterprises—Korea (South) 2. International business enterprises—Taiwan. 3. Investments, South Korean—Europe. 4. Investments, Taiwan—Europe. 5. Electronic industries—Korea (South) 6. Electronic industries—Taiwan. I. Title. II. Series.
HD2908.H64 1999
338.8′895124904—dc21 98-41198
CIP

ISBN 0-415-19488-1

Contents

List of figures

List of tables

Foreword

John H. Dunning

This is a well crafted and lucid monograph, and is a worthy addition to a long line of studies on the determinants and impact of outbound foreign direct investment. Its contents contain a wealth of empirical data, all of which the author carefully interprets in the light of the mainstream theories of MNE activity. Though there have been several previously published volumes on the growing phenomenon of the internationalisation of Third World firms, none, to my mind, has documented, in such extensive and fascinating detail, the *raison d'être* for, and the extent and composition, of such internationalisation. Roger van Hoesel's study is also especially intriguing for the comparisons he makes between the post-war development of the Korean and Taiwanese economies, and how this has fashioned the ability and propensity of their own firms to invest abroad. While there are some similarities, e.g. in the industrial composition of their foreign operations, the differences, due *inter alia* to the cultural and institutional structures of the two economies, are the more revealing.

The time profile of Korean and Taiwanese foreign direct investment is also examined. Roger van Hoesel distinguishes between two waves of investment. The first, viz that in the 1960s and 1970s, he shows, was directed primarily at adjacent countries in Asia; and this fits in well with the internationalisation model of the Uppsala economists. The second wave occurred in the 1980s and early 1990s, and was directed at the more developed regions of the world. As country specific advantages of the Taiwanese and Korean countries have become more sophisticated, and their firms broadened and intensified their international networks, the need to both exploit the markets of more advanced countries and to tap into their intellectual assets has become more apparent.

Roger van Hoesel had largely completed his empirical work before the third wave of Taiwanese and Korean investment really got under way, and before the current Asian economic crisis took its grip. For in the mid-1990s there was a renewed upsurge in intra-Asian fdi, particularly among the Chinese ethnic communities. The slowdown in Asian economic growth has also led to some retrenchment of Korean and Taiwanese outward fdi, particularly in the first part of 1998. Flows of new investment into the two countries have, however, continued to remain vibrant; in 1997, in dollar terms, for example they were the highest they had ever been.

This monograph has many excellent attributes and is to be highly commended to all scholars interested in fdi and the strategies of MNEs. Among other things, it emphasises the importance *context* in testing specific explanations of MNE activity. Roger van Hoesel well demonstrates that several well renowned partial theories have contributed to our knowledge about the investment development path of Korea and Taiwan; but he is also careful to point out that a variety of country, industry and firm specific characteristics need to be explicitly identified if the internationalisation process of the firms from the two countries is to be fully understood. The extent to which, in future years, there will be greater convergence or divergence between the industrial and geographical pattern of fdi by Korean and Taiwanese MNEs is, indeed, an intriguing question; and one which, no doubt, future generations of scholars will address.

John H. Dunning
Rutgers & Reading Universities

Acknowledgements

This book is based on research for a dissertation which I defended at Erasmus University Rotterdam, the Netherlands, in December 1997. Writing a book is a long-term commitment and involves time-consuming effort. Many people have helped me to make my effort a successful one for which I am very grateful.

Crucial has been the kind and co-operative attitude of many people in the two countries I studied in particular, namely Korea and Taiwan. It is impossible to name all those who have helped me and I know that some are not very eager to find their names in this preface. Nevertheless, I would first of all like to thank all the companies that have allowed me to interrupt their daily work and who received me in their offices. Special thanks, of course, go to those working for the four companies I bothered most, in their home countries as well as in Europe: Samsung Electronics, LG Electronics, Acer and First International Computer. The representatives of these companies have been so kind to collect a lot of material for me and time and time again answered my quest for clarifications, opinions, and more information.

My fieldwork in Korea and Taiwan would have been impossible without having a base to work from. During my first round of field activities this base was provided by the offices of the Netherlands Foreign Investment Agency (Ministry of Economic Affairs) in Seoul and Taipei. Especially I would like to thank Mrs K.J. Hahn in Seoul and Mr Guido Tielmans (currently working for the Ministry of Foreign Affairs) in Taipei for their kind help. For the second round of field activities I moved to the academic world in Korea and Taiwan. In Seoul, at Chung-Ang University, Professor Yongwook Jun was an excellent host who among others provided invaluable introductions that made my entry into Korean business circles much easier. Moreover, the help of Hyun-Chang Park and Byoung-Jun Kim in collecting data for the empirical part of the study has been invaluable. In Taipei, I had the privilege to be a guest of the Chung-Hua Institution for Economic Research. The willingness of the scholars working at the Institute to share their data of and insights into the Taiwanese economy was unique. Special thanks go to Professor Tain-Jy Chen (currently at the National Taiwan University) and Dr Weng-Jeng Kuo who have also taught me a lot more about Taiwan in general. In addition, Mr Meng-Chun Liu helped me greatly with the census data. Moreover, Mr Yi-Cheng Chien collected important statistics for the econometric work.

Notwithstanding the impression given to some, during most of the time I worked on this book I lived in the Netherlands. Also here, many have supported me in various ways. I would like to thank my supervisors, George Waardenburg of my own university, and Danny van den Bulcke of the University of Antwerp, for their comments and continuous encouragement. A special word of gratitude should be dedicated to Rajneesh Narula who has greatly helped me in advancing my academic activities beyond 'merely' writing a dissertation. Thanks also to Kees van Opijnen, Thijs de Ruyter van Steveninck, Rudi Verburg, Caroline van Ruyven, and Laraine Visser who all helped me in various ways. Without the financial support of the Erasmus University, the 'Trustfonds' of the university, the Tinbergen Institute, the Netherlands Organization for Scientific Research (NWO) and the Netherlands Foreign Investment Agency (Ministry of Economic Affairs) this research project would not have been possible.

Last, but certainly not least, I would like to express my sincere gratitude for the interest and patience that my parents, friends and, of course, Mila, have shown and without whose encouragement this venture would not have materialised.

Roger van Hoesel

Abbreviations

A/C	Air Conditioner
A/V	Audio/Visual
AMG	American Military Government
ASIC	Application-Specific Integration Circuit
Bipa	Brand international promotion association
BOK	Bank of Korea
CAD/CAM	Computer Aided Design/Computer Aided Manufacturing
CBC	Central Bank of China (Taiwan)
CCIS	China Credit Information Service
CCL	Computer and Communication research Laboratory
CD-I	Compact Disk-Interactive
CE	Consumer Electronics
CEPD	Council for Economic Planning and Development
CIECD	Council for International Economic Cooperation and Development
CIS	Commonwealth of Independent States
CPU	Central Processing Unit
CTV	Colour TeleVision
DC	Developed Country
DCC	Digital Compact Cassette
DGBAS	Directorate-General of Budget, Accountancy and Statistics
DRAM	Direct Random Access Memory
DTCI	Division on Transnational Corporations and Investment
EIAK	Electronics Industry Association of Korea
EIE	Early Industrialised Economy
EO (industrialisation)	Export Oriented (industrialisation)
EPB	Economic Planning Board
EPFZ	Export Processing Free Zone
EPIP	Eclectic Paradigm of International Production

EPZ	Export Processing Zone
ESCAP	Economic and Social Commission for Asia and the Pacific
ERSO	Electronic Research and Service Organisation
EU	European Union
FDI	Foreign Direct Investment
FOREX	FOReign EXchange
FT	Financial Times
GDFC	Gross Domestic Fixed Capital
GDP	Gross Domestic Product
GNP	Gross National Product
GSP	Generalised System of Preferences
GTC	General Trading Company
H-O-S	Heckscher-Ohlin-Samuelson
HCI	Heavy and Chemical Industries
HDTV	High-Definition TeleVision
IC	Integrated Circuit
IC	Investment Commission (Taiwan)
IDI	Inward Direct Investment
IDIC	Industrial Development and Investment Center
IDP	Investment Development Path
IFC	Investment and Finance Company
III	Institute for Information Industry
IMF	International Monetary Fund
IS (industrialisation)	Import Substitution (industrialisation)
ITC	Investment Trust Company
ITRI	Industry Technology Research Institute
KAIS	Korean Advanced Institute of Science
KFTA	Korean Foreign Trade Association
KITA	Korea Industrial Technology Association
KMT	KuoMinTang
LAN	Local Area Network
LCD	Liquid Crystal Display
LDC	Less Developed Country
LI (paradigm)	Late Industrialisation (paradigm)
LIE MNE	Late Industrialising Economy MultiNational Enterprise
LIE	Late Industrialising Economy
LSL	Labour Standards Law
LTC	Large Trading Company
M&A	Merger and Acquisition
MFE	Ministry of Finance and Economy
MIC	Market Intelligence Centre
MNE	MultiNational Enterprise
MOEA	Ministry Of Economic Affairs

MWO	MicroWave Oven
NAFTA	North American Free Trade Agreement
NCP	Net Commodity Product
NDP	Net Domestic Product
NIE	Newly Industrialised Economies
NOI	Net Outward Investment
NSO	National Statistical Office
NTB	Non Tariff Barrier
NT$	New Taiwan Dollar
OBS	Own Brand Sales
ODI	Outward Direct Investment
ODM	Original Design Manufacturing
OECD	Organisation for Economic Co-operation and Development
OEM	Original Equipment Manufacturing
OLI (advantages)	Ownership Locational Internalisation (advantages)
PC	Personal Computer
PCB	Printed Circuit Board
PRC	People's Republic of China
R&D	Research and Development
RD (ratio)	Relative Development (ratio)
REF	REFrigerator
RI (ratio)	Relative Investment (ratio)
RID (ratio)	Relative Investment Development (ratio)
ROC	Republic of China
ROK	Republic of Korea
SAIT	Samsung Advanced Institute of Technology
SME	Small and Medium sized Enterprise
TFT-LCD	Thin Film Transistor-Liquid Crystal Display
TTC	Technology and Technological Change
UN	United Nations
UNCTAD	United Nations Conference on Trade And Development
UNCTC	United Nations Centre on Transnational Corporations
UNDP	United Nations Development Programme
USAID	United States Agency for International Development
VCP	Video Cassette Player
VCR	Video Cassette Recorder
VLSI	Very Large-Scale Integration
W/M	Washing Machine

1 Introduction

The research objectives and why it matters

South Korea and Taiwan are among the most prominent examples of the 'Asian economic miracles' that are characterised by almost continuous high economic growth since the 1960s combined with a relatively equal income distribution (World Bank 1993).[1] Unlike the bulk of the literature dealing with these economies (cf. Balassa 1988, Cuyvers and van den Bulcke 1989, Singh 1995, Smeets 1991, World Bank 1993), this book does not aim at explaining the overall economic success of Korea and Taiwan.[2] Instead, we are particularly interested in one important aspect of both economies, namely their cross-border economic activities.

For a long time the promotion of manufactured exports has been regarded as one of the most important engines of the Korean and Taiwanese economy (Linnemann *et al.* 1987, World Bank 1993). At present, however, their outward orientation is no longer merely confined to exports. Since the 1980s, companies from both countries are also increasingly active abroad by means of foreign direct investment (FDI). As will be discussed in Chapter 2, the emergence of FDI from (by that time) developing countries in the beginning of the 1980s received considerable academic attention.[3] In the beginning, however, investment projects were limited in size and most research focused on activities in other developing countries, which were most prominent then.

With annual FDI outflows exceeding US$ 1 billion, Korea and Taiwan at present are among the most important new sources of overseas direct investment (see Chapter 5). This new development appears to justify an adjustment of the conventional image of Korea and Taiwan which in the past – as far as cross-border economic activities are concerned – fully depended on *exports* from their home economies. Such modification is needed especially since companies from both economies have not confined their investment activities to developing countries (activities that are relatively easy to explain), but are increasingly active in more advanced industrialised nations.

Until now, very little empirical research has been done into Korean and Taiwanese investments in the industrialised world. This study aims to fill this gap. Some authors did attempt to explain the emergence of these new multinational enterprises (MNEs) from a more theoretical perspective. Especially the Investment Development Path (IDP) (cf. Dunning 1981a, 1981b; Dunning and Narula 1994), and more recently the Stages Theory of Industrial Upgrading and Overseas Investment (Ozawa 1992, 1996) tried to capture the relationship between the development level of economies and (inward and) outward investment patterns. The insights provided by these theories will serve as a guideline for the present study. In addition, however, the emergence of Korea and Taiwan as outward investors is examined here against the background of the fact that both belong to the group of late industrialising economies (LIEs). For these economies, the starting-point of their industrialisation and the path they have followed since then differs considerably from the typical industrialisation pattern of early industrialised economies (EIEs). This unique industrialisation path also had a clear impact on the way companies originating in these economies have developed (cf. Amsden 1989, 1991; Amsden and Hikino 1993; Hikino and Amsden 1994). As will be discussed in Chapter 2, the strengths ascribed to their leading companies usually differ from the competitive edges that are characteristic for multinational enterprises from early industrialised economies. Typically, for instance, the strategic focus of LIE companies has been on shopfloor operations and not on R&D activities or the exploitation of new technologies. The aim of this study is to gain a deeper insight into how this has influenced the internationalisation pattern of their companies. More specifically, the objective of this investigation is to get a better understanding of outward direct investment activities by leading companies from Korea and Taiwan in industrialised nations.

In order to serve the aim of this study, a central hypothesis will be formulated here. On the basis of this hypothesis, which merely acts as a general guideline for this book, a number of more specific research questions will be derived.

The central hypothesis reads as follows:

> As a result of the typical (late) industrialisation path of Korea and Taiwan their leading companies historically have not developed the competitive advantages typically ascribed to MNEs from early industrialised economies. Consequently, their internationalisation (*namely* outward investment) pattern will differ from that of MNEs originating in early industrialised economies.

To 'test' this hypothesis, the following research questions will be addressed:

a. *How does the industrialisation pattern of Korea and Taiwan look and which characteristics may influence their outward direct investment pattern?*

To examine the central research question of this study, it is important to obtain an insight into the characteristics of the actual industrialisation patterns of both Korea and Taiwan and the way in which economic activities are in fact organised. Special attention will be paid to those aspects that, according to the theory, may influence outward direct investment from these economies.

b. *To what extent is received MNE theory applicable to outward FDI from LIEs?*

Since outward FDI from countries like Korea and Taiwan only recently gained momentum, we cannot expect received MNE theory to have fully captured the specific nature of their operations. As a result, not only conventional MNE theory has to be discussed, but also the insights provided by concepts that analyse industrialisation patterns of countries like Korea and Taiwan. On the basis of both strands of literature, a number of theoretical questions will be addressed that hopefully make the theoretical framework more geared towards our subject matter. Based on these questions and the (empirical) insights gained in this study, we shall assess the appropriateness of received MNE theory to explain the internationalisation of Korean and Taiwanese companies. Moreover, new 'building blocks' will be suggested to improve our theoretical insights into this phenomenon.

c. *What is the actual size and nature of Korean and Taiwanese outward direct investment flows?*

To assess the absolute and relative importance of Korea and Taiwan as new sources of FDI,[4] it is important to analyse the available FDI statistics thoroughly. One relevant exercise is to compare Korean and Taiwanese outward FDI with other ('old' and 'new') home countries of MNEs. In addition, the geographical and industrial diversification of their outward investments will be studied. Moreover, we will examine to what extent Korea and Taiwan are not only late industrialisers, but are also late as outward investors.

d. *What are the characteristics of Korean and Taiwanese MNEs and what is the nature of their internationalisation strategies?*

These questions aim at getting a better picture of the characteristics of companies that represent the investment flows analysed under research question c. In addition, we intend to increase our knowledge of their internationalisation strategies, such as the reasons as to why they invest abroad, on the basis of what strengths, and what entry modes are chosen to internationalise further.

e. *What problems do companies face in operating in industrialised economies and to what extent are these hurdles related to their late industrialisation background?*

In view of the fact that, according to theory, MNEs from LIEs such as Korea and Taiwan have not, historically, focused on building up competitive strengths similar to that of leading companies from early industrialised economies, obstacles may have emerged in operating in industrialised nations. It is therefore imperative to examine the nature of these hurdles. Subsequently, the question can be posed as to which of these problems can be attributed to their typical industrialisation history, as described in the theoretical discussion.

f. What differences in the outward investment pattern of Korea and Taiwan can be discerned and how can they be explained?

Although Korea and Taiwan are both labelled as 'late industrialising economies' and are often put in the same category, according to theory, national differences are also expected to have an impact on the nature of outward FDI. Therefore, the actual differences between Korean and Taiwanese overseas investment have to be analysed. Subsequently, possible explanations have to be brought forward to explain possible divergences.

The research strategy

As can be derived from the questions formulated earlier, the broad aim of this book is twofold. On the one hand, we try to gain a deeper understanding of a relatively new phenomenon, namely Korean and Taiwanese investment abroad – especially in industrialised economies. In this context, we also hope to 'discover' new relationships and intend to give an impulse to new 'building blocks' to complement existing theoretical explanations. In this respect, the investigation can be labelled as exploratory. On the other hand, it is investigated to what extent *received* (MNE) theory is able to explain the phenomenon studied here. For this purpose – *inter alia* – hypotheses are derived from existing theoretical insights that are tested on the basis of a 'new' sample.

This double aim is also reflected in the research techniques used in this study. A tobit model, for instance, is used to test the hypotheses related to received MNE theory in an econometric manner whereas the case study method is used to understand more of the typical nature of the internationalisation patterns of Korean and Taiwanese companies and the problems encountered herein. Both techniques are concerned with the analysis of micro level data. In addition, however, data are also analysed at the meso and macro level. More specific features of the various research techniques used are discussed in the respective chapters.

Part of the investigation took place in Korea and Taiwan. As a guest of the local offices of the Netherlands Foreign Investment Agency (Ministry of Economic Affairs) (in 1993), Chung-Ang University in Korea (in 1995) and Chung-Hua Institution for Economic Research in Taiwan (in 1995 and 1997) I was able to collect a lot of information that was not (or only partly) available in Europe. Moreover, from these localities some 50 interviews with Korean and Taiwanese managers of parent companies were arranged in which their internationalisation strategies were discussed. Because of (too) large variations between the sectors in which these companies operate, the wealth of information and insights that these discussions yielded is not systematically processed in this study. With the exception of the case studies it is used only for illustration purposes.

As with most research, this study is also limited in scope. First of all, our aim is only to gain a deeper understanding of *industrial* Korean and Taiwanese companies investing abroad. Although Boddewyn *et al.* (1986) argue that in principle, existing MNE theories can be used to explain the international expansion of the service sector, at the same time they admit that it is already difficult to compare one group of service MNEs with another – let alone service with non-service firms. It therefore appears an inappropriate research strategy to include MNEs from sectors that cannot be compared.

Moreover, although the investments in the developing world will not be left out of consideration, the emphasis is on the operations of Korean and Taiwanese companies in industrialised nations. As will be shown in Chapter 2, these investments are more difficult to explain than FDI in developing countries and raise the most interesting questions.

Also in terms of industries covered, the study is limited. The econometric analysis concerning the determinants of FDI is confined to the electronics industry and the case studies deal with operations of firms in subsectors of the electronics industry, namely consumer electronics (Korea) and the computer industry (Taiwan). As Yip (1994) among others, has illustrated, substantial differences exist between branches of industry as far as their 'propensity to globalise' is concerned. By limiting the study to (parts of) the electronics industry, these variances are not taken into account.

Some definitions

Since various, distinct, definitions are used in the literature on multinational enterprises, it is useful to dwell on this possible source of confusion. The UN (UNCTC 1992a: 45) define *foreign direct investment* as an:

> investment involving a long-term relationship and reflecting a lasting interest of a resident entity in one economy (direct investor) in an entity resident in an economy other than that of the investor. The direct investor's purpose is to exert a significant degree of influence on the management of the enterprise resident in the other economy.[5]

A very general definition of *multinational enterprises* is proposed by the UN saying that this is 'an entity controlling foreign assets' (UNCTC 1992a: 46). Often, a more limited category of companies is included in the definition. Dunning (1993: 3) defines a multinational or transnational enterprise as an 'enterprise that engages in foreign direct investment and owns or controls value-adding activities in more than one country'.[6] *Global companies* refer to 'enterprises that engage in value-adding activities in all of the major regions in the world and possess an integrated strategy towards these activities' (Dunning 1993: 11). *Internationalisation* refers to the process of extension of the domestic value-added chain of economic activity. In the internationalisation literature, the focus is more on the strategic management of the resources and competencies owned or acquired by individual companies to achieve their goals related to value-added activities outside their

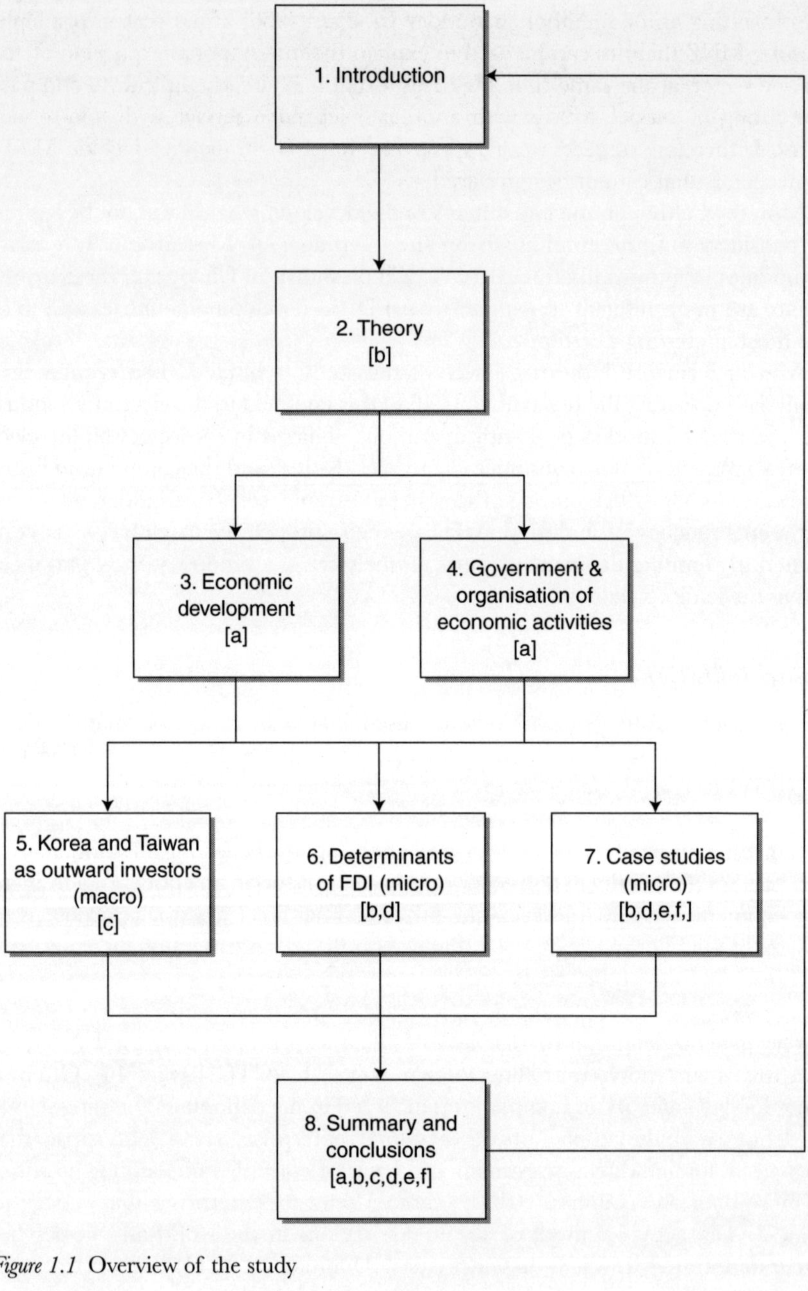

Figure 1.1 Overview of the study

national boundaries (Dunning 1993: 186). Unless stated otherwise, these definitions reflect the meanings of these terms throughout this study.

How this book is organised

In Figure 1.1, an overview of the study is provided. In the figure, not only the chapters are indicated but also the research questions that were formulated. In Chapter 2, received MNE theory as well as the concept of late industrialisation is discussed. Both strands of literature are subsequently confronted to derive new theoretical questions addressed in the remainder of this book. Chapters 3 and 4 provide a contextual analysis that is needed to gain a better understanding of the context in which outward FDI from Korea and Taiwan emerged. In Chapter 3, a concise overview will be given of the economic development of Korea and Taiwan in the post-war period. In Chapter 4, more specific attention is paid to the role of the government in organising economic activities.

Chapters 5–7 discuss the empirical exercises conducted for the research. Chapter 5 is mainly concerned with an empirical analysis of the importance and nature of Korean and Taiwanese outward investment on an aggregate level. In Chapters 6 and 7, we will 'descend' to the firm level. In Chapter 6, the determinants of outward investments in the electronics industry will be tested by means of an econometric analysis. To further deepen our understanding of the internationalisation of Korean and Taiwanese firms, in Chapter 7 case studies are conducted of leading companies from both countries operating in the electronics industry. Finally, in Chapter 8, conclusions are drawn with regard to the questions posed in Chapters 1 and 2. Moreover, suggestions are made to adapt received theory to explain the internationalisation pattern of leading Korean and Taiwanese companies more adequately.

2 Multinational enterprises and late industrialisation

A discussion of the literature

Introduction

Only since the 1960s have distinct theories dealing with MNEs been developed (Dunning *et al.* 1986). Even at present, however, mainstream neo-classical (international) economics does not fully acknowledge the relevance of these theoretical concepts.[1] Dunning (1993: 78) explains this lack of interest as follows:

> The lack of interest by traditional trade economists with ownership or governance questions arises because they have tended to assume . . . implicitly rather than explicitly that firms engage only in a single value-added activity. The effect on trade patterns of the vertical integration or horizontal diversification of firms or their reaction to uncertain markets or government intervention are rarely discussed in the literature. Since the option of internalising domestic markets for intermediate products within a country has not generally interested trade economists, it is hardly surprising that they have been relatively unconcerned with issues of international production.

In spite of this lack of interest by mainstream neo-classical economics, during the last three decades an impressive amount of literature has been published addressing the emergence and behaviour of MNEs. In this chapter, we will discuss the theoretical insights brought forward in the literature to understand the emergence of multinational enterprises from late industrialising economies. First, general theories on multinational enterprises are discussed.[2] Subsequently, the typical characteristics of late industrialising economies (LIEs) are investigated. Next, we will examine to what extent received MNE theory is able to explain outward direct investment from LIEs. Finally, on the basis of the insights rendered by these exercises, a number of research questions are formulated that will be addressed in the remainder of this study.

General MNE theories

Micro theories

In this section, the most important MNE theories which focus on the micro level are discussed, i.e. those theories that primarily explain the existence and behaviour of companies having value-added operations abroad.[3]

The origin of modern MNE theory: Stephen Hymer

The doctoral dissertation of Hymer (defended in 1960, and posthumously published in 1976) is usually regarded as the starting-point of modern MNE theory. In his work, Hymer focused on the typical characteristics of FDI as a modality for international business. By the time of his study, the prevailing conceptualisation of international capital movement was based on a neo-classical theory of portfolio flows. If perfect competition is assumed, capital moves across borders in response to interest rate or profit differentials. In such a concept, there was no role for the MNE.

Hymer argued that because of market imperfections such as risks and uncertainty, volatile exchange rates and costs of acquiring information, this theory did not relate to real-life conditions. In addition, he indicated that FDI represents more than a flow of international financial capital, and consists of a package of resources such as technology, managerial capabilities, and entrepreneurship. Perhaps the most fundamental contribution of Hymer concerns the distinction he made between foreign direct investment (where no change of ownership of the transferred sources or rights takes place) and portfolio investment (where a change of ownership does take place). The organisation modality of the sources and value-added activities of both forms of investments are different.

To explain foreign direct investment (or international production), Hymer used a typical industrial organisation approach. To own and control value-added activities abroad, a company should possess some oligopolistic advantages that outweigh the disadvantages that exist in competing with indigenous firms abroad. Based on the operations of US MNEs abroad, Hymer created an image of companies that erect entry barriers for other companies by means of specific knowledge, scale economies, distribution networks, product diversification, etc. Dunning and Rugman (1985) drew attention to the fact that Hymer thus only acknowledged structural market imperfections, which are so-called Bain-type advantages, meant to enhance the asset power of MNEs. Williamson-type transaction costs imperfections that are assumed to be exogenous to the MNE, however, are not explicitly taken into account by Hymer. In the latter case, the MNE bypasses the market as a response to transaction costs.

Internalisation theory

At present, the internalisation theory is widely accepted as a key element of MNE theory. The roots of this theory can be traced to Coase (1937) who was primarily interested in the existence of firms *per se*. He did not pay any special attention to cross-border value-added operations of MNEs, however. In his article 'Nature of the firm' Coase showed that because of transaction costs and market failure, it may be more efficient for a firm to 'internalise' market transactions than incur the considerable costs involved in using an external market. The most obvious example of such internalisation is vertical integration in which a firm decides to control either its own products of supply or destination of output. In both cases, the functions of independent material suppliers or of wholesale and retail merchandise are absorbed – internalised – within the firm.

Coase never used the word 'internalisation' as such; this notion was used much later by Buckley and Casson (1976) who introduced it for their explanation of MNE behaviour.[4] The core hypothesis of present-day internalisation theory is that a multinational hierarchy can in some cases be a substitute for the market to organise cross-border value-added activities. Whether a firm decides to internalise (or replace) markets across national boundaries – and thereby create an MNE – is governed by the relative costs and benefits to the firm of such internalisation. Buckley and Casson argue that in particular, the costs and benefits of intermediate items such as managerial, technical and other skills that are specific to firms but easily transferable between countries, are relevant. In the internalisation theory (or market imperfection approach), a distinction is made between natural (or structural) and unnatural (or transactional) market imperfections that induce firms to internalise (Dunning and Rugman 1985). Structural market imperfections (à la Hymer) give rise to monopoly rents as a result of the presence of entry barriers that companies seek to erect or increase. Transactional market failures (à la Williamson), on the other hand, concern the inability of markets to organise transactions in an optimal manner. The latter may be the result of (Dunning 1993: 78): (a) the fact that buyers and/or sellers have incomplete, asymmetrical information or face uncertainty; (b) the market cannot take account of the benefits and costs of external effects of a transaction (e.g. those related to international exchange and capital markets); or (c) the demand for a particular product, while infinitely elastic, is insufficient to allow the producing firms to fully capture economies of size, scope and geographical diversification.

More recently, Buckley (1989) suggested that the internalisation concept preferably should be labelled a 'paradigm' instead of a 'theory' since a wide divergence exists with regard to the kind of market failures that determine the mode of organisation of intermediate-good transactions.

Closely related to the internalisation theory is the transaction costs approach that also explicitly takes into account the notion of markets and hierarchies as alternative routes of transactions. Whereas Williamson (1975, 1979) has contributed considerably to this notion within the national context, others scholars [such as Hennart (1982, 1989)] extended the concept to the international business environment. According to this approach, FDI consists in the extension of hierarchical modes of organisations across national boundaries. This form is chosen because it reduces the costs of international co-ordination between parties located in different countries.

Although these theories, as we noticed earlier, have become part of received MNE theory, an important criticism concerns the negligence of the role of location-specific variables in explaining the direction of FDI (cf. Dunning 1993).[5]

Eclectic Paradigm of International Production

At present, the Eclectic Paradigm of International Production (EPIP) is the most widely accepted theoretical framework to analyse multinational enterprises. The concept was introduced by Dunning in 1976 and since then has been refined a number of times (cf. Dunning 1977, 1988, 1993, 1995).[6] As opposed to the ruling theoretical concepts at the time of the introduction, he opted for an 'eclectic' approach. The meaning of the word 'eclectic' was twofold (Dunning 1988); it suggested the use of various strands of economic theory and, at the same time, it indicated that international production is only one possible mode of international economic activity. The choice for a particular mode of international business is influenced by a common set of factors that will be discussed hereafter.

The Eclectic Paradigm primarily draws upon three strands of received economic literature, namely, industrial organisation, conventional trade theory and the internalisation theory. In accordance with the industrial organisation theory, EPIP underlines the importance of ownership (O) specific advantages. Since overseas production involves additional costs to a firm arising from legal, linguistic, cultural and political differences between the home and host countries as well as increased costs of transportation and communication, a company needs to own certain advantages over national firms in the host country.[7]

Internalisation (I) advantages refer to the perceived advantages of a hierarchical control of value-added activities to exploit or circumvent market failures. As we have seen earlier, a distinction is made between structural and transactional market failures. Internalisation advantages may arise from exploiting structural or circumventing transactional market failures to improve a company's position.

Traditional Heckscher-Ohlin-Samuelson (H-O-S) trade theory, finally, has drawn the attention to the importance of locational (L) advantages. L advantages refer here to the specific attractions of a host country as a site of investment ('pull factors') as well as to the disincentives in the home country ('push factors').

The basic hypothesis of the Eclectic Paradigm of International Production is as follows. The extent, form and pattern of international production is determined by the configuration of these three sets of advantages:

1 Companies need to possess O advantages to be able to compete with companies in another country;
2 It must be in the best interest of enterprises that the possessed O advantages are transferred across national boundaries within the organisation itself instead of selling them or their rights to use by others (namely, I advantages have to be apparent);
3 Assuming that conditions (1) and (2) are satisfied, L advantages determine whether export from the home economy or local production in the host market will take place.

The nature and value of the ownership–location–internalisation advantages is also referred to as the OLI constellation. Recently, a fourth condition was added which deals with the strategy of companies (Dunning 1993). Given the OLI configuration of a firm, the question whether a company will produce abroad also depends on the consistence of such a choice with its long-term strategy.

It is needless to say that the paradigm as presented in the foregoing merely represents a very general explanatory framework. The actual factors that are relevant in each specific case may vary considerably. One important determinant of the sort of factors that apply is the motivation to invest abroad. Dunning (1993) distinguished four types of international production (namely, natural resource seeking, market seeking, efficiency seeking, and strategic asset seeking) that are each characterised by their own OLI constellation. Examples of the variables relevant in each category are provided in Appendix 2.1.

An important explanation for the popularity of the Eclectic Paradigm of International Production is that it – in a straightforward manner – categorises the many factors that influence the complex phenomenon of foreign direct investment. This does not mean, however, that the paradigm itself renders hypotheses that can be tested. For that purpose, additional theoretical concepts have to be introduced that deal with more specific questions. Notwithstanding Dunning's claim that his paradigm is 'eclectic', another limitation concerns the strong bias towards the explanation of international production. Furthermore, the EPIP is rather static in nature (Vernon 1985). Although Dunning, for instance in the framework of the Investment Development Path, argues that the OLI constellation changes over time, the mechanisms underlying these changes – especially those related to the O advantages – are not revealed convincingly. Moreover, many other modes of international business exist that in the EPIP are all headed under the category of 'portfolio resource transfer'.[8] Closely related to this, since the 1980s is the rapid growth of the importance of strategic alliances – a development that Dunning (1995) has only recently formally addressed. These alliances, which are neither hierarchically controlled nor arm's length transactions are often formed to gain access to new or complementary technologies, speed up innovatory or learning processes or to upgrade the efficiency of particular activities. According to Dunning, the advent of 'alliance capitalism' has a number of implications for the Eclectic Paradigm of International Production. First, the role of innovation in sustaining and upgrading competitive advantages of firms and countries has to be better recognised. Second, the Eclectic Paradigm has to recognise that strategic alliances can be a viable alternative to internalisation to reduce some kinds of market failure – particularly those alliances that have to do with opportunism and information concentration by participants in the market. Third, the EPIP has to acknowledge that the traditional assumption that capabilities of the individual firm are limited to its ownership boundaries is no longer acceptable whenever the quality of a firm's efficiency-related decisions is significantly influenced by the collaborative agreements they have with other firms.

Dynamic aspects

The foregoing MNE theories are all basically static in nature. An early attempt to develop a more dynamic concept was made by Scandinavian scholars in the 1970s (cf. Johanson and Vahlne 1977, Luostarinen 1979, Welch and Luostarinen 1986). Models were developed to describe the internationalisation of individual firms.[9] Johanson and Vahlne, for instance, focus on the gradual acquisition of knowledge about foreign markets and operations and on the incrementally increasing commitment in foreign markets. The latter, according to their model, is reflected in an internationalisation path that firms follow starting with export to a country via an agent, followed by the establishment of a sales subsidiary and, eventually, in some cases, production in the host country. Luostarinen's research (1979) revealed a process of evolutionary development not only in terms of the depth of international mode, but also in terms of the diversity of modes used, as well as product offerings and the range of markets penetrated.[10]

Later empirical investigations (cf. Sullivan and Bauerschmidt 1990) have shown that such models are not necessarily applicable today. Zander (1994) mentions a number of possible explanations why firms might not follow the expected internationalisation path, such as: increasing international competition requiring faster establishment in leading markets, more general experience and sharing of knowledge about international operations and the increasing use of acquisitions as a means of foreign entry, which has made it possible to achieve substantial operations in foreign markets more rapidly.[11] As will be shown later, the companies investigated in the present study pursue catch-up strategies that also imply a deviation of their internationalisation patterns from the patterns described by the conventional Scandinavian models. Notwithstanding these counter-arguments, some of the insights emphasised by these Scandinavian authors have remained popular in later theories. 'Psychic distance' (such as differences in language, business practices, culture, and industrial development), for instance, is supposed to influence the time order of establishments abroad. Also important is the idea that internationalisation is a learning process, which explains its incremental character. Learning here does not only relate to acquiring information and reducing risks regarding a particular market, but also to the process of internationalisation as such. As will be illustrated in later chapters, the initial stages of 'going abroad' imply organisational changes, extensive training of employees, and a general change in their attitudes. Besides these more specific elements, the internationalisation path described by the Scandinavian School is also a useful frame of reference to analyse the internationalisation strategies of 'new' MNEs, such as those from Korea and Taiwan.

More recent 'dynamic' theories that have attracted substantial attention offer a deeper insight into the growth and technological competence of firms (cf. Cantwell 1991, Cantwell and Tolentino 1990, Lall 1990, 1992, Tolentino 1993, Zander 1994). In these theories on 'technology and technological change (TTC)', technological competence is seen as the regulator of competitive success or failure.[12] Cantwell (1991: 4) defines technological competence as 'all changes to the immediate conditions of production which, over time, raise the productivity of inputs and provide for new products'. The TTC approaches are strongly rooted in evolutionary economics. As opposed to neo-classical theory, evolutionary theory has incorporated endogenous technological change. The opportunity to develop unique technology combined with the imitation efforts by other firms creates the endogenous drive for technological change that makes an equilibrium impossible. Moreover, it is based on the premise that technological opportunities of firms depend on what has been achieved in previous periods which underlines the dynamic nature of these concepts.[13] The cumulative nature of firm-specific capabilities generally prevents firms from developing new capabilities quickly. Instead, they are locked into existing fields of business, a situation that has been termed 'path dependent'. Evolutionary models emphasise the proprietary nature of technology. When firms invest in the development of new technologies, they usually make use of some sub-set of publicly available knowledge, but even more of highly specific and tacit technological knowledge which resides in the firm (Zander 1994).

The TTC theories acknowledge that firms can only build up technological capabilities if the country of location is endowed with well-functioning factor markets (e.g. the availability of skilled workers and financial sources) and supporting technology institutions. Lall (1990, 1992) labels these necessary prerequisites 'national technological capabilities'. In addition, the incentive structure in which a firm has to operate is also an important determinant of the actual development of technological capabilities. An outward-oriented industrialisation strategy and a competitive environment are supposed to stimulate firms to develop their capabilities more. The more technologically advanced a sector is in which the firm operates, the larger the skill endowments and the more focused the domestic technological support efforts will have to be.

As is also the case with the theory of the firm and the internalisation theory, the TTC theories were not specifically developed to explain MNE behaviour. Nevertheless, in recent years their insights have been linked up with MNE theory. It thus provides us with additional insights into the mechanisms behind the accumulation of (technological) O advantages. In addition, a discussion has arisen about whether MNEs are constrained by their technological trajectories in their country of origin ('home-based MNEs') or that technological development should be seen as an international process (Zander 1994). According to the latter view, MNEs – as opposed to national firms – can broaden their technological search by tapping in different specialisations in various locations. Ghoshal (1988) in this respect refers to 'environmental scanning' whereas Ozawa (1992) speaks of 'internationalised learning'. To optimally use the existing local competencies in other countries, the partially firm-specific nature of technology forces MNEs to carry out research and development activities abroad. Cantwell and Tolentino (1990) and Tolentino (1993) further developed the relation between the stage of technological capability of the firm from LIEs and the nature and stage of FDI. The thrust of this relation will be discussed later in this chapter.

Meso theory

Although the distinction between macro, meso and micro theories is somewhat arbitrary (none of the concepts deals exclusively with just one level of analysis), the revised version of the so-called Product Cycle model is the only major theory that takes the industry as the central level of analysis. The Stages Theory of Industrial Upgrading and Overseas Investment also draws considerable attention to the development of industries. Since its primary goal, however, is the development of countries, this theory is addressed in the next ('macro') section.

The Product Cycle model (Vernon 1966, 1974) received considerable attention in the 1960s when it seemed to present a generally applicable explanatory framework. The theory focuses on the dynamic nature of international trade and investment and was primarily based on the experience of US firms in that period. The original version of the model distinguishes three stages in the life cycle of a product. The first stage is the innovation of a product. The basic model states that new products are developed and introduced in response to cost and demand conditions in the innovator's home market. The high average income levels in the US supported new wants and the relative labour scarcity and high unit labour costs greatly stimulated the development of labour-saving consumer goods and industrial products. Innovational development work is mostly done in the home market because of the presence of engineers and scientists with the required skills and the possibility to interact with potential customers. In addition, close communication with the headquarters is needed in the initial stages. Furthermore, by bringing together an innovational team in one location, additional benefits can be gained, such as subdividing tasks among specialists and maintaining efficiency of communication between the research specialists. Usually innovating firms also prefer to locate the actual production facilities in the home country. Because of the unstandardised nature of the production process, the producer attempts to improve production technology by experimenting with alternative inputs. The characteristic inelasticity of demand of early users of new products in the beginning makes the innovator rather indifferent to the relatively high production costs in the (US) home country. Moreover, because communication between production and marketing needs to be frequent, and communication costs increase with distance, it is important to locate production close to the market.

The second stage is that of the maturing product. The demand for a product increases, and a certain degree of standardisation takes place. As a result, the importance of flexibility declines whereas the possibilities of economies of scale through mass production increase. Buyer knowledge increases, and demand becomes more price-elastic, so that the firm becomes increasingly sensitive to routine production costs. Because technology has stabilised it has become less important for the company to be close to the home market. Although initially the main market will be the home country, demand in other relatively advanced economies will grow rapidly. Depending on labour cost differentials between home and overseas, it eventually becomes economic to invest overseas. Apart from cost considerations, the risk that local firms abroad may start manufacturing similar products can be an important stimulus to the initial investment of a firm.

The final stage of the original Product Cycle model refers to the standardised product when competition between producers is based solely on price. Market information is easily available and the prime concern is the lowest-cost source of supply. The most labour-intensive stages of production are thus transferred to developing countries.

Not long after Vernon published the first version of his model in 1966, it was rendered out of date by new developments. Other countries (first Europe and later Japan), for instance, caught up with the US in terms of per capita income and overall factor conditions. Increasing cross-investments between powerful home countries (e.g. American firms investing in Europe and European firms investing in the same sector in the US) made it difficult to explain the investment relationship between the US and these other countries by means of the Product Cycle model.

As a result of these changes in the international economic conditions, a revised version of the Product Cycle model was developed in which Vernon (1974) shifted the emphasis of his analysis from the firm to the meso level. A central theme became oligopolistic behaviour and the desire of firms to maintain an oligopolistic market structure by erecting barriers to entry. The first stage of the model, renamed innovation-based oligopoly, acknowledges that different national environments generate different kinds of innovations. While US firms tend to generate innovations consistent with high incomes and high labour costs, European companies, for instance, might specialise in innovations that are consistent with high land and material costs. The second stage of the modified model, the mature oligopoly, differs more considerably from the second stage in the first version of the concept. As opposed to the first stage, in which the oligopoly was based on product innovation as a barrier to entry, in this stage economies of scale in production, transportation or marketing constitute effective entry barriers. Behind these entry barriers rival firms nullify aggressive strategies initiated by the others, by matching them move for move. To minimise the risk of destabilising the oligopoly, leading firms produce in each other's principal markets and exchange threats in the form of price cutting ('taking hostages'). The final stage is the senescent oligopoly when economies of scale become less effective as barriers to entry and also attempts to erect other barriers (e.g. through advertising) fail. Whereas some companies will leave the industry altogether, others stay on. The firms that continue will be subject to some form of price competition in which case the question of costs takes on added importance. If cheaper materials or components can be secured from other countries, the pull to those locations is strong.

Notwithstanding these adaptations, at the end of the 1970s the Product Cycle model had lost an important part of its explanatory power – a conclusion also drawn by Vernon (1979). Innovations started to originate in different overseas branches of the MNE instead of in its home country. In addition, particular production processes, also for the newest products, were detached from the home base and could even be located in developing countries. A general limitation of the concept apparent from the beginning was that it primarily explained market-seeking investment and did not apply to resource-based, efficiency-seeking or strategic-asset-acquiring FDI. When firms established global networks throughout the world the Product Cycle concept became irrelevant as a general explanatory framework for the operations of most MNEs (Giddy 1978).

One reason for including the model in this overview is that it has, however, not become totally useless. A case in which the model still seemed to hold was in explaining the operations of MNEs originating in developing countries that invest in other less developed countries (LDCs). As will be addressed later on in this chapter when discussing FDI from late industrialising economies, Wells (1977) viewed the internationalisation of firms from developing countries as a stage in the product life cycle.

A second reason to pay attention to the work of Vernon is that other interesting studies built on his approach. Especially noteworthy for the present research is Knickerbocker's (1973) dissertation in which he empirically investigated the strategic responses of companies to the behaviour of competitors. Knickerbocker showed that in oligopolistic markets, in trying to avoid destructive competition, companies normally follow each other into new (foreign) markets. As we shall see later on in this book, this 'follow-the-leader' hypothesis also applies to Korean MNEs.

Macro theories

Most MNE theories primarily deal with the existence and/or behaviour of the firms involved. Much less attention is paid to the interaction between FDI and the economic development of nations. In this section, we shall discuss two theoretical concepts that specifically try to explain this relationship, namely, the Investment Development Path and the Stages Theory of Industrial Upgrading and Overseas Investment.

Investment Development Path

The Investment Development Path (IDP) was initially developed by Dunning (1981a, 1981b, 1986, 1988) and was elaborated by Dunning and Narula (1994, 1996) and Narula (1996a). The building blocks for the IDP are provided by the Eclectic Paradigm of International Production. The IDP relates a country's net outward investment (= outward direct investment minus inward direct investment) position to its stage of development relative to that of the rest of the world. In principle, two types of catalysts for change are distinguished, non-FDI and FDI induced changes. Non-FDI induced changes are exogenous to firms, but endogenous to countries and mainly reflect the influence of government policy and the economic system of a country.[14] Both determine how L specific natural and created assets are organised. FDI induced changes refer to the activities of domestic MNEs abroad and foreign MNEs in the domestic economy and the way in which they interact with the national economy. At any point in time, country-specific characteristics influence the nature of O advantages possessed by firms (e.g.

through export subsidies) and L advantages (e.g. import restrictions); they may also affect the propensity of firms to internalise O advantages across the border. The IDP also states that as changes occur in the OLI configuration, the net outward investment (or NOI) position of a country changes.

First, we present the 'average' IDP. The IDP distinguishes the following five stages of development as far as the net outward investment position of countries is concerned:

Stage 1 Because of insufficient L advantages (such as an inadequate domestic market or a poor physical and human infrastructure), potential inward foreign investment is discouraged, except for those due to the possession of natural assets. Since only little indigenous technology accumulation has taken place, domestic firms do not yet have the O advantages needed to invest abroad.

Stage 2 As domestic markets enlarge (in size and/or purchasing power) and/or the local infrastructure is improved, inward direct investment begins to become profitable. Import-substituting investment in particular, is often stimulated by host governments by imposing tariff and non-tariff barriers. As far as investments in export-oriented industries is concerned (at this stage primarily in natural resources, primary commodities and simple, labour-intensive low technology manufacturing), the quality of the existing physical and human infrastructure is very important.[15] Due to the development of supporting industries clustered around primary industries, O advantages of indigenous firms are gradually developed. In addition, the emergence of local semiskilled and moderate technology-intensive consumer goods is observed. Some outward direct investment may take place either in neighbouring countries at lower stages of the IDP ('resource seeking') or in industrialised economies aimed at acquiring strategic assets.

Stage 3 Net inward investment declines as a result of a decreasing growth of inward direct investment and an increased growth of outward FDI. The original O advantages of MNEs begin to erode when local firms have accumulated their own competitive advantages to compete with them in the same sectors. New L advantages based on enlarged markets and improved domestic innovatory capacity will encourage more technology intensive manufacturing. Outward investment in countries in lower stages of the IDP increase, partly as a result of a loss of comparative advantages in labour-intensive industries. Next to investments in export platforms, investments in these countries are partly also market seeking. In Stage 3 and 4 countries, market seeking as well as strategic asset seeking investments (to upgrade the firm's O advantages) are made.

Stage 4 In this stage, the country has become a net outward investor and the growth rate of outward FDI is still higher than that of inward FDI. Domestic firms now use state-of-the-art production processes while the cost of capital is lower than that of labour. To maintain their competitive advantage, these firms shift conventional industries abroad and respond to trade barriers erected by major trade partners. In view of a high propensity to internalise their O advantages, the latter is primarily done through international production instead of export.[16] Inward investment is increasingly sequential in nature and directed

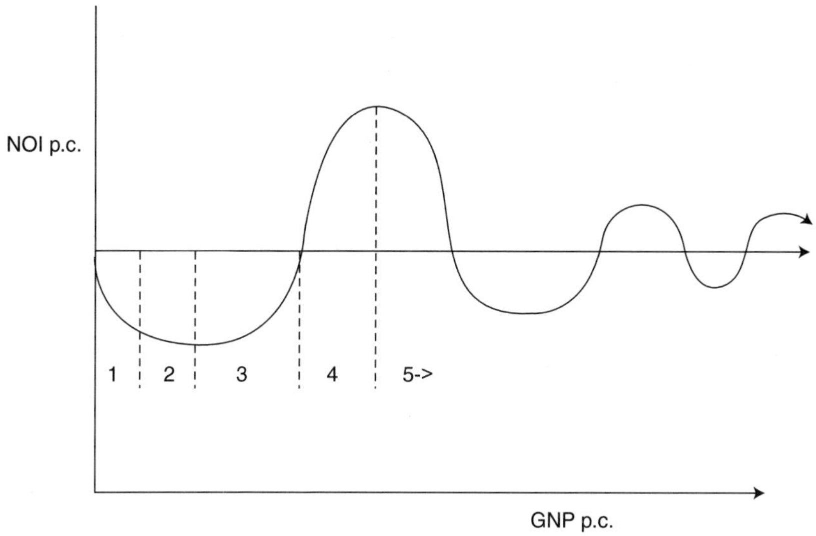

Figure 2.1 The 'general' Investment Development Path
Source: Dunning and Narula (1994)

towards rationalised and asset-seeking operations undertaken by firms from Stage 4 countries. The government increasingly gives more attention to the structural adjustment of the economy by fostering technology accumulation in infant industries and phasing out declining industries.

Stage 5 In the final stage, which was added to the model later,[17] a reconvergence of outward and inward investment flows takes place. This is a result of a steady process away from those MNEs that base FDI decisions primarily on factor endowments towards those that instead do so more on the basis of their ability to acquire assets and their ability to efficiently organise their advantages and exploit the gains of cross-border common governance. In the (near) future, this phase may apply to the 'Triad' (USA, Japan and EU), in which L advantages – with the exception of natural resource-based activities – will no longer exist.

The pattern of these stages is presented in Figure 2.1. As was indicated earlier, the figure illustrates an 'average' pattern and therefore in most cases will not match reality. In later versions of the IDP it is acknowledged that countries may follow different paths, depending upon country-specific characteristics, such as (1) resource availability, (2) market size, (3) industrialisation strategy and (4) government policy and organisation of economic activity (Dunning and Narula 1996, Narula 1996a). In the following, some effects of these country-specific conditions – *ceteris paribus* – will be illustrated:

1 *Resource availability* In a case where a country possesses substantial primary commodities, it is likely to receive considerable inward investment from MNEs that wish to internalise the supply of raw materials for their upstream

activities elsewhere. As other L advantages develop (e.g. skilled manpower, infrastructural facilities), the size of inward investment will rise further, which may lead to vertical investment in upstream activities by both domestic firms and MNEs. In this case, the NOI position may remain negative for a longer-than-average period, while the reverse case (namely, a lack of natural resources) *ceteris paribus* will imply an 'early' positive net outward investment position.

2 *Market size* Small countries have a limited attraction in terms of market size. According to the IDP the lack of economies of scale will constrain inward FDI in the early stages. As the country's human capital and infrastructure improves, some inward investment may occur, for instance for export processing purposes. For domestic firms, the limited size of the home market implies a high propensity to do business abroad to realise scale advantages. Domestic investors who are involved in export activities at a relatively early stage will seek overseas locations to compensate for the shortage of low wage labour for labour-intensive production. Countries characterised by a limited market size therefore will reach (and remain at) a positive NOI position at an early stage of development. Large home markets on the other hand will show early inward FDI (because of the attractions of their large markets) and relatively late outward investment (because of the lack of incentive for domestic firms to seek overseas markets) which implies a longer development period characterised by a negative NOI position.[18]

3 *Industrialisation strategy*[19] A third refinement concerns the industrialisation strategy of a country. Broadly speaking, a distinction can be made between import substitution (IS) and export-oriented (EO) industrialisation. The aim of IS is to protect a country's infant industries so that the overall industrial structure can be developed and the dependence on foreign technology and capital reduced. Often, very high tariffs were imposed on those sectors chosen for protection. Import quotas and other devices (e.g. licences, multiple exchange rates) were also used to encourage domestic protection.[20] Given the many pitfalls associated with IS strategies (cf. Singer 1989), it is no surprise that EO strategies in recent years have become much more popular.[21] In this case, exports are seen as generating greater productivity growth as a result of: greater capacity utilisation in industries in which the minimum efficient size of a plant is large relative to the size of the domestic market; increasingly familiarity with and absorption of new technologies; greater learning-by-doing insofar as this is a function of cumulative output, and exports permit greater output in an industry; and the stimulating effects of the need to achieve internationally competitive prices and quality (Adriaansen *et al.* 1989).

How does the choice for an IS or EO policy affect the IDP? It is argued that an EO policy is likely to achieve faster economic growth and structural upgrading. Economic reality in countries with an inward orientation instead shows that such policy often implies limited competition in the home market – which may lower incentives to operate abroad. The most obvious example has been India, where the lack of heavy competition in many markets (ranging from soft drinks to cars) did not stimulate them to accumulate the O advantages needed in international markets. Even after a shift to a more outward-oriented policy, this will necessarily delay the entrance of many producers from such countries on the international market, first by means of exports and later by means of foreign direct investment. In other words, countries that have pursued an IS policy will show a negative NOI position for a longer period than export-oriented economies. To move beyond the second stage of the IDP a reorientation towards exports is needed which, according to the IDP, will lead to faster structural adjustment and economic growth.

4 *Government policy and the organisation of economic activity* Finally, it is observed that the nature of government policy can vary between countries with basically the same (e.g. export oriented) industrialisation strategy and which are at the same stage of development. The aspects of government policy that directly and indirectly may impinge on the IDP pattern of individual countries cover a wide range of instruments.

Unfortunately, the IDP in its present form remains rather vague about the most important factors, nor does it provide us with a deeper insight into the underlying relationships between these factors and inward and outward FDI.[22] In Chapter 4, some of these factors will be discussed in more detail in the case of Korea and Taiwan.

Empirical analyses have suggested that, although important exceptions have been noted, by and large the relationship between net outward investment and the level of economic development as described by the IDP indeed applies (Dunning 1981b, Narula 1996a, Tolentino 1993). Such aggregate empirical exercises, however, fail to provide us with deeper insights into the underlying mechanisms at work here. Recently, an evaluation of the concept was carried out through a set of studies on eleven countries at various stages of development (Dunning and Narula 1996). According to Lall (1996), the studies render general support of the IDP analysis, although important deviations exist depending mainly on the pattern and efficacy of government interventions. Since countries show long-term departures from the 'predicted' average path, Lall concludes that the theory may need to be extended and modified to take these subpatterns into account. The latter conclusion brings us back to the four deviations that have already been suggested within the IDP framework. It seems that more research is needed for a more detailed understanding of the existing subpatterns, especially with regard to the influence of the fourth category of factors leading to a deviant IDP pattern.

Stages Theory of Industrial Upgrading and Overseas Investment

An interesting alternative conceptualisation of the relationship between FDI and economic development has recently been put forward by Ozawa (1992, 1995,

1996). His 'Stages Theory of International Upgrading and Overseas Investment' (hereafter: Stages Theory) draws largely from the Japanese experience that '. . . delineates the close interrelatedness between structural upgrading, dynamic comparative advantage and FDI, along paths of its physical/human-capital-intensive factor endowment and technological progress' (Ozawa 1992: 39).[23] Although a general applicability of the concept is not ruled out by the author, Ozawa initially has developed it to explain the economic developments as observed in the East and Southeast Asian region. The basic premises are that the superiority of an advanced nation in industrial structure and technological progress is transmitted to 'lower-echelon' nations and that inward and outward investment play a major role in this explanation. Investment is not limited to conventional direct investments but also includes the 'new forms' of investment as defined by Oman (1984). This is an important extension since in Japan, for instance, the annual payments (royalties, fees) for technology imports were for a long time much larger than the amount of inward direct investment (Ozawa 1996: 160).

Important for Ozawa's analysis is a hierarchy of countries characterised by four sequential stages (or 'tiers') of industrialisation. These four stages are reflected in Table 2.1. Although the examples of the industries are derived from Japan's experience, the various stages of industrial advancement are supposed to be the same for all countries. Or, as Ozawa puts it (1995: 224) Japan shows the rest of Asia 'the image of its own future'.

According to the Stages Theory, an important underlying mechanism that explains shifts from one stage to the other is factor incongruity. Factor incongruity is defined as the incompatibility that appears over time between the factor intensity of a good and the factor endowments of an economy in which the good is produced. This phenomenon occurs under two different circumstances. In the case where the technological nature of the product rapidly becomes standardised (and less human capital intensive) and at the same time the factor endowments of the country change only gradually, the firm which innovated a new product cannot retain its competitive advantage in the home economy. As a result, production has to be transplanted abroad to restore appropriate factor congruity. Such development is basically the scenario described by the Product Cycle model (Vernon 1966). The other case refers to the situation that the factor endowments of a country become more capital abundant and labour scarce so that a product which was compatible with the initial factor endowments cannot be produced in a cost efficient manner any more. In this case, production of these goods also has to be moved abroad. The latter, so-called structural upgrading mechanism, is especially relevant in the case of the Asian economies developing at high speed.

In the first stage, industrialisation is primarily based on labour-intensive manufacturing ('Heckscher–Ohlin') of such goods as textiles (making use of cheap, abundant labour) and/or on the primary sector. This stage only lasts, however, as long as wages are kept low. When unemployment decreases and real wages go up, factor incongruity increases. When, as a result, profit squeeze occurs, labour

Table 2.1 Industrial hierarchy and type of investment

Tier	*Examples*	*Inward investment (sources)*	*Outward investment (sources)*
4 Innovation-intensive ('Schumpeterian')	Biotechnology, optoelectronics, new materials, super-semiconductors	Hosting FDI in chemicals, pharmaceuticals, machinery	Innovation-enhancing, strategic alliances type of overseas investment in R&D intensive, client-interactive and flexible-manufacturing-based industries
3 Assembly-based ('differentiated Smithian')	Automobiles, electronics, machine tools	Continuous adaptation of and improvement on imported technology and stepped-up domestic R&D	Assembly-transplanting, market-securing FDI in such products as automobiles, electronics
2 Heavy and chemical industries ('undifferentiated Smithian')	Steel, shipbuilding, petrochemicals	Adoption and adaptation of advanced foreign technology through licencing and inward FDI (joint ventures) in petrochemicals, machinery, chemicals	Initially resource-seeking FDI, later joined by 'house-cleaning' FDI in resource-extractive and resource-processing industries
1 Labour-intensive light ('Heckscher-Ohlin')	Textiles, toys	Selective imports capital goods and licencing agreements or inward FDI	Low-cost seeking FDI in standardised, labour-intensive light industries in neighbouring countries

Source: Based on Ozawa (1995, 1996)

intensive production will be transplanted to low-wage (often neighbouring) countries (namely, 'natural labour resource seeking' FDI).

As a result, a shift to the second stage will take place, the resource-intensive heavy and chemical ('undifferentiated Smithian') industrialisation phase. In this stage of development, countries become more dependent on overseas resources, especially if it concerns a resource-poor nation. If prices of these resources go up and profit squeeze again (threatens to) occurs, FDI in 'natural resource-seeking' activities takes place to secure a stable overseas supply. Profit squeeze may also occur as a result of increased international competition or an appreciation of the home currency. Ozawa labels the threat that such subsequent profit squeeze leads to a stagnation at the macroeconomic level as 'Ricardian bottlenecks', which, according to the author, can be solved by (among others) overseas investment. Increasing environmental costs may also encourage 'house cleaning' investments in other, less strict countries.

In the third, assembly-based ('differentiated Smithian') stage of industrialisation, competitiveness is based on economies of scale and scope and learning-by-doing with regard to technological progress. Companies increasingly undertake R&D activities. To secure large volumes of production, the maintenance and expansion of market share is crucial. Hence, threats of protectionism in export markets induce market seeking or retaining investment in export markets. Labour-intensive components, parts and accessories of the assembly-based industries at the same time will be shifted to low-labour costs countries. Moreover, Ozawa (1992) argues that 'localised learning' (namely, the situation-specific improvement of imported technologies) turns to 'internationalised learning' if companies start to invest in technology seeking, info-gathering and trade-supportive investment in industrialised economies.

In the fourth, innovation-intensive ('Schumpeterian'), stage, which has not yet been fully developed theoretically, the 'Smithian growth *élan*' (namely, market expansion and organisational and technological efficiency) becomes even more important. In this stage, MNEs actively co-operate to set up joint research, development and engineering ventures as well as marketing arrangements. 'Strategic alliances' play an important role in this stage of development.

Ozawa argues that, until now, in Asia, only Japan has successfully reached the most advanced tier. In graduating from one stage to another, Japan transplanted (through FDI) industrial activities that no longer were factor congruent to Asian newly industrialised economies (NIEs). With increasing factor incongruity in these Asian NIEs, they also started to shift production to countries lower on the industrial ladder. According to Ozawa, this 'intra-Asian tandem investment' explains the rapid upgrading of the countries in the region through the various stages of industry.

At present, the Stages Theory is still in its infancy. Nevertheless, it is an intuitively appealing concept that provides us with important additional insights into the role of FDI in the dynamic process of upgrading and fading out of industries in national economies. Important is also the explicit inclusion of the 'new forms' of investment. As is shown in Table 2.1, different sources are needed in different industrial stages. As Ozawa (1995: 241) has already indicated, however, the role of governments as a facilitator of structural upgrading is still under-exposed. The strong reliance on the Japanese experience may also imply that other country-specific elements (such as the industrial structure) are not yet adequately taken into account. Further theoretical conceptualisation as well as empirical testing are needed to advance this theory.

Characteristics of late industrialisation

Before exploring the extent to which the general MNE theories are able to explain the emergence of multinational enterprises from late industrialising economies (LIEs), we shall first examine some characteristics that the home economies of these companies have in common. The typical features discussed in this section are put forward by the so-called 'Late Industrialisation' (LI) paradigm. This paradigm is part of the 'Revisionist View' that since the second half of the 1980s has emerged as an alternative for the neo-classical explanation of the rapid growth of some (especially East Asian) LIEs.[24] For a long time, the neo-classical interpretation of East Asia's success has dominated (World Bank 1993), arguing that a central role was reserved for the market and that only a minor role was played by governments. Although later explanations were nuanced, the thrust has remained that governments merely provided a relatively stable macroeconomic environment with generally low inflation levels and no considerable appreciation of real exchange rates. According to this view, inputs flow to sectors following static comparative advantage, whereas international competition led to cost discipline and technological upgrading.[25] The trade neutrality advocated by proponents of this view implies that policies should not selectively discriminate among industries but that specialisation of an economy instead should follow from existing comparative advantages. Careful studies of countries (for Korea: see Amsden 1989; and for Taiwan: Wade 1990) have shown, however, that this view seriously lacks factual validity, which led to alternative, revisionist, explanations. A major benefit of the LI paradigm discussed in this section is that it explicitly relates the industrialisation path of LIEs to the development that (leading) companies from these economies underwent – this as opposed to, for instance, Wade's analysis of the role of the government in LIEs, which primarily focuses on the macro level (Wade 1990).

According to the Late Industrialisation paradigm (Amsden 1989, 1991, Amsden and Hikino 1993, Hikino and Amsden 1994), a common characteristic of economies that started to industrialise only this century is that even their leading companies had to grow without the competitive asset of a pioneering technology. This as opposed to the First (Britain – end of the eighteenth century) and the Second Industrial Revolution (especially Germany and the United States – about one hundred years later) where inventions and innovations were the major forces behind industrialisation. In their case, companies became world leaders either because they realised a major invention or innovation or because they were the first to commercialise new products or processes on a large scale. In LIEs 'learning from others' (or 'borrowing' technology) in the beginning was the only input for their technological enhancement. As opposed to 'new trade' and 'new growth' theories that also emphasise the importance of technological development, the LI paradigm refutes the notion that LIEs can freely draw on the 'world stock of knowledge' (Lo 1995). The imperative to industrialise almost exclusively on the basis of learning is responsible for many characteristics that (former) developing countries such as Korea, Taiwan, Brazil, India, Mexico and Turkey have in common; economies that otherwise are very diverse in terms of resource endowments, history and culture.[26] The LI paradigm argues that this learning path has influenced various aspects of the industrialisation of the countries: the role governments play in promoting industrial development; the competitive focus of (leading) companies and firm strategies and their structure.

Role of the government

Also during the First and Second Industrial Revolution governments played an important role in building up an institutional and physical infrastructure. In LIEs, however, their task has been far more widespread and crucial. Given the absence of pioneering products and processes, low wages in most cases seemed to be the most important competitive asset shared by firms in these countries. Studies have shown (e.g. Clark 1989) that – if not already offset by higher productivity levels in other countries – low wage levels are only a temporary advantage. Active government intervention is often observed in these economies. By allocating subsidies to targeted industries and/or firms or by manipulating foreign exchange rates, prices are often deliberately distorted.[27]

We can safely assume that in general the interventionist nature of government policy in LIEs will also affect cross-border investment flows – both inward and outward – thus influencing the shape of the IDP. In case of a protectionist policy, foreign MNEs may be excluded from the domestic economy. In other cases, inward investment is perceived as an important instrument to upgrade a domestic industry. In addition, a government's strict control over foreign exchange reserves may severely restrict the outflow of capital for overseas investment.

Competitive focus of companies

Leading indigenous companies from LIEs typically use mid-level technologies. Frontier technologies developed by leading companies from economically advanced countries are usually not available. In addition, as was argued by the TTC approaches, the incremental nature of technological learning prevents leapfrogging through increased efforts within the firm. Focusing on the bottom end market, on the other hand, would imply fierce competition with firms from low-wage countries. Therefore, it is argued that the only possibility for leading companies from LIEs to compete with companies from early industrialised economies (EIEs) is to continuously sharpen their managerial and organisational skills, shorten their learning period and particularly by realising incremental improvements in the quality of their products and the productivity of their production processes. In other words, their strategic focus is on the shopfloor, and not on R&D activities or the organisation of the exploitation of new technologies.

How this difference in strategic focus between leading companies from EIEs and LIEs affects their O advantages and internationalisation behaviour is an important research question addressed in this study. In the next section, we shall discuss the insights that have been brought forward in the literature.

Firm strategy and structure

The strategic focus on mastering and improving existing products and production processes bears a continuous risk of becoming completely obsolete as a result of the introduction of a revolutionary new product. As far as LIE firms are not capable of innovating around a core technology, they typically protect themselves by diversifying into technologically unrelated mature product markets. This explains the large number of business groups found in LIEs. As will be illustrated in later chapters, Korean conglomerates are among the most well known examples of such a strategy. According to Hikino and Amsden (1994), in Taiwan these conglomerates also would have been more important if the authorities had not been strongly opposed to private economic concentration. The trend of diversification, it is argued, is strengthened by the bandwagon effect found in oligopolistic markets. Companies typically follow each other into new markets to maintain their overall national position. As was pointed out earlier in this chapter, this strategic behaviour was first hypothesised by Knickerbocker (1973).

The way in which the conglomerate structure influences outward direct investment by these companies is not yet very clear. The comparison between the internationalisation strategies of Korean and Taiwanese companies made later on in this study may shed more light on this issue.

The Late Industrialisation paradigm provides very useful additional insights into the unique characteristics of the industrialisation of Korea and Taiwan that are highly relevant for our investigation into outward investment from these economies. Nevertheless, Kang (1995) argues that this perspective needs to be complemented with insights derived from New Institutional Economics, such as those dealing with the importance of: the politics behind economics (i.e. extend the analysis beyond economic outcomes); the historical origin of Korean and Taiwanese capitalism (e.g. the influence of Japan); and the importance of international systems (e.g. security concerns and the influence of the US). In Chapter 3 and especially in Chapter 4 such aspects – as far as they can be related to outward FDI – will be discussed in more detail.

Late industrialisation and outward direct investment: theoretical implications

The question arises to what extent outward direct investment from LIEs can be explained by received MNE theory. We start here with the macro-oriented

Investment Development Path. According to Dunning (1993: 89): '. . . the (IDP) concept . . . is very relevant in explaining the growth of outward investment from Third World countries, especially from Korea, India, Singapore, Hong Kong and Taiwan'. As was pointed out earlier, according to the IDP a relationship exists between (inward and) outward investment and economic development. Although some outward FDI may already take place in stage 2, it is argued that investments abroad especially gain momentum in the third stage (cf. Dunning 1981a, 1981b, 1986, 1988). Investments in economies at a lower stage of the IDP are partly motivated by an erosion of comparative advantages in 'traditional' indigenous industries and are partly market-seeking in nature. Investments in more advanced economies, on the other hand, are market and/or strategic asset seeking. The IDP suggests that while 'real' developing countries (such as India) are in stage 2, newly industrialised economics typically can be found in stage 3. Earlier it was suggested that many countries will divert from the average NOI path. It can be assumed that this will certainly hold with regard to LIEs; for the LI paradigm suggests that the strong role of the government in these economies will affect the *allowance* of (inward as well as) outward FDI flows. LIE authorities also typically interfere with the market structure of certain industries. The deliberate creation of an oligopoly, for instance, leads to the emergence of large-sized firms, which is usually perceived as an important stimulus of outward FDI (Horst 1972b).

Fully in line with the Eclectic Paradigm, the IDP remains indeterminate with regard to the precise nature of the various advantages relevant in this case. It only suggests that in stage 2, O advantages of indigenous firms are gradually developed, whereas in stage 3 these companies have accumulated sufficient O advantages to compete with MNEs that initially invested in relatively less advanced industries. The precise *nature* of the ownership-specific advantages, however, remains unclear. The same holds for the reasons to internalise these O advantages and – to a lesser extent – for the L advantages that apply here.[28] Like the IDP, Ozawa's Stages Theory also does not provide a more detailed explanation for the emergence of outward direct investment from LIEs – especially as far as firm level characteristics are concerned. A deeper insight into this phenomenon therefore requires more specific theories. First, the theoretical discussions that resulted from the 'first wave' of outward direct investment from LIEs are reviewed, followed by the deliberations on the 'second wave'.

The emergence of LIE MNEs: the 'first wave'

The emergence of companies from LIEs that invest abroad (LIE MNEs) for the first time received academic attention in the literature on 'Third World multinationals' published by the end of the 1970s and the 1980s. Notwithstanding some notable exceptions (such as India), most home countries belonged to the NIEs (Lall 1983, Wells 1983). Although the size of these investments was relatively small (e.g. in 1984 a scanty 1.7 per cent of world-wide FDI flows originated in developing economies), this new phenomenon nevertheless received considerable academic attention. Important topics that were dealt with – among others –

concerned the effects of these LIE MNEs on the balance-of-payments of the home countries (cf. Agarwal 1985a); the possibilities LIE firms offered to realise collective self-reliance among developing nations (e.g. Green 1986); and other typical 'developmental effects' such as employment generation, technology transfer, and the appropriateness of goods produced (cf. Agarwal 1985a, ESCAP/UNCTC 1988, Wohlmuth 1986).[29] Moreover, questions were addressed with regard to the theoretical implications of the emergence of LIE MNEs. Since received MNE theory had mainly been developed on the basis of western companies investing abroad, it is no wonder that a debate arose on the applicability of these theories for LIE MNEs.

Wells (1977, 1986) and Vernon (1979) put the shift from export activities to overseas production by LIE companies in the framework of the Product Cycle model. Countries manufacturing a particular product first export it down the 'pecking order' to those countries which are not yet able to produce it themselves. In a later stage, when demand abroad is sufficient and the threat of local competition arises (or, for instance, import restrictions are imposed), production may be shifted to that particular country. While this pattern was limited initially to advanced nations, later, according to Wells, the more advanced developing countries also appeared as exporters of capital to countries further down the line. Probably the most severe competition between LIE MNEs and EIE MNEs (multinational enterprises from early industrialised economies) occurs at the tail of the product cycle (Wells 1986). But what particular advantages did these companies from LIEs possess?

Although a generally accepted explanation could not be found, it appears that during the early stages of outward investment their O advantages resulted from the typical characteristics attributed to companies from LIEs. Usually the O advantages of LIE MNEs were not based on some kind of innovational lead, as is characteristic for EIE MNEs (Agarwal 1985b, Kumar and McLeod 1981, Lecraw 1977, 1981, Wells 1977, 1983). The technologies used by LIE MNEs were considered to be more suitable for the local environment of developing nations. These smaller scaled, less advanced, less expensive, and more labour-intensive technologies most often were the result of an adaptation of an existing innovation or consisted of a renovated technology that had been forgotten or discarded in developed economies. In addition, it was argued that LIE MNEs on a large scale make use of local (instead of imported) inputs that are not only cheaper, but also increase their popularity with local host governments (Lecraw 1977). Other O advantages concerned the lower overhead and lower expatriate costs of LIE MNEs. As opposed to EIE MNEs, these companies generally made very low investments in expensive office buildings and attractive work facilities; they also paid relatively low salaries to their managers and technicians.[30] Furthermore, the familiarity with the conditions and problems of developing nations was often seen as an advantage in comparison with western MNEs (Kumar 1982). A special case are the so-called 'ethnic investments' made in order to serve a local community that is related to a specific ethnic group of the investor's home country.[31] Moreover, the, mostly non-branded, goods produced by LIE MNEs tended to be less advanced and luxurious and more appropriate for local mass markets. In other words, it appeared that these companies exploited particular skills and cost advantages rather than a monopolistic hold over new technology. The foregoing arguments illustrate the O advantages of LIE MNEs in developing markets *vis-à-vis* EIE MNEs. In comparison with local firms in developing countries, Wells (1983) noted that the technologies used by LIE MNEs embody an amount of know-how not readily available to local firms. In addition, one may imagine that LIE MNEs often had a longer corporate tradition (and thus more management experience) than local firms and that, in comparison with local firms, economies of scale frequently occurred.

Lall (1983: 15–16) drew attention to the fact that O advantages may differ considerably between LIE MNEs from different countries. In addition, he distinguished between process industries and consumer goods and engineering industries (Lall 1983: 260–1). His study shows that in many process industries (such as paper and pulp, synthetic fibres, textiles, oil refining, and paint production) LIE MNEs use technologies that are identical to those in use in EIEs. In these cases, only marginal technological advantages can be offered by these firms, for instance following from the ability to use different raw materials. Consumer goods and engineering industries may be amenable to more substantial adaptations. As opposed to the rather static explanations provided by Wells and others, Lall also emphasised the importance of genuine indigenous technological capabilities of the companies involved. Although he acknowledged the relevance of (most of) the O advantages put forward by the other authors,[32] he argued that these are the outcome of a process of conscious accumulation of technological knowledge, which, as we will see in the next section, has important implications for their future potential.

What made LIE firms internalise these O advantages (Agarwal 1985b, Wells 1981)? As we have seen earlier, advantages owned by firms such as skills, know-how or reputation are often intangible. Intangible assets can be transferred abroad without reducing the supply of assets at home. In the case where these advantages are sold to others (e.g. by means of a licence agreement), they can be used by the new owners in various ways and may even be passed on to other enterprises – in which case their value is diminished (Wells 1981). In principle this danger is also present in the case of traditional MNEs and often contractual arrangements can cover many of these risk issues. In developing countries legal systems provide little protection in case of disputes, however. Because of these hazards, LIE MNEs – especially those that have built up considerable expertise themselves – may establish their own affiliates abroad. In addition, potential buyers (namely, local firms) may not have sufficient information about the manufacturing technologies of LIE MNEs because the latter tend to be less well-known than EIE investors. This may lead to a lack of confidence of the potential buyer in the reliability of what is being sold. Establishing a subsidiary again allows the circumvention of this problem. Besides, Agarwal (1985b) argued that local entrepreneurship in host developing countries is often not able or not prepared to take up the complete risk of a new enterprise using the adapted (LIE MNE-) capital because it lacks managerial know-how.[33] In those cases, the exploitation of O advantages is more profitable through internalisation (namely, FDI) than through direct or indirect sale of the advantages to local entrepreneurs.

The L variables that induced LIE MNEs in the early stages not to export but instead invest in other developing countries are for the greater part similar to those relevant for EIE MNEs (Kumar 1982). One of the motives frequently referred to is a threat to existing or potential markets. Import substituting policies and protectionist measures taken by host country governments regularly induce firms to shift activities to countries that introduce or consider such measures. Arguments that are more related to the specific situation of a particular country or firm are: the need to secure reliable future supplies of raw materials and a lack of knowledge about their products in the host markets ('buyer uncertainty'). Other L variables, that are more specifically relevant for LIE MNEs, are the limitations to the growth of the home market with which LIE MNEs are confronted. In addition, some governments restrict the growth of their own, indigenous firms, for instance by antitrust legislation or by the requirement that importing firms earn their own foreign exchange; the latter may be accomplished by the establishment of foreign subsidiaries which earn their foreign exchange locally. Besides, it is suggested by some authors (e.g. Kumar 1982) that their presence in other less developed countries (LDCs) is perceived as less threatening to the host country than the appearance of MNEs from early industrialised economies.[34]

A last remark concerns the limited scope of the foregoing discussion. It should be realised that the initial research into LIE MNEs has focused primarily on market seeking investment in developing countries. Investments meant to set up export platforms elsewhere were not yet very relevant, since low cost labour was still available in the home countries (Wells 1977). The importance of investments abroad to secure raw materials (e.g. investments by Korean and Philippine wood products manufacturers in Indonesian forestry projects; see Wells 1977), were acknowledged but not explicitly included in the theoretical discussions. Moreover, although FDI in developed countries was also noticed, it was perceived as 'fairly exceptional' (Kumar and McLeod 1981, Lall 1983). Therefore, Monkiewicz's (1986) complaint that the phenomenon is much more heterogeneous and complex than portrayed in the literature appears to be justified.

Spread and transformation of LIE MNEs: the 'second wave'

We saw that the first stream of research on LIE MNEs was carried out by the end of the 1970s and the first half of the 1980s. During the second half of the 1980s this subject received much less academic attention. As a result of new developments that have taken place since then, a renewed research interest has manifested itself (e.g. Cantwell and Tolentino 1990, Dunning *et al.* 1997, 1998, Lall 1991a, 1991b, Lecraw 1991, Tolentino 1993, Wells 1993). First, as will be illustrated in Chapter 5, the importance of outward direct investment from LIEs as a whole has increased substantially, both in absolute and in relative terms. Second, outward FDI is facilitated by the much more liberal policy that many LIE home authorities have recently established with regard to capital outflows. During the first wave, government policies towards outward FDI in most developing countries mainly consisted of capital export restrictions to keep investments at home (UNCTAD 1995). Third, as will also be shown in Chapter 5, in terms of their regional origin, a significant shift towards a more pronounced dominance by Asia as a home region has occurred. Fourth, as far as FDI in developing countries is concerned, we observe that it is no longer (almost) entirely market seeking in nature. Steep rises in real wages and the appreciation of local currencies, among others, has caused deterioration in the L conditions, especially in the richer NIEs. As a result, an important part of the more labour-intensive production activities was moved to cheaper locations from where the goods are exported to other regions in the world. Fifth and most interesting, an investigation into the destination of FDI outflows from Asian LIEs (van Hoesel 1992) revealed that the regional pattern has changed a lot since scholars observed that most LIE MNEs invested predominantly in developing countries. Especially for the more advanced newly industrialised economies, such as Korea and Taiwan, the developed economies have become important host countries. (In Chapter 5, the exact size and composition of FDI flows from both economies will be investigated in more detail.) These so-called 'upward investments' gave rise to new deliberations. What advantages may be relevant in terms of the Eclectic Paradigm? The O advantages used to explain the operations of LIE MNEs in the developing world do not seem to be relevant in developed host economies. On top of that, the lack of knowledge about the conditions in the newly penetrated developed markets is even an outright disadvantage for these LIE firms.

As far as I advantages are concerned, the risks of proprietary knowledge being spread further than agreed upon with the transferee because of failing protective legal systems, appears much less relevant in developed host countries than in LDCs. Furthermore, it is unlikely that a lack of managerial know-how in developed countries (DCs) will be a motive for LIE MNEs to internalise their O advantages instead of selling them. Nevertheless, a lack of information about the manufacturing technologies owned by LIE firms and a general distrust of knowledge which comes from, or at least, has been modified in LDCs may also be present in DCs. In addition, the unwillingness to share know-how with others may also stimulate LIE MNEs to internalise their assets.

Low wage levels constitute one of the most important L advantages attributed to operations in developing countries – a condition that obviously does not apply to industrialised nations. In addition, the discrepancies in cultural and economic environment ('psychic distance') between most developing and developed countries can be quite high. The negative perception in host countries about unknown foreign firms intruding in their (often increasingly protectionist) markets may form another L disadvantage. In principle, however, the higher labour cost in DCs and other L disadvantages may (at least partly) be offset by advantages that the presence in advanced economies may offer, such as higher productivity, access to large markets with high purchasing power and the new opportunities (e.g. in terms of technology accumulation) that will emerge from being present in such markets.

Directly related to the inquiry into the precise nature of O advantages is the analysis of how the firms involved can acquire competitive advantages needed to invest in industrialised economies. According to the Product Cycle model, the O advantages of LIE MNEs are based on a rather passive absorption of mature technologies from EIEs that are adapted to local circumstances to produce in a more labour-intensive, small-scale manner. According to the concept, the – mainly undifferentiated – products are sold in countries, where indigenous firms have not yet been able to master these technologies. In following the Product Cycle model, the future of LIE MNEs would not look very bright, since, as Wells (1982: 40), put it '. . . the market advantages that . . . these firms have are too tenuous for long-term survival. Local businesses will eventually learn to do the same as they do'. But more importantly, these kind of advantages cannot explain the operations of LIE MNEs in developed economies, especially with regard to the manufacturing of differentiated products with more advanced scale and technological requirements (such as consumer electronics and automobiles; see also Lall 1991a). This rather pessimistic attitude appears to be shared by the proponents of the Late Industrialisation paradigm. Hikino and Amsden (1994) observe that especially in the case of foreign markets in which technologically advanced domestic firms are dominant, LIE firms typically rely on exports and delay (greenfield) overseas production. The reason is that their O advantages (based on incremental process and product improvements) are often too intangible to transfer overseas easily. Only in the case of trade barriers, may these companies be provoked into starting overseas operations as a defensive measure. A typical LIE strategy that does occur, however, is the establishment of overseas R&D laboratories to scan advanced technologies or the acquisition of ailing high tech companies in advanced economies. In other words, both the PC model and the LI paradigm do not offer adequate explanations for *greenfield* operations of LIE MNEs in the industrialised world; the latter does, however, acknowledge the importance of mergers and acquisitions (M&A) strategies in developed nations.

As compared to the foregoing concepts, from a TTC perspective the future of LIE MNEs looks much brighter. As we have seen earlier, technological competencies result from a cumulative process in which companies build upon their existing and past achievements. The explanations of the operations of LIE MNEs in other LDCs that follow from the PC model and the TTC approach are for an important part similar (namely, based on mastering of existing technological know-how with some adaptations) (Lall 1991a). Yet, important differences occur with respect to the more advanced activities of LIE MNEs in developed countries. Contrary to the Product Cycle model, the TTC perspective asserts that mastering technologies is not a passive process of absorbing mature technologies that will be eroded over time. Although these theories argue that technological leapfrogging based on own efforts is not possible, it is however assumed that at some stage LIE companies will have accumulated sufficient technological capabilities to compete with firms from EIEs, eventually also in the latter's home economies.

Cantwell and Tolentino (1990) and Tolentino (1993) found a relation between the stage of technological capability of LIE MNEs and the nature and direction of their FDI activities. First, as domestic firms in LIEs upgrade their domestic production activities, resource-based and less technologically sophisticated activities are transferred to countries that are at an early stage of development and have lower levels of technological capabilities and production costs. As learning-by-doing and production experience accumulates, firms get increasingly involved in their own R&D thereby generating new or adapted product designs, process technology and machinery. During this period, more efforts from the individual firm are required, together with more support from national institutions in matters of finance, technology and marketing. In these later stages, more sophisticated manufacturing activities (consisting of the production of higher quality, more research-intensive products and an increased involvement in product differentiation strategies) are transferred abroad, even to developed countries. These investments in DCs increasingly take place in order to gain access to the more sophisticated and advanced forms of manufacturing technologies used there. It is expected that these R&D investments will result in an upgrading of the industrial structure in the home base, i.e. a changeover from production based on unskilled, intensive labour towards higher skilled and more research-intensive activities.

An important implication of this explanation is that it shows that – following a path of accumulating knowledge and experience – LIE firms can build up considerable O advantages of which the opportunities for exploitation are not restricted to their activities in LDCs. Unfortunately, no distinction is made between the various sources of technology accumulation (e.g. own R&D, licensing agreements, M&A) and their impact on the speed of the learning process. Another shortcoming of this approach relates to the process of internationalisation itself. Accumulation of knowledge and experience is not only needed to build up technological capability, but also for the internationalisation process the company undergoes – a notion emphasised by the Scandinavian school. Although Cantwell and Tolentino (1990) do link TTC with the 'complexity' of FDI, the latter is perceived as a result of the increasing technological capability itself. Although both are certainly interrelated, internationalisation of firms (namely, LIE MNEs) can be considered as a separate process which, as with the accumulation of know-how, is characterised by a process of accumulating knowledge and experience with relation to activities abroad. Even if the latter notion was to be incorporated into the present concepts of TTC, the approach still does not reveal which specific O advantages enable LIE MNEs to compete with Western firms on the latter's home ground; the concept remains rather vague and does not elaborate on the outcomes of the (accumulation) mechanisms at work here. More primary data are seriously needed here.

Discussion and research questions

The aim of this chapter has been twofold. First, we discussed the most important theories that may contribute to a deeper understanding of the emergence of new multinational enterprises from Korea and Taiwan. Our main sources have been MNE theories (dealing with the various levels of analysis) and the Late

Industrialisation paradigm. Second, we looked for the relation between the two: Is it possible to derive from received theories in what ways the internationalisation pattern of Korean and Taiwanese firms (especially their outward investment) is influenced by the fact that they originate in late industrialising economies?

In general, we observe that a considerable overlap exists between the theoretical concepts discussed in this chapter. For instance, the Stages Theory put forward by Ozawa regarding economic development and FDI shows important similarities with Dunning's Investment Development Path; an assessment that is confirmed by Ozawa (1992: 36). The same holds for the TTC approach as used by Cantwell and Tolentino to 'explain' outward direct investment. On the (partial) overlap of competing concepts Dunning (1988: 26) notes the following: 'It is one of the irritating characteristics of academic researchers (and the present author is no exception!) for them to invent their own nomenclature for concepts that are familiar under different names to other researchers'. The existence of these distinct concepts nevertheless seems to be justified by their, in part, different focus. Ozawa's Stages Theory brings the interaction between trade and international production to the fore. In addition, he emphasises more than the IDP, the changes that take place at the industrial level. The Technology and Technological Change perspective is primarily based on a firm level analysis of the growth of companies, an approach that – at least to some extent – has been neglected by 'conventional', more static, MNE theories. Moreover, this concept perceives technological capabilities as the sole determinant of growth of the firm, including its investments abroad. One of the most interesting characteristics of the Investment Development Path is the direct link with the Eclectic Paradigm of International Production connecting a macro economic phenomenon like national FDI flows to micro level variables. Typical of the IDP is also the notion of the 'net outward investment' position used as the dependent variable.

It appears that a combination of the various concepts would enrich our insights into the phenomena studied. The process of technology accumulation as described by the Technology and Technological Change theories perhaps could be integrated into the IDP. Moreover, an analysis of NOI patterns at the industry level as emphasised by Ozawa should be illuminating. The same may be true for the element of accumulating experience in doing business abroad as brought forward by the Scandinavian internationalisation school. If such notions were to be integrated in a formal way, however, a more fundamental theoretical problem might occur. For these concepts do not all share the same theoretical assumptions. The IDP and the Stages Theory, for instance, clearly show traces of traditional (neo-classical) trade theory, whereas the Late Industrialisation paradigm as well as the Technological Accumulation approach are primarily based on insights derived from institutional and evolutionary economics. Yet they all provide us with building blocks for a more comprehensive theoretical explanation of the emergence and operations of LIE MNEs. What is seriously lacking at present, are new empirical findings that would enable us to make theoretical statements and hypotheses more concrete.

From the discussion of the theories in this chapter it has become clear that there still exist considerable knowledge gaps regarding LIE MNEs. As a guideline for the collection and analysis of empirical material, the following research questions will be addressed in particular in this study:

What is the nature of the O advantages possessed by LIE MNEs as far as FDI in industrialised economies is concerned?

According to the Eclectic Paradigm, companies investing abroad need certain O advantages to overcome the disadvantages of being a non-indigenous firm. We have noted that the O advantages of LIE MNEs that are relevant in LDCs, do not hold for their operations in industrialised nations. According to the LI paradigm, leading companies from these countries have historically focused on shopfloor operations and not on innovational leads. Given the fact that an increasing number of companies from LIEs invest in industrialised nations, the question arises whether LIE MNEs nevertheless have been able to accumulate O advantages similar to those of EIE MNEs or have developed a different set of advantages. If the latter is the case, the following question is of particular relevance:

To what extent can the O advantages possessed by LIE MNEs be related to the typical Late Industrialisation nature of their home economies?

The aim of this question is to find out whether the ownership-specific advantages accumulated by LIE firms typically match the industrialisation pattern of their home economies – a hypothesis suggested by the LI paradigm.

Closely related is the following question:

Are strong technological capabilities necessary and sufficient to invest in industrialised economies?

In most theoretical concepts discussed in this chapter, and especially in the LI paradigm and the TTC approach, the crucial role of technology is underlined. Is this strong focus on technological capabilities justified or should other important explanatory variables be taken into account as well? An example of the latter is the possible influence of the firm (conglomerate) structure on outward investment.

What is the role of the home country government in building up the O advantages needed to be active in DCs by means of FDI?

As is suggested by the theories discussed in this chapter, home country governments play an important role in the accumulation of the O advantages LIE firms need to invest abroad. The LI paradigm indicates that their influence is even stronger in LIEs than in EIEs and that LIE firms initially primarily derive their strength from L advantages created by the authorities. Examples mentioned, for instance, by the Investment Development Path or the Stages Theory vary widely ranging from the establishment of national R&D institutions (contributing to the technological capabilities of individual firms) to industry level policies admitting only a limited number of companies (indirectly influencing the structure of firms).

What is the importance of accumulating experience in going abroad for the activities of LIE MNEs?

LIEs have been characterised by a rapid transformation from an agricultural or raw-material-based to an industrialised economy and, consequently, a fast upgrading of various industries. The companies themselves have accordingly developed at a rapid pace. As was suggested by the Scandinavian school, internationalisation, however, is a stepwise learning process that is difficult to speed up. The question therefore arises whether and how the incremental character of this process itself shapes the internationalisation strategies of LIE MNEs.

What is the importance of 'reverse technology transfer' for LIE MNEs in industrialised economies?

The TTC approach, the Investment Development Path and Ozawa's Stages Theory all point to the importance of investments by LIE MNEs in industrialised nations that are driven by the desire to tap into technological know-how that is locally available. At present our insight into the importance of 'strategic asset seeking' or 'internationalised learning' for LIE MNEs in DCs is still very limited.

What is the role of alliance capitalism in the internationalisation of LIE companies?

As was indicated, strategic alliances have rapidly gained importance in global business. The question therefore emerges to what extent this development is of relevance to the emergence of LIE MNEs.

Although this study pays particular attention to the questions formulated in the foregoing, they primarily serve as a guideline. The aim is not to render generalisable answers for all of them. That is not feasible because of the exploratory nature of most questions, which requires a case study research strategy. Consequently, in some cases the end result will be the sharpening of questions and the formulation of more concrete hypotheses.

Appendix

Appendix 2.1 Types of international production: some determining factors in the case of MNEs from EIEs

Types of international production	*O advantages*	*L advantages*	*I advantages*	*Strategic goals*	*Illustration of activities that favour MNEs*
Natural resource seeking	Capital, technology, market access, size	Possession of natural resources, transport and communication infrastructure, tax and other incentives	Ensure stability of supplies at right price, control markets	Gain privileged access to resources *vis-à-vis* competitors	(a) Oil, copper, bananas, hotels (b) Export processing, labour-intensive products
Market seeking	Capital, technology, management skills, surplus R&D, economies of scale, brand image	Material and labour costs, market size, government policy (e.g. import controls)	Wish to reduce transaction or information costs, buyer ignorance, or uncertainty	Protect existing markets, respond to behaviour of competitors, preclude (potential) rivals	Computers, pharmaceuticals, cigarettes, airline services
Efficiency seeking (a) products (b) processes	As above, but also market access, geographical diversification, international input sourcing	(a) Economies of product specialisation and concentration (b) Low labour costs, incentives to local production by host governments	(a) As for second category plus gains from economies of common governance (b) Economies of vertical integration and horizontal diversification	As part of regional or global product rationalisation and/or gain advantages of process specialisation	(a) Motor vehicles, electrical appliances, business services (b) Consumer electronics, textiles and clothing, cameras, pharmaceuticals
Strategic asset seeking	Any of the first three offering synergy with existing assets	Any of the first three offering technology, markets and other assets in which firms are deficient	Economies of common governance, improved competitive or strategic advantage, reduce or spread risks	Strengthen global innovatory or production competitiveness, gain new product lines or markets	Industries with high ratio of fixed to overhead costs and that offer substantial economies of scale or synergy

Source: Dunning (1993)

3 High-speed economic development of South Korea and Taiwan

An overview

Introduction

It is not without reason that the economic success stories of Korea and Taiwan are so often cited and investigated. Indeed, only very few countries in the world have been able to come up with such remarkable records, that are particularly extraordinary if the short period needed to accomplish them is taken into account. Being aware of the absence of one single indicator that captures the diversity of the overall welfare situation of a country, a number of frequently used indicators are presented in Table 3.1. It convincingly illustrates the socio-economic achievements of both Korea and Taiwan.[1]

The countries in the table are listed in four categories. In the first category, the four Asian newly industrialised economies (NIEs) can be found – next to Korea and Taiwan are the two city states of Singapore and Hong Kong. The latter two economies already had a substantially higher welfare level in 1960.[2] Nevertheless, they also have progressed in a very fast manner and by 1994 had reached welfare levels comparable to those of many early industrialised economies. It should be noted, however, that the character of their development process shows important deviations from that of Korea and Taiwan. In the case of Singapore and Hong Kong one cannot speak of a transformation process from a predominantly agricultural or raw material base, a pattern that is usually associated with the 'industrialisation' process (Amsden 1991).

The second category consists of the Asian so-called 'second tier' NIEs. Although these countries have also performed quite well, they, especially the Philippines and Indonesia, clearly still lag considerably behind the Asian NIEs. What is remarkable is that the welfare level (measured in per capita GNP) in 1960 was much higher in the Philippines and Malaysia than in Korea and Taiwan.

The third category of countries belong to the group of Latin American NIEs. The table shows that in 1960 Argentina, Brazil and Mexico possessed a higher per capita GNP level than Korea and Taiwan. In 1994 this position had changed completely. In that year, only Argentina realised a welfare level comparable to that of Korea, the 'poorer' of the two countries studied here; Brazil and Mexico had fallen considerably behind.

Table 3.1 Socio-economic indicators of Korea and Taiwan vis-à-vis other countries

	GNP/per capita (current US$)		*Population (million)*		*Life expectancy at birth*		*Adult literacy rate*	
	1960	*1994*	*1960*	*1994*	*1960*	*1994*	*1960*	*1994*
Korea	156	8,508	25.0	44.5	54	72	71	98
Taiwan	141	11,597	11.2	21.1	64	75	54	94
Hong Kong	322	21,681	3.1	6.0	66	79	71	92
Singapore	437	23,435	1.6	3.3	65	77	69	91
Malaysia	275	3,480	8.1	19.7	53	71	53	83
Thailand	97	2,382	26.4	57.8	52	70	68	94
Philippines	251	950	27.6	66.4	53	67	72	94
Indonesia	70	886	96.2	194.5	41	64	39	83
Argentina	589	8,110	20.6	34.3	65	72	91	96
Brazil	337	2,970	72.6	156.9	57	66	61	83
Mexico	329	4,180	36.9	89.6	58	72	65	89
India	72	320	442.3	913.5	43	61	28	51
Pakistan	81	430	50.0	132.7	44	62	15	37
Colombia	268	1,670	15.9	35.2	53	70	63	91
Egypt	160	720	27.8	60.9	46	64	26	51
Turkey	349	2,500	27.5	59.9	51	68	38	82

Sources: CEPD (various issues), UNDP (1997), World Bank (various issues)

The countries in the fourth category are not usually labelled as 'new industrialised economies' or 'second tier NIEs', but at the beginning of the 1960s had welfare levels comparable to those of Korea and Taiwan. Three decades later the welfare gap has become tremendous. This even holds for the best performer in this category, Turkey.

Although less pronounced, a gap also exists between the Asian NIEs and the other countries with regard to the social indicators (namely life expectancy at birth and adult literacy rate) presented in the table. This holds especially for countries in the fourth category, which show considerably lower literacy rates and life expectancies at birth – India, Pakistan and Egypt.

By all these standards, the performance of Korea and Taiwan since 1960 has been quite exceptional among the developing countries. In the remainder of this chapter, we will look more closely at the characteristics of their development patterns. The emphasis is on the post-World War II period, when the transformation of both economies occurred at an unprecedented pace. First, an overview of the economic development of Korea will be given, followed by Taiwan. This chapter ends with some conclusions regarding the nature of the transformation of the Korean and Taiwanese economy.

South Korea

Japanese colonisation

Although officially Korea was a Japanese colony from 1910–45, Japan's influence was already felt much earlier. In 1876 the presence of a Japanese fleet forced the Koreans to sign a commercial treaty, which – *inter alia* – opened up their ports.[3] Japan's supremacy in the region was increased by its victories in the Sino-Japanese War in 1895 and the Russo-Japanese War in 1905. In 1905 Japanese primacy over Korea was officially acknowledged and with the fall of the Chosun Kingdom in 1910 the country was formally annexed by Japan. Before the official annexation, however, Japan had already invested in a railway network, power generation, extractive industries and in the Korean banking system.

The role of Japan during the colonisation of Korea clearly changed over time. From 1910–20 Japan focused on setting up an administrative and legislative framework and on developing the agricultural sector. The latter was done mainly to cater for the needs of the Japanese home country.[4] During this first decade of Japanese rule, agricultural exports (especially rice) made up more than two-thirds of total Korean exports, while the remainder primarily consisted of raw materials. Industrial activities took place only on a very limited scale and were even discouraged by the 'Corporation Law' enacted in 1911 (Kuznets 1977).[5] Nevertheless, during this period Korea's infrastructure was improved considerably: the railway system was extended substantially, the port facilities were improved and a communication network was developed.

The 1920s showed a steady expansion of manufacturing ventures (especially in foodstuffs and textiles) in Korea, which was stimulated by the repeal of the Corporation Law in 1920. The share of manufacturing in total net commodity product (NCP) increased from 7 per cent in 1910–12 to 22 per cent in 1929–31 (annual averages) (Kuznets 1977). The emergence of entrepreneurship was accompanied by a shift away from handicraft and cottage industries. In 1926–28 57 per cent of factory output was in food and textiles, while one quarter originated in 'new' industries such as chemicals and metals.[6]

During the third decade of colonisation (1930–40) the importance of manufacturing in the Korean economy further grew to 29 per cent in 1939–41.[7] The emphasis in the manufacturing sector, however, clearly shifted from light to heavy industrial activities such as chemicals (especially fertilisers), metal production and hydro-electric power generation. In 1930 heavy industry accounted for 23 per cent of total factory output; in 1940 this share had gone up to 50 per cent. Most of these activities were carried out by Japanese financial-industrial conglomerates ('zaibatsu') which increasingly invested in Korea. These companies were 'pushed abroad' because of rising wage levels at home and the strict regulation of competition in Japan which the government had initiated after the Depression. To attract Japanese businesses the colonial government offered various incentives, such as low taxes, guaranteed dividends in selected industries and subsidies (Ho 1984). In addition, the abolition of trade barriers further integrated the Japanese and Korean economy. After the outbreak of the Sino-Japanese War in 1937 Korea's industry was also utilised for Japan's war efforts, especially during the 1940–45 period. This, among others, meant increased government control, a further squeezing out of agriculture and even more emphasis on heavy industry.

During Japan's rule over Korea, trade for the first time became an important engine of economic growth. From 1911–15 to 1936–38, exports grew from a relatively small base nearly seventeen times, while imports grew more than five times. Next to this steep rise of export volume, the composition of exports also changed significantly during this period (Ho 1984). The share of foodstuffs in total exports decreased from an annual average of 72 per cent in 1911–15 to 49 per cent in 1936–38, while the importance of manufactures increased from only 6 per cent to 18 per cent in the same period. Moreover, as a result of Korea's abundant minerals, the share of industrial raw materials in total exports grew from 22 per cent to 32 per cent in this period.

Making a sound evaluation of colonial rule is an impossible task. In many respects the Japanese dominated Korean society. Although economic growth had never been so high (3.1–3.5 per cent annually from 1910–40), it was characterised by an enclave structure of the economy (Jones and Il 1980). Almost all larger factories were Japanese owned and produced primarily for the Japanese market. In 1941, for instance, 59 per cent of all factories and 91 per cent of total paid-in capital was in Japanese hands. Most managerial positions were also held by Japanese.[8] In addition, the Japanese controlled the transportation and communication infrastructure in Korea.

Nevertheless, during the occupation death rates dropped (from 35 per 1,000/year in 1910 to 23 per 1,000/year in 1935–40) and the health situation improved considerably. Also, an increasing number of Korean children received some (mostly technical and primary) education.[9] Real wages also went up considerably. After the defeat of Japan in 1945 Korea was left with about 2,500 functioning companies and a well developed physical and communication infrastructure. In addition, the Japanese had introduced modern technologies and management skills. And although most crucial positions were held by Japanese, the number of Korean workers employed in industry increased from 23,000 in 1910 to 440,000 in 1940 (Jones and Sakong 1980). Next to the exposure to modern technology and organisation, Das (1992: 17) notes that 'Exploitation and assumption of superiority created indignation and a firm determination to catch up with the Japanese.' According to Das, 'Many observers of the Korean scene take this as an important source of motivation for the subsequent economic growth'.

Turmoil

Korean euphoria about its regained independence did not last long. Although the 'industrial heritage' of the Japanese occupation may have caused optimism in the beginning, soon a number of serious setbacks took place which unsettled the economy completely. The end of the war resulted in a partition of the Korean peninsula. North of the 38th parallel the former Soviet Union took over control while in the South an American Military Government (AMG) was installed. A direct consequence of this partition was a disruption of the economic complementarity that had existed between the North and South. The northern part of Korea produced mining products, heavy manufacturing goods and most electricity, while the South had only limited mineral deposits but owned most arable land (especially paddy fields) and produced light industrial goods (Michell 1988). Furthermore, many factories were stripped from their assets, first by the Japanese and later by the AMG. Moreover, most well trained personnel and managers left with the Japanese. In addition, the Japanese and Manchurian markets in which Korea had had a comparative advantage because of subsidies and favourable trade tariffs as a result of colonial policy, were lost. These factors let to a serious drop in industrial output. At the same time, Korean refugees and emigrants were repatriated to the peninsula.[10] Instability was further fuelled by serious food shortages and high inflation rates. Although the US authorities attempted to get the Korean economy going again,[11] the drop to extremely low levels of industrial and agriculture output could not be reversed. In 1948 industrial output was still only 14 per cent of the 1938 level (Jones and Sakong 1980). Disagreement between the AMG and the Korean authorities about economic policy further aggravated the situation. More importantly, tensions with North Korea rapidly worsened and in June 1950 resulted in the outbreak of the Korean War.

Although the actual war lasted only one year (after this, a standstill was reached which lasted until 1953) its effects were devastating. Estimates suggest that there were about 1 million civilian and some 400,000 UN casualties (Kuznets 1977). Besides human suffering, the destruction of the industrial infrastructure was tremendous: it is estimated that this amounted to Korea's 1953 GDP. Indirectly, the war also had long-term implications such as the primacy of the military in the new Republic of Korea and the sizeable share of GDP that was to be spent on the military in the future.

Recovery and import substitution

In the mid-1950s South Korea was still recovering from the partition and the destruction due to the Korean War. In 1953, the economy was primarily agriculture based: 49 per cent of GDP was realised by the agricultural sector while industry contributed only 9 per cent (Michell 1988). The emphasis during this period was on the reconstruction of the infrastructure and industrial facilities.

Like many other developing nations at that time Korea followed an import substitution (IS) policy, after the War, focusing primarily on the domestic market; however, this move was not only inspired by ideological motives. As noted earlier, the independence from Japan had meant the loss of the Japanese and Manchurian market, by far the most important export destinations for Korean goods in the past. During the IS period, final consumer non-durables and other light industry, in particular, were heavily protected.[12] During the 1950s the exchange rate of the Korean won was highly overvalued because of a shortage of essential goods and associated inflation. As a result of serious balance-of-payment problems a complex system of multiple exchange rates was developed which further hampered export growth.

Although the recovery from the War was complete by 1958, economic growth remained rather low until 1962. The average annual per capita real GNP growth during 1953–62 was only 1.7 per cent. Exports were still of very limited importance to the Korean economy: as a percentage of GNP their value in 1962 amounted to only 2.5 per cent. Raw materials (iron ore, tungsten, raw silk, graphite) represented 36.5 per cent of total exports, followed by fish products (9.0 per cent) and agricultural products (6.3 per cent).

During this period Korea was strongly dependent on foreign aid. Estimates indicate that about US$ 2 billion (amounting to some 8 per cent of GNP) was poured into the Korea economy in the 1950s, mostly being official US assistance. According to Das (1992), a substantial part of the assistance came in the form of either non-essential consumer goods or US agricultural surpluses. Foreign aid financed no less than 70 per cent of Korea's imports during 1953–60. By the beginning of the 1960s balance-of-payment problems had not disappeared, however, and it became clear that the ceiling of first stage import substitution had been reached. A second IS stage replacing, in addition, imports of consumer durables, machines, and intermediates did not look promising because of unsuitable factor endowments and the limited size of the home market. Expansion of agricultural exports also was not an attractive alternative because of Korea's limited arable land. Moreover, international prices were depressed by large American agricultural surpluses (Kuznets 1977). Furthermore, foreign aid started to decrease by the end of the 1950s (Cho 1994). These considerations more or less automatically led to a new policy of export oriented (EO) industrialisation.

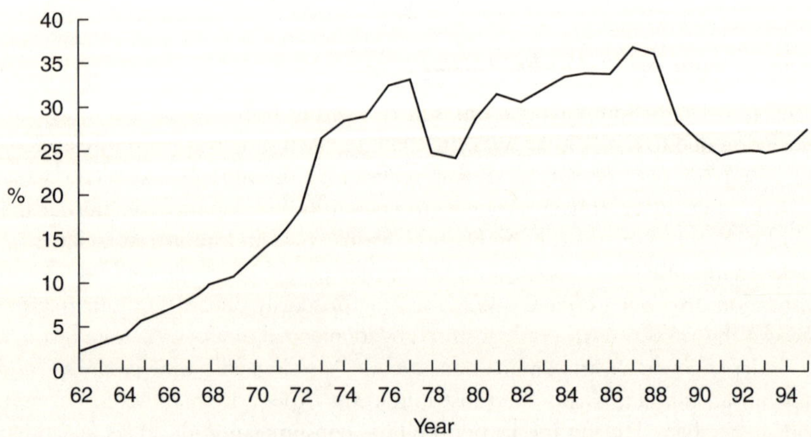

Figure 3.1 Korean exports as a share of GNP
Source: NSO (various issues)

Export orientation: take-off

During this period important institutional changes were carried out that have unmistakably shaped the later development of Korea (Haggard 1988). Characteristic of this transformation was the strengthening of the state. New instruments of economic control and planning were introduced, such as the Economic Planning Board (EPB) from which most policy reforms originated.[13] In addition, as will be discussed in Chapter 4, the banking sector was nationalised.

The new focus on EO industrialisation did not imply that import substitution was thrown overboard completely . The Korean government continued to selectively promote import substitution in cases where an efficient scale of operations was realised.[14] Nevertheless, in the first Five-Year Plan launched in 1962, export promotion was promulgated as the central premise of economic policy.[15] An extensive package of policy measures was advanced step-by-step to promote export performance (Choi and Lee 1990). A uniform (devalued) exchange rate was introduced in 1964 pegging the Korean won to the US dollar.[16] Later, the won was periodically devalued to neutralise its tendency to overvaluation. Exporters obtained tariff exemptions on imported intermediate inputs used in export production (1959) and indirect tax exemptions on imported and domestically purchased intermediate inputs (1962). Furthermore, indirect taxes on export sales were abolished. In addition, income taxes on foreign exchange earnings were reduced by 50 per cent. In 1965 the export incentive schemes were extended to producers who supplied intermediate inputs to exporters (Linnemann *et al.* 1987). From 1966 onwards the imports of capital goods and equipment utilised for export production were also exempted from tariffs. Other incentives introduced to relieve the burden of companies involved in exports were: accelerated depreciation schemes, loans against preferential rates (mostly lower than the market rate), reduced prices for electricity and railway transportation and the wastage allowance subsidy.[17] The government also participated more actively in the promotion of exports, for instance by setting export targets for individual firms by product and regional destination.

Next to production for export, basic industries (such as steel, fertilisers, cement, oil refining and electricity) were built up during the first two Five-Year Plans. In order to mobilise domestic savings the nominal interest rate was raised to 30 per cent. The Foreign Capital Inducement Act of 1966 encouraged private enterprises to borrow abroad by providing guarantees. In this way the gap between the domestic savings rate (14.6 per cent) and the investment to GNP ratio (23.4 per cent) could be closed. Until 1965, when the diplomatic relations with Japan were normalised, hardly any FDI had flowed into Korea (Sakong 1993). Also in the second half of the 1960s, inward FDI remained quite low. Between 1962 and 1971 a total of US$266 million (on approval basis) was invested, which is a modest level (Das 1992).

That the new policy in the 1960s resulted in a substantial increase in the importance of exports for the Korean economy is clearly illustrated by its share of GNP going up from 2.5 per cent in 1962 to 13.3 per cent in 1970 (see Figure 3.1).

The composition of these exports had changed completely in this period. In 1970, textiles and garments represented 40.8 per cent of total exports, followed by plywood (11.0 per cent) and wigs (10.8 per cent). The importance of iron ore had gone down to 5.9 per cent whereas electronic products had captured 3.5 per cent of all exports. The success of export production, the improved infrastructural facilities and emergence of basic industries boosted economic growth which reached an average 6.7 per cent per year (in per capita real GNP) between 1962–72.

The heavy and chemical industry drive

Industrialisation in the 1970s was characterised by a so-called 'Heavy and Chemical Industries (HCI)' drive, the central theme of the third Five-Year Plan. In 1973 the 'Heavy and Chemical Industry Declaration' was proclaimed, designed to improve heavy engineering and petrochemical processing industries, such as iron and steel, non-ferrous metals, shipbuilding, industrial machinery, electronics, and petrochemicals. This change in policy implied a replacement of the earlier industry-neutral incentives for industry-specific and sometimes even firm-specific measures (see also Chapter 4). What were the reasons behind this policy shift? First of all, the international economic (trade) environment had become less favourable (Das 1992). Protectionism in industrialised countries, the oil crises and a world-wide slump again deteriorated Korea's balance-of-payments which led to renewed emphasis on import substitution in intermediate goods and heavy manufactures. Because of increased competition from other developing countries and domestic real wage rises, the traditional comparative advantage of Korean exports, based on labour-intensive production, started to erode (OECD 1994). This underlined the need to transform the economy towards more skill and knowledge intensive production activities. In addition, it was assumed that labour productivity in HCI would be higher than in light manufacturing. Moreover, it was felt that the time was ripe for a 'deepening' of Korea's industrial structure. Next to economic motives, political considerations also played a major role in the reorientation of industrialisation. Improved American–Chinese relationships and Nixon's remarks regarding a possible withdrawal of US troops intensified Korea's efforts to gain (economic) independence.

Since the government was convinced that the investments and risks of the intended industrial transformation could not be borne by the private sector alone, it reserved an important, if not dominant, role for itself. A major instrument was the establishment of the National Investment Fund in 1974, which lent no less than two-thirds of its portfolio to the HCIs. Since HCIs were designated as priority sectors they also received a major share of regular bank finances, often against preferential (negative real) interest rates. External borrowing was also supported (Das 1992). As a result, the debt–GNP ratio rose from 25 per cent during the 1960s to 40.5 per cent in 1975. Other means to promote HCI were: protection under the label of infant industries by means of prohibitive tariff and other import restrictions and tax benefits through exemptions and reduction of corporate taxes (Choi and Lee 1990).

Although most emphasis during the third Five-Year Plan was on the HCI drive, some other new developments should not go unnoticed (Linnemann *et al.* 1987). The EPFZ (export processing free zone) programme was introduced, intended to attract foreign direct investors. EPFZs were established in Massan in 1970 and Iri in 1973. These zones were explicitly developed to attract foreign direct investment in export industries. This implied a departure from the 1960s policy, during which exports were almost exclusively produced by domestic firms. FDI in the export sector – in and outside EPFZs – was disproportionately directed to electronics (see also Chapter 7), followed by textiles and clothing. Between 1972 and 1981 a total US$1.6 billion of foreign investment in Korea was approved. Although the share of FDI in fixed capital formation reached its post-war peak in the second half of the 1970s (2.4 per cent), this was still rather low.

Another new instrument to promote exports was the introduction in 1975 of the so-called 'General Trading Companies'. As will be discussed in more detail in the next chapter, the government started to promote the grouping of companies with their own trading houses that would function as purchasing and sales agents for their (and others') products.

The HCI drive was realised against enormous costs (Sakong 1993). The aforementioned policy had been an important source for inflation which went up from an average yearly 11.1 per cent between 1965–73 to 19.3 per cent between 1973–80 – adversely influencing the effective exchange rate of the won. The economic power of a restricted number of major corporations increased further because of the preferential credit and subsidies they received. In addition, the disproportionately large share of funds allocated at HCIs implied that other goals, such as the development of smaller and medium-sized enterprises and the improvement of the social infrastructure received less attention. Although nobody will deny the importance of a restructuring of the Korean economy, it is felt that the approach and implementation of the policy had been too ambitious and had caused too many undesirable side-effects (cf. Das 1992).

Between 1973–80, per capita real GNP nevertheless grew 6.9 per cent per year. Notwithstanding the heavy emphasis on the promotion of heavy and chemical industries,[18] the importance of exports for the Korean economy continued to grow. In 1980, the value of exports amounted to 28.6 per cent of GNP. The gradual upgrading of the economy is reflected in the composition of the export items. In 1980, the share of textiles amounted to 28.6 per cent, whereas the share of plywood in that year contributed a modest 2.0 per cent to total exports. The importance of electronic products, on the other hand, had gone up to 11.4 per cent. The growing importance of HCI is among others illustrated by the increased importance of steel products in total exports; their share had increased from 1.5 per cent in 1970 to 10.6 per cent ten years later. The new policy with regard to foreign direct investment also clearly left its mark. By 1978 three-quarters of all electronics and 10 per cent of textiles and clothing exports was realised by foreign firms (Linnemann *et al.* 1987).

Stabilisation and liberalisation

The imbalance that emerged as a result of the privileges granted to HCIs and to large groups in general at the expense of small and medium-sized companies was only one of the many problems Korea faced as it entered the 1980s (Sakong 1993). The construction boom in the Middle East of the mid-1970s had increased the pressure on the skilled labour market at home and, consequently, pushed up general wage levels.[19] The activities of Korean companies and workers in the Middle East also resulted in an inflow of liquidities which improved the balance-of-payments but also implied further inflationary tendencies. Combined with existing fixed exchange rates the international competitive position of Korea worsened rapidly. At the same moment, some developing countries (e.g. China and Southeast Asian countries) rapidly started to industrialise, focusing on low-skilled labour-intensive industries. This trend was further strengthened by a world-wide recession and the second oil shock in 1979 which led to increased protectionism in the industrialised world. A definite low was the negative GDP growth which Korea first experienced in 1980. Other serious setbacks were a disastrous rice harvest in 1980 and growing political instability.[20]

All these developments led the government to drastic reforms of Korea's economy. The 'growth at any costs' policy was gradually replaced by a 'stability at any costs' strategy. During the first half of the 1980s attaining price stability had top priority and had to be realised by restrictive fiscal and monetary policies. The salaries of civil servants and employees of publicly owned enterprises were frozen, with the hope that the private sector would follow. The government also introduced a 'zero-based budgeting system' in the beginning of the 1980s. Soon these measures started to bear fruit: in 1982 the increase in wholesale and consumer prices had dropped to 2.4 per cent and 4.7 per cent, respectively.

Next to stabilisation policies the 1980s also witnessed the prudent start of a general liberalisation of the Korean economy. As will be discussed in Chapter 4, important financial liberalisation efforts were initiated (World Bank 1993). In addition, measures were taken to promote market competition and eliminate factors that hinder a competitive environment.[21] In 1984 major reforms were also introduced for the public enterprise sector, which had played an important role in Korea's economy.

Major political changes also took place in this period. In 1987, for the first time a president was elected directly by the population. In the wake of this political transformation, labour laws were changed, resulting not only in a sharp growth in the number of unions and members but also in a rapidly increasing number of labour disputes.

Also in the international context a policy of liberalisation was gradually introduced. The share of automatically-approved import items increased from 52.7 per cent in 1977 to 96.3 per cent in 1990. Bilateral trade disputes, especially with the United States, clearly speeded up this trend. In the second half of the 1980s, after substantial pressure from its major trading partners, Korea – by that time generating a balance-of-payment surplus – decided to further liberalise manufactured imports (Sakong 1993). In addition, a tariff reduction programme was introduced. Also in the 1980s markets for technology and foreign direct investment were partly liberalised, as well as the market for services.[22] As will be discussed in more detail in Chapter 4, in the second half of the 1980s restrictions on outward FDI were eased. These liberalisations resulted in a rather strong growth of foreign presence in Korea as is illustrated by the total amount of inward FDI flows going up from US$ 1.8 billion in 1982–86 to US$ 4.2 billion in 1987–90. These measures, as well as the improved protection of intellectual property rights were all targeted at 'bringing Korea in line with international conventions'.

How did Korea perform economically during this decade of change? As compared to the difficult second half of the 1970s (the average per capita GNP growth from 1977–81 was 'only' 4.2 per cent), the 1980s showed a remarkable recovery. Between 1981–90, on average per capita real GNP grew 8.0 per cent per year. Although export as a percentage of GNP did not go up any more and even decreased (from 31.2 per cent in 1981 to 25.8 per cent in 1990), its growth in absolute terms was still considerable. Electronics (including semiconductors, consumer electronics and telecommunication equipment) by now had become the most important export category, realising 23.3 per cent of total exports. Textiles ('woven fabrics' and 'apparel') in 1990 represented 18.7 per cent. In that year, automobiles and ships realised 2.8 per cent and 4.3 per cent of Korean exports, respectively.

Present and future challenges

Although the first half of the 1990s again showed high economic growth, even before the crisis in 1997 it had already become apparent that the country still faced major challenges. In becoming a member of the OECD, Korea has committed itself to abolish important fundaments of its successful policy, such as the liberalisation of capital flows, the opening up of strategic sectors such as financial services and telecommunication, and the elimination of low interest subsidised loans for selected industries (FT 06.06.96, OECD 1994). With regard to the financial sector, for instance, by now it has become obvious that only drastic consolidation and refining will render Korean banks in any shape to meet the challenge of market opening and competition. Closely related to a restructuring of the financial sector is the call for diminishing the economic power of the chaebol – among others by means of antitrust legislation and fair trade laws. As Sakong (1993) noted, this implies that the nature of the whole government–business relationship will have to change (see Chapter 4). Another weakness in this respect is the strong economic nationalism which is rather persistent in Korea. Up to now, the attitude towards take-overs of companies, especially if foreigners are involved, has been quite hostile.

At the same time, however, to prevent Korea getting squeezed between low-cost producers from cheaper countries and the economically more advanced countries, it is clear that the technological and innovative capabilities have to be further strengthened. Such transformation is also needed to improve Korea's balance-of-payments position which is constantly threatened by substantial capital goods imports.

Until now, Korea's reform process has been very gradual and is not characterised by major changes. As will be discussed more elaborately in the remainder of the book, the crisis which hit Korea at the end of 1997 may trigger more substantial measures to address these challenges. We will also show that not only the authorities, but also many companies will have to make major efforts to safeguard their survival in the longer run. Our investigation of their internationalisation behaviour will reveal important challenges faced at the firm level.

Taiwan

Japanese colonisation

Taiwan was the first colony of Japan. Its colonisation was a direct consequence of Japan's defeat, in 1895, of China, of which Taiwan had officially been a part since 1683. On the eve of Japanese rule Taiwan had an agricultural economy based on small peasant-owned and tenant holdings, similar to that of Fukien and Kwantung where most of the population originated. Unlike the situation in Korea, Japan had not been involved in Taiwan's economy prior to colonial rule. Apart from this and other (sometimes minor) differences, the role of Japan in Taiwan is in many ways comparable to that in Korea.

As in Korea, the first period of Japan's colonisation of Taiwan was characterised by efforts to consolidate its control of the island. Similarly, the Japanese saw Taiwan primarily as an agricultural appendage of the home country. Two crops were of prime importance in this respect: sugar cane and rice (Amsden 1985). Next to considerable investments in the transportation (roads, railroads and harbours) and communication infrastructure, the Japanese authorities invested heavily in the development of the agricultural sector.[23] In these years, 20 per cent of Taiwan's rice and 90 per cent of its sugar were exported, nearly all to Japan.

Taiwan's industrial start is closely linked to its agricultural activities (Fei *et al.*, 1979).[24] Japanese enterprises, attracted by similar benefits as in Korea, were invited to invest in sugar refining factories in Taiwan (Gold 1986). In 1923 the rules for establishing companies were liberalised, which facilitated the entry of Taiwanese companies. Nevertheless, most factories remained in Japanese hands. This can be accounted for not only by the requirement for modern joint-stock companies to have Japanese partners, but also by the lack of accumulated Taiwanese wealth. The first elaborations of industrial activity were mainly restricted to the food processing industry. By 1930 64 per cent of all registered factories, 55 per cent of all factory employment and 76 per cent of the gross value of industrial production was still realised in this industrial sector. From 1920–24 till 1930–34 the manufacturing production index rose from 64.98 to 124.18 (Ho 1984; 1925–29 = 100).[25]

Gradually, Japanese *zaibatsu* also started to invest in other industries. This was not only a response to rising wage levels in Japan itself and domestic policy on competition, in the 1930s. Japan was also increasingly seeking economic self-sufficiency. Next to safeguarding the supply of rice and sugar cane this implied investments in heavy industries that could process raw materials (which were imported into Taiwan from Southeast Asia) for stock-piling and shipment for Japan's war operations. Industrial chemicals, ceramics, aluminium, machine tools and textiles were some of the industries in which investments took place in Taiwan at that time. From 1930–34 till 1935–39 the manufacturing production index grew from 124.19 to 177.59 (Ho 1984). The output composition nonetheless did not change as radically as in Korea. Throughout the colonial period foodstuffs remained by far the most important production item, reflecting some 70 per cent of total manufacturing output.

Although not as spectacular as in Korea, trade volume in Taiwan also grew considerably during Japan's colonisation of the island: exports increased more than ninefold between 1905 and 1940 while the import volume went up six times. Also the export composition changed less drastically than in Korea. Foodstuffs (especially rice and sugar) remained dominant in exports; its share even grew from 80 per cent in 1906–10 to 89 per cent in 1936–38. In 1936–38 the remainder of exports were made up by consumer goods (5 per cent), industrial raw materials (5 per cent)[26] and producer goods (1 per cent). As in Korea, World War II for Taiwan meant an interruption of its economic development.

The hesitance expressed at evaluating the costs and benefits of Japan's occupation of Korea also applies for the Taiwanese case. Five decades of Japanese rule caused substantial changes in Taiwan's socio-economic situation. Between 1912 and 1937, Taiwan's net domestic product (NDP) for instance increased at an annual compound rate of 4.1 per cent. It should be clear however that, as in Korea's case, this was realised under a 'dependent capitalism' structure (Gold 1986). This among others implied that by the end of World War II a class of Taiwanese (industrial) entrepreneurs was almost non-existent. Notwithstanding this observation and despite the hardships which occurred during colonial rule, unmistakably a foundation was laid for the industrial advance Taiwan has realised since then. Examples of the elements of this foundation are manifold: an adequate infrastructure, the presence of factories fairly distributed over the island, basic education for many,[27] an improved public health[28] and sanitation system, and standardised units of weights and measures.

Interregnum

After Japan was defeated in 1945, Taiwan became a province of the Republic of China (ROC). Preoccupied with power struggles on the mainland, the nationalist government (Kuomintang) paid little attention to the island during the early years. Taiwan's start therefore did not look very promising. World War II had left Taiwan with a seriously damaged industrial capital stock and infrastructure, a lack of qualified human resources and a production level which in 1945–46 was below one half of the peak in the colonial period. Even worse was the tremendous inflation reflected by a wholesale price index increasing 260 per cent in 1946, 360 per cent in 1947, 520 per cent in 1948 and 3,500 per cent in 1949![29]

On the other hand, Taiwan had a number of important assets (Ho 1978). Despite wartime destruction, the infrastructure was still relatively extensive. In addition, the agricultural sector – after Japan – was the most advanced in the Far East. Moreover, the literacy rate, as a result of Japan's educational policy, was unusually high. The well developed institutional system (such as the banking system) also helped Taiwan's post-war start.

In the late 1940s the position of the Kuomintang on the mainland had become untenable. As a result, more than one million mainland refugees – under the leadership of Chiang Kai-shek – fled to Taiwan. Among the group of military and civilians was a relatively small number of managers, technicians, and entrepreneurs who could partly neutralise the human resource gap that emerged after the Japanese had left.

Stabilisation and import substitution

The turbulent economic situation in Taiwan by the end of the 1940s led to a strict stabilisation programme. An important constituent was a conservative fiscal policy of maintaining balanced budgets. The realisation of this target was supported by the enormous aid rendered by the United States. Up to 1965 the USA provided US$1.5 billion economic aid and US$2 billion military aid. Of the economic aid, 38 per cent went to finance the import of intermediate goods (such as cotton, yarn, ores, metals, and fertiliser), 30 per cent was spent on consumer goods (mainly food), and 19 per cent was used for capital goods (machinery and tools). The US supplied 35–45 per cent of Taiwan's imports and took 5–10 per cent of its exports. In addition, the US economic and military advisers exercised considerable influence over the government's policy (Wade 1990).

The stabilisation policy turned out to be highly successful as inflation went down to an annual average of 30 per cent in 1951–52 and less than 10 per cent during the rest of the 1950s. As a response to inflation experiences, balance-of-payments problems and the loss of former target markets Taiwan, as did Korea, instituted an import substitution policy (Clark 1989). Although not as extreme a situation as in Korea, Taiwanese agriculture at that time was much more important than industry.[30] In this period, the domestic market was especially sealed off for non-durable consumer goods. This was accomplished by high import tariffs, quotas for specific consumption goods (especially luxury items), and multiple exchange rates, keeping the domestic New Taiwan dollar (NT$) overvalued. The foreign exchange controls obviously acted as a disincentive to export performance. Industrial companies which were oriented towards the domestic market, however, benefited from the overvalued exchange rate in the form of lower costs of imported inputs, as well as from the trade controls through higher prices for products sold on the Taiwanese market (Wade 1990). In this decade, the private sector gradually outgrew the public sector. Important new sectors in which

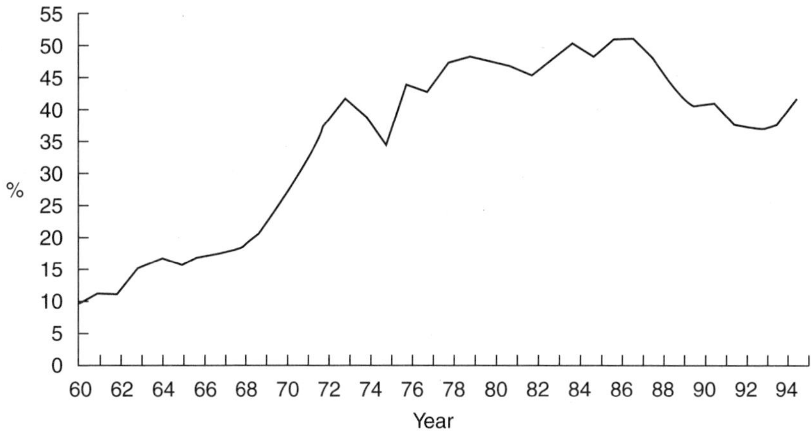

Figure 3.2 Taiwanese exports as a share of GNP
Source: CEPD (various issues)

private companies started their business were textiles, clothing, wood and leather products and bicycles.

The average annual growth of per capita real GNP between 1951–60 was a modest 4.5 per cent. Exports were still very limited in importance reflected by a share of 9.6 per cent of GNP in 1960. In the same year, no less than 56 per cent of these exports was accounted for by agricultural products whereas textiles represented 14 per cent of the total. By the end of the 1950s it had become clear that the growth potential of first stage import substitution was exhausted. The domestic market became saturated with Taiwan produced goods, while the heavy import demands (in terms of capital goods and industrial raw materials) of light industry caused balance-of-payments problems. As a result, the government decided to refocus towards industrial promotion through the export of labour-intensive goods (Clark 1989).

Export-led surge

The decision to switch to an export-oriented (EO) based industrialisation was not an easy one. Many leading figures doubted whether Taiwan's light industries could become international competitors. Moreover, this strategy caused opposition from the military, who wanted to develop a defence-related heavy industry, as well as from the bureaucracy who had to delegate a substantial part of their authority (Clark 1989). Despite these objections, between 1958–60 USAID (US Agency for International Development) and the Taiwanese government developed a reform package which was incorporated into the Third Four-Year Plan (1961–64).

The most encompassing element of the reform package was the implementation of the 'Nineteen-Point Program of Economic and Financial Reform'. The programme consisted of a wide range of measures intended to liberalise controls on trade and industry, promote exports and establish a business climate to stimulate private local and foreign investments (Ho 1978). Gradually, the New Taiwanese dollar was devalued and a uniform exchange rate was introduced. The allocation of foreign exchange for permissible imports was simplified. Quantitative restrictions on the import of intermediary goods and capital goods for exports were practically removed. The list of forbidden 'controlled' imports was also shortened step-by-step. Export promotion schemes were expanded considerably. Loans to exporters were provided at preferential rates. A rebate system originating in 1955 was expanded and simplified to include import duties, commodity and other taxes on export products. Other measures to promote exports included income tax exemptions, cheap export insurances and marketing support. The government also approved and supported the formation of export cartels, in which case the production of certain goods was regulated by the allocation of export quotas.

Another change in policy concerned inward foreign investment. The reduction of US aid in the 1960s encouraged the Taiwanese authorities to attract foreign investors (van Hoesel 1996). For this purpose, the Statute for the Establishment and Management of Export Processing Zones (EPZs) was promulgated in 1965 followed by the establishment of the first EPZ in Kaohsiung in 1966. This EPZ offered modern port facilities, tax incentives, no duties on equipment and parts as long as they are used for export production and a considerably limited red tape. Later, other EPZs were established in Taichung and Nantze.[31] As a result of this new policy, approved inward FDI flows went up from a scant US$35.7 million in the 1950s to US$523.6 million during the period 1961–70. Most inward direct investment in the 1960s took place in labour-intensive operations, especially in electronics and, to a much lesser extent, chemicals and textiles and garments.

As in Korea, institutional reforms coincided with these changes in economic policies (Gold 1986). New institutions such as the Industrial Development and Investment Center (IDIC) and the China Development Corporation were established to promote private investment. The newly formed Council for International Economic Co-operation and Development (CIECD) was responsible for Taiwan's Four-Year Plans and regularly improved the investment climate.[32]

The 1960s witnessed impressive economic records in Taiwan. Real per capita GNP, for instance, grew on average 6.7 per cent per year. Taiwan's balance-of-payments difficulties rapidly diminished as a result of fast export growth. This development demonstrates the growing dependence of Taiwan's economy on export performance in this period: the share of exports in GNP grew from 9.6 per cent in 1960 to 26.2 per cent in 1970 (see Figure 3.2). Although most industries were still labour intensive, a shift towards more sophisticated operations could be observed as, among others, is illustrated by the export composition. By 1970, the share of agricultural products had dropped to a mere 13 per cent. The relative importance of electronic goods, on the other hand, had gone up from 0.5 per cent in 1960 to 12 per cent ten years later. Textiles by now had become the most important item sold overseas, realising 32 per cent of total exports.

Industrial consolidation

At the beginning of the 1970s Taiwan became a victim of its own success. Its export performance was more and more threatened by protectionist measures of its trading partners, especially the United States. New competition in light manufacturing came from other developing countries with low labour costs, while in Taiwan full employment and rising living standards pushed wages up (Gold 1986). Moreover, the existing transportation and communication networks and electricity provisions did not match the fast growth of industry any more – most of it originated from the colonial period. The derecognition of Taiwan by the United Nations in 1971 followed by Japan in 1972 and the stepwise approach of the United States towards the People's Republic of China (PRC) were a major political shock to Taiwan which also had economic implications such as serious capital flight and the emigration of 'human capital' to the United States.[33] Moreover, inflation was fuelled by the rapid expansion of money supply resulting from large trade surpluses that Taiwan had built up in the beginning of the 1970s (Clark 1989).[34] The first oil crisis, resulting in a quadrupling of oil prices, also meant a big setback for Taiwan. In 1974 real GNP growth was only 1 per cent, inflation went up to 40 per cent and with a negative real export growth of 7 per cent the trade deficit amounted to more than US$1,000 million. It was obvious that the authorities had to launch measures to counteract these negative developments.

A major component of the stabilisation programme was a sharp increase in interest rates (Gold 1986, Clark 1989).[35] Other measures taken to dampen inflationary pressures were a substantial cut in real government expenditures[36] and price increases managed through the monopolies exercised by government corporations, especially the price of oil and electricity rates. By the end of 1974 prices were under control again.

After taking advice from the American consulting firm Arthur D. Little, the Taiwanese authorities shifted their policy towards more self-reliance, without giving up their export orientation (World Bank 1993). New plans were worked out in the Sixth Four-Year Plan (1973–76). Ten major development projects were formulated to modernise the infrastructure and initiate the development of heavy and chemical industries (HCIs) ranging from building highways to the construction of nuclear power plants. Total costs of these projects amounted to about US$8 billion. One of the new sectors to develop was the petrochemical industry, seen as a means to integrate two important industries, namely synthetic fibre and plastics.[37] The role of public enterprises increased, not only in infrastructural projects, but also in HCI; the share in gross domestic fixed capital (GDFC) formation grew from 28 per cent in the 1960s to 33 per cent in the 1970s (Wade 1990), reflecting the major role the state assumed in supporting future economic growth. By the mid-1970s, Taiwan's competitiveness had improved considerably. In 1976 exports shot up 46 per cent in real terms (over twice the increase in imports), restoring thereby a healthy trade balance.

The second half of the 1970s witnessed a continuing labour shortage and a sharp rise in wage levels, which clearly increased export prices substantially. In addition, industrialised countries adopted neo-protectionist non-tariff barriers (NTBs) to restrict imports, especially of such Taiwanese staples as textiles, footwear, mushrooms, and television sets. Notwithstanding these negative developments, approved FDI flowing into Taiwan increased to more than US$2.2 billion in the period 1971–80.

Notwithstanding these difficulties, for the whole decade, per capita real GNP averaged an annual growth of 7.1 per cent. The economy's dependence on exports grew further: its share of GNP went up from 31.3 per cent in 1971 to no less than 47.9 per cent in 1980. Although labour-intensive industries remained relatively important, the structural transformation of Taiwan unmistakably continued. By 1980, the importance of electronics exports had grown to 18 per cent of total exports. Textiles, however, was still the most important export item realising about 23 per cent of the total. The share of food processed goods, on the other hand, had gone down to less than 6 per cent.

Modernisation and liberalisation

The second oil crisis of 1979 was considerably milder for Taiwan's economy than the first one. Resulting inflation was relatively low (namely 14 per cent in 1979 and 22 per cent in 1980), while the trade account managed to stay positive. Interest rates and electricity rates were only gradually raised and a tightly controlled floating exchange rate system was introduced. The increased integration of Taiwan into the world economy revealed serious structural weaknesses, however (World Bank 1993). Perhaps the most obvious example was the obsolete financial system that was not able to channel the considerable savings to an optimal use for investments and trade purposes. In addition, the enormous trade surplus which Taiwan had accumulated (by the end of the 1980s this amounted to about US$80 billion!) led to growing protectionism against Taiwanese products, especially in the United States.[38] The strong appreciation of the New Taiwan dollar during the second half of the 1980s (under strong pressure of the US government) and the rapidly rising real wage levels further squeezed Taiwan's traditional labour-intensive manufactures between low wage countries in the region (namely Southeast Asia and the PRC) and high-tech goods from industrialised economies.

These developments led to a further restructuring of the Taiwanese economy. In the future, the focus would be on non-energy intensive, non-polluting and technology-intensive industries instead of heavy or capital-intensive ones (Gold 1986). Since the government was convinced that the predominantly small-scale industries were not able to realise these goals on their own, they founded and funded R&D institutes in strategic areas and provided incentives for the development of high-tech industries (Wade 1990). Another landmark was the realisation of the Hsinchu Science-Based Industry Park where foreign and domestic high-tech firms operate in close proximity to laboratories of the Industrial Technology Research Institute.[39]

In 1984 new tax incentives were introduced for companies carrying out R&D activities. Other measures taken to advance Taiwan's technological capabilities were: a government sponsored programme to diffuse labour-saving automation equipment, the promotion of venture capital to access the capital for high-tech firms, increased efforts to repatriate more Taiwanese graduates (in technical fields) from abroad and increased protection for intellectual property (Wade 1990, World Bank 1993). In addition, in 1985 the government launched fourteen major infrastructural projects, including the expansion of the energy, telecommunication and transportation networks. In 1990, the Statute for Industrial Upgrading was enacted, offering firms incentives to upgrade their technology and enhance their research and development. The government identified ten emerging industries as targets.[40]

Next to all these efforts to diminish the dependence on external technology sources, other changes took place as well. Imports of goods and financial markets were considerably liberalised, while many state-run enterprises were privatised. Moreover, restrictions on inward capital flows were removed. As a result, total approved inward FDI went up substantially and amounted to US$10.5 billion for the period 1981–90. As will be discussed at length in Chapter 4, outward capital flows were also liberalised considerably in the second half of the 1980s.

The average annual per capita real GNP growth during the 1980s amounted to 7.0 per cent. While in the first half of the 1980s growth expectations – because of sluggish export growth – stayed somewhat behind expectations (5.8 per cent per year), the second half of the decade again witnessed an impressive growth record (7.3 per cent per year). Although the export volume continued to grow, its relative importance declined somewhat: the share of exports in GNP went down from 47.2 per cent in 1981 to 41.0 per cent in 1990. The industrial transformation is again reflected in the continuing change in the composition of exports. In 1990, textiles and garments represented only about 15 per cent of the total. The share of electronics (including information and communication products), on the other hand, had gone up to more than 22 per cent. Also machinery had surpassed textiles, realising about 17 per cent of total exports in that year.

Present and future challenges

The 1990s have also shown rapid economic growth in Taiwan. Nevertheless, the country still faces important challenges. One of the biggest uncertainties regarding its future is in its relationship with the People's Republic of China. Although business contacts across the Taiwan Strait flourish as never before, the hostilities which now and then take place remind us of the fragile economic and political situation which needs a more stable solution for the future. Closely related to the 'two China's problem' are Taiwan's attempts to become a member of international organisations such as the United Nations and OECD. Not only from a political, but also from an economic point of view these efforts represent another major challenge which will definitely shape Taiwan's economic achievements in the coming years.

Although the damage caused by the Asian crisis of 1997 was much less serious for Taiwan than in the case of Korea, it has clearly underlined the need to restructure the banking sector and the financial sector in general. One of the steps to be taken, as already pointed out by Ranis (1992), is the further privatisation of the banking system. Moreover, as Liu (1992) argued, capital accounts have to be liberalised further, permitting foreigners to purchase domestic equities and reduce central bank interventions in the foreign exchange markets. In addition, major infrastructural investments (including telecommunications) are needed to keep up with the expansion of economic activity. Another major challenge concerns the shortage of skilled labour in Taiwan (San 1992). Increasingly important is also the environmental protection movement which for many (especially smaller) entrepreneurs in Taiwan is a new complication they have to deal with.

Furthermore, it is unclear how the increasing globalisation of industries will affect the many small and medium sized enterprises that lack strong brand names and possess only limited technological capabilities. As will be examined further in the remainder of the book, it appears that in the future, more 'critical mass' is needed to build up O advantages enabling firms to compete in the global market. As in the case of Korea, the investigation of the further internationalisation of Taiwanese companies will yield important new insights on these matters.

Summary and conclusions

The high-speed economic development of Korea and Taiwan are not only rather unusual in recent history, they also show a remarkable resemblance. Both in Korea and Taiwan the foundation of post-war successes was laid during Japanese occupation when important investments in locational determinants (e.g. infrastructure and education) were made. After their independence both economies experienced a period of severe political and military instability after which an import substitution policy was introduced in the 1950s. Being faced with relatively small home markets and poor resource endowments, both Korea and Taiwan refocused their attention to overseas markets in the 1960s – by that time almost solely through the export of goods produced in a labour-intensive manner.

Since then, we have witnessed continuous efforts to transform the economies in line with their comparative advantages. Although some periods have been more successful than others, the 'outward' orientation has remained. The fact that, as a result of this policy, companies had to compete with other firms throughout the world, does not imply that these economies can be labelled as 'liberal'. In this chapter many examples were presented which illustrated the highly important (intervening) role that the authorities have played in the economic development of Korea and Taiwan. The financial and economic crisis in Korea also revealed important drawbacks of such a dominant role of the authorities in the domestic economy.

We have also seen in this chapter that inward FDI in Korea has been relatively low. By the end of 1995, total approved inward FDI amounted to US$14.5 billion whereas in Taiwan US$22.3 billion had been invested. On a per capita basis this implies that in the latter economy 3.2 times more had been invested than in

Korea. According to Das (1992: 68) 'the belief that FDI from the US and Japan immensely benefited the Korean economy and exports is erroneous.' Sakong (1993) notes that an important explanation for the relatively minor role FDI has played is the general fear of foreign domination in Korean industries. Closely related is the strategy of Korean companies to acquire foreign technology through non-equity arrangements. As will be illustrated in Chapter 7, however, an important exception to the rule is the electronics industry where foreign MNEs no doubt have contributed considerably to the accumulation of O advantages by indigenous companies. The presence of foreign companies in the Taiwanese economy is not only more elaborate than in Korea in terms of FDI. Elsewhere (van Hoesel 1996) we have argued that inward direct investment also more in general has played a significant role in keeping the Taiwanese economy competitive. In Chapter 7, we will illustrate in more detail how conventional and non-conventional FDI has contributed to the upgrading of the Taiwanese PC market. It will also be shown that, in accordance with the arguments brought forward by the Late Industrialisation paradigm, the government played a very active role in this process.

According to the IDP framework discussed in Chapter 2, each country follows its own investment path. It was argued that four main variables influence this path which determines the nature and extent of the O advantages of (foreign companies and) local firms, namely: its resource structure, market size, industrialisation strategy and the role of the government in the organisation of economic activities. We have seen that both Korea and Taiwan are poorly endowed with natural resources and have relatively small home markets. These characteristics no doubt have influenced their broad strategies of economic development (such as their export orientation), which, as we have seen, show many similarities. The role of the government in the organisation of economic activities has not yet been sufficiently addressed. Since there are strong indications that in that respect important differences between Korea and Taiwan can be discerned, this fourth factor influencing the investment pattern is discussed separately in the next chapter.

4 Governments and the organisation of economic activities

Introduction

As was discussed in Chapter 2, according to the Investment Development Path (IDP) individual countries show idiosyncratic (outward) investment patterns whose shape is influenced by their (natural) resource structure; home market size; industrialisation strategy; and governments and the organisation of economic activities. Chapter 3 illustrated that with regard to the first three factors – grosso modo – Korea and Taiwan are quite similar. However, there are a priori reasons to assume that more significant differences may exist with regard to the role of the government and the organisation of economic activities in both economies. As pointed out earlier, especially in late industrialising economies, governments have played a very profound role in the formation of capital and enterprises thus strongly shaping the structure of companies and institutions as a whole (cf. Fields 1995, Hikino and Amsden 1994). We therefore expect important distinctions to exist in this respect between Korea and Taiwan which also influence the outward investment pattern of both economies.

We noted earlier that the IDP concept does not unfortunately really specify the role of the government and the organisation of economic activities. The question then arises how these characteristics can be investigated more in detail so that a comparison between Korea and Taiwan can be made in a systematic manner. Without intending to initiate a debate on various strands of economic theory, it appears that well received economic concepts based on neo-classical ideas or conventional transaction cost economics do not offer the insights nor the instruments needed to make such comparison (Whitley 1994). According to Whitley, despite considerable differences between these approaches, they all believe that competitive markets select efficient forms of business organisations and destroy inefficient ones. If they are to survive, every firm has to follow the same basic 'recipe' within each industry. In recent years, however, new contributions have been rendered by authors who address these issues more explicitly by perceiving the role of the state as an independent variable and taking into account factors beyond economic rationality, such as 'the dynamics of familial or peer approval, community status, political power, and implicit and explicit coercion' (Fields 1995: 17).

Based on these new insights, often referred to as 'New Institutionalism' (Powel and DiMaggio 1991), 'New Institutional Economics' (Kang 1995), or 'New Economic Sociology' and 'New Political Economy' (Fields 1995), a number of idiosyncratic features of the Korean and Taiwanese economic systems are addressed in this chapter. First, the most important business actors in both economies are typified as well as the background of their characteristics. Subsequently, the role of trading companies is discussed in more detail. Next, the most important sources of finance for Korean and Taiwanese companies are examined. In the following section, we investigate how the markets are organised. Subsequently, government policy in Korea and Taiwan with regard to outward investment is addressed. In the final section of this chapter, conclusions are drawn and an attempt is made to relate these characteristics to the possible impact they may have on outward investment.

Business actors in Korea and Taiwan

To get a first impression of the variations that exist between the most important business systems in Korea and Taiwan, the differences in firm size are shown in Table 4.1. Unfortunately, these (census) figures are based on individual establishments and therefore do not take into account the links that may exist between companies, i.e. the existence of conglomerates. Nonetheless, the table suggests that in Korea larger-sized companies play a more important role in the domestic economy than in Taiwan.

In Table 4.2, some average characteristics of the ten biggest companies (in terms of total turnover) in Korea and Taiwan are listed. With regard to almost every aspect the Korean companies easily outnumber their Taiwanese counterparts. Again, however, the numbers for Korea are underestimated because most chaebol do not publish consolidated figures. Another interesting observation is that among the 500 biggest companies in the world, twelve Korean and only one Taiwanese company can be found (Fortune, 05.08.96) These figures strongly suggest that the company characteristics in Korea and Taiwan are quite different. Who

Table 4.1 The size of manufacturing enterprises in Korea and Taiwan (1991 data)

	Share of firms (%)		*Share of workers (%)*	
Number of employees	*Korea*	*Taiwan*	*Korea*	*Taiwan*
<50	86.8	94.0	33.7	43.7
50–99	7.4	3.6	12.6	12.8
100–499	5.0	2.1	23.9	21.3
500+	0.8	0.3	29.8	22.2
Total	100	100	100	100
(Total number)	(72,213)	(140,572)	(2,918,015)	(2,665,435)

Sources: DGBAS ROC (1993), NSO (1993)

Table 4.2 Average characteristics of the ten biggest companies in Korea and Taiwan (1995 figures)

	Korea	*Taiwan*
Employees/firm	43,740	19,191
Sales/firm (US$ m)	23,958	5,260
Share of total workforce (%)	2.1	2.1
Share of GNP (%)	53.0	19.8

Sources: CCIS (1996), *Fortune 500* (05.08.96)

Note:
The Korean companies included in the sample, are: Daewoo, Samsung, Ssangyong, Sunkyong, Samsung Electronics, Hyundai, Samsung Life, Hyundai Motor, LG International and Korea Electric Power. The Taiwanese companies are: Chinese Petroleum, Taiwan Power, Cathay Life Insurance, DGPRSB, DG of Telecommunications, Shin Kong Life Insurance, NanYa Plastics, Taiwan Tobacco & Wine Monopoly Bureau, China Steel and Hotai Motor.

are the most important business actors in both economies and what are the reasons behind these differences?

Korea

When the main actors behind Korea's economic successes are investigated, most attention usually goes to the chaebol and their relationship with the authorities. This is not without reason: as Kang (1996: 9) for instance points out, 'there is little dispute that . . . the economic engines behind the real economic growth have been the . . . chaebol'. Here too we therefore focus primarily on the development and role of these conglomerates.

Most definitions of 'chaebol' in the literature share a number of characteristics (cf. Fields 1995, Steers *et al.* 1989).[1] Chaebol are typically owned and controlled by a family and managed in a paternalistic manner. Moreover, although the companies belonging to a chaebol are legally independent of each other,[2] the membership of a conglomerate is unambiguous. In addition, the product lines and industries in which the chaebol are active are typically highly diversified. At least until recently, the chaebol had no control over important financial institutions and were therefore heavily dependent on external sources. And finally, an important characteristic of these conglomerates are the close links that have always existed with the Korean government.

How important is the role that these chaebol play in the Korean economy? In Table 4.3, the weight of the thirty largest chaebol is shown. From this, we observe that the five biggest chaebol (namely Samsung, Hyundai, LG, Daewoo, Sunkyong) are very dominant in the mining and manufacturing sector, accounting for no less than 24.6 per cent of shipments, 11.1 per cent of employment, 25.6 per cent of assets and 24.0 per cent of value added. The activities of the chaebol are not limited to these sectors, however. Commerce and trade, for instance, are also important sources of income for these conglomerates in which they are even more dominant. According to Kang (1996), 88.6 per cent of sales in this sector is realised by the thirty largest chaebol.

Table 4.3 The weight of the thirty largest chaebol in the mining and manufacturing sector (%)

Chaebol	*Shipments*	*Employment*	*Assets*	*Value added*
Top 5	24.6	11.1	25.6	24.0
Top 10	32.1	13.7	35.1	30.2
Top 15	35.0	15.2	39.8	32.7
Top 20	36.8	16.2	41.3	34.1
Top 25	38.3	17.0	43.0	35.7
Top 30	39.6	17.7	45.0	36.9

Source: Lee and Lee (1996)

In order to understand why in Korea these conglomerates have become such a dominant actor in economic life, it is important to trace their origins. Most chaebol were founded at the end of the 1940s and in the 1950s (Fields 1995).[3] During the import substitution regime in the 1950s companies with strong political connections were able to gain various important favours – in return for political support – which stimulated a rapid take-off.[4] One important benefit concerned former properties of the Japanese rulers (such as plants, land, infrastructure, and inventories) that were distributed for very low prices while accompanied by a number of preferential conditions such as cheap bank loans and protection of monopoly or oligopoly positions. In addition, the import licences and necessary foreign exchange that were allocated by the government provided extremely profitable business opportunities. Moreover, the acquisition of aid, whether in the form of dollars or raw materials, was an important factor in the early days of the chaebol.

Notwithstanding their rapid take-off, it was only after president Park Chung Hee came into power in 1961 that these companies really started to build up their present dominant position. To direct the new policy of export-oriented industrialisation, the economic power was concentrated in a new super ministry, the Economic Planning Board. Using 'sticks' (e.g. tax penalties, revoked import licences, calling in of loans) and 'carrots' (e.g. preferential loans, access to foreign exchange, tax concessions) Park directed the chaebol to realise export goals and other goals, such as the development of new industries (Cho 1994). The financial instruments that were used are discussed later in this chapter.

The present-day position of the chaebol was largely shaped in the 1970s. Especially important in this respect was the initiation of the Heavy and Chemical Industries drive in 1973. The authorities strongly believed that either a monopoly or an oligopoly would be more desirable than a competitive structure (because of the small size of the home market, the importance of economies of scale, and the high capital requirements) for the selected strategic industries (namely petrochemicals, non-ferrous metals, general type machinery, electronics and shipbuilding) to be commercially successful (Kang 1996).[5] The small number of chaebol that were given a licence to participate in these industries were protected by the government not only from domestic competition, but also from imports. In addition, incentives were granted to the selected companies such as preferential financing, tax exemptions and reductions. Next to the support of such public banking institutions as the Korea Development Bank and the allocation of loans from commercial banks to the selected companies, the National Investment Fund was established in 1974 to expand the available financial resources for them (Koo 1994). The penetration into the Middle Eastern construction markets following the first oil price hike in 1973–74, also meant an important source of expansion for the chaebol. The government strictly controlled the allocation of the licences needed to construction overseas.[6] Another important source of chaebol growth was the creation of 'general trading companies' (GTCs) which will be discussed in greater detail in the next section.

Notwithstanding various attempts to curb their economic power, in the 1980s the dominance of the chaebol increased further. Unlike earlier times, the names of the players hardly changed suggesting that the same groups have consolidated and strengthened their positions in the Korean economy (Kang 1996). The chaebol started to establish non-bank financial institutions, not only to mobilise funds but also to expand their structure by circumventing government regulations. It appears that their dominance has continued to grow in recent years (Kang 1996, Lee and Lee 1996). In 1995, in no less than 64.6 per cent of the manufacturing sector, the three biggest companies had a combined market share of at least 70 per cent for a specific item (*Business Korea Yearbook* 1996). Moreover, notwithstanding government attempts to force the chaebol to focus on a limited number of industries, Lee and Lee (1996) note that between 1993 and 1996 the number of subsidiaries of the top 30 chaebol went up from 604 to 669.

With the gradual introduction of democracy in Korea, the perception of the public about the chaebol has become increasingly critical.[7] The attitude of the Korean authorities *vis-à-vis* the chaebol has, however, always been rather ambiguous. It appears that the government has always been dependent on these companies achieving high economic performance (*Business Korea Yearbook* 1996, Kang 1996). Whereas in earlier periods these companies depended very much on the goodwill of the authorities, it has even been suggested that nowadays, as a result of their mere dominance in the Korean economy, the chaebol could hold the government hostage by refraining from new investments in cases where genuinely inconvenient measures are launched against them (*Business Korea Yearbook* 1996). Without new investments made by these conglomerates the Korean government would certainly be confronted with a sharp decline in economic growth.

Notwithstanding their dominant position, the conglomerates are, of course, not the only actors in the Korean economy. Although since the 1960s attempts have been made to promote the growth of small and medium sized enterprises (SMEs),[8] the core idea has remained to realise growth and export targets through the support of large-scale companies (Cho 1994). Since SMEs are usually seen as an important source of innovation, their weak technological, financial and commercial position in Korea remains one of the biggest worries. Next to the private sector, Korea also has a substantial number of state-owned enterprises. Although the importance of publicly owned companies in the Korean economy has gone down since the 1980s,[9] in the first half of the 1990s their role in the financial sector as well as in mining and transportation and communication was still considerable. Particularly in manufacturing, state-owned companies play a substantial role in iron and steel.[10] In Korea, however, privatisation programmes have also commenced to rapidly reduce the size of the public sector.

Taiwan

To understand the firm structure in Taiwan it is important to study the institutional context in which it has been shaped. In fact, according to Amsden (1992), this rather unique context makes it much harder for other countries to follow Taiwan's policy model than is often assumed.

The figures in Tables 4.1 and 4.2 suggested that, certainly in comparison with Korea, the Taiwanese economy is dominated by small and medium sized enterprises. According to Fields (1995), in 1990, firms employing less than 300 persons realised 60 per cent of total exports. Wade (1990), however, cautions against exaggerating the small size of Taiwanese companies. As we noted already before, data such as listed in Table 4.1 do not take into account the role of business groups. The Taiwanese business groups (or '*guanxiqiye*') began to be formed in the early 1970s by successful entrepreneurs and individuals who invested in new companies or each others' firms (Gold 1986).[11] This trend was not only stimulated by the need to spread company risk, but also by capitalists' fears of growing too large and thereby attracting government attention in combination with laws restricting the amount of capital one firm could invest in another; tax incentives favouring new firms; and the evasion of creditors by moving capital about rapidly. In the beginning it was particularly companies with Kuomintang (KMT) connections that entered into the forming of such business groups, but increasingly other firms have followed. FIC, one of the companies studied in detail in Chapter 7, belongs to the Formosa Plastics Group – the largest industrial conglomerate in Taiwan (see also Box 4.2). According to Gold (1986: 89), however, the *guanxiqiye* are only 'but a shadow of their East Asian cousins, Japan's zaibatsu and Korea's chaebol, in terms of capital, scale, control of their own trade, and government connections'. The average group contains only 7.76 firms which, on average, are active in four different industries (Orrù *et al.* 1991). Although smaller in size, the highly diversified nature of the business groups is quite similar to that of 'conglomerates' in other LIEs (Amsden 1992). Although no figures are available on Taiwanese conglomerates, it is generally accepted that the *guanxiqiye* are much less dominant in the domestic economy than the Korean chaebol. In fact, most companies in Taiwan are single-unit operations (Wade 1990); this also holds for many leading firms in their sector such as Acer, one of the computer companies discussed in Chapter 7 (Amsden 1992). According to Orrù *et al.* (1991) of the 500 largest manufacturing firms in Taiwan, only 40 per cent belongs to one of these business groups.

The seemingly dualistic industrial structure in Taiwan of smaller-sized exporters and big enterprises that are primarily domestically oriented is in effect highly interconnected. The larger (private or public) enterprises are often important suppliers of inputs (e.g. petrochemicals, steel) to the exporting SMEs (Wade 1990). In addition, they frequently serve as a source of trade and working capital credits. Moreover, employees that leave the larger-sized firms and start up on their own account are important for transfer of management and technological knowledge (Amsden 1992, Wade 1990).

Various explanations for this rather divergent firm structure have been brought forward in the literature. According to Fields (1995: 91) '.. the fact that the regime automatically suspects any enterprise with sales exceeding US$40 million in an economy with annual GNP of over US$200 billion dramatises the state's penchant for control and bias against the concentration of private capital.' It is beyond doubt that the authorities have played a significant role in preventing (or actively discouraging) the economic system to be dominated by a limited number of (private) companies. Taiwanese banks for instance, which for a major part are government-controlled, have been very hesitant to lend large sums to private enterprises thus erecting a major hurdle to their expansion.

How can we explain this attitude? From an ideological perspective, the adoption of Sun Yat-sen's 'Principles of Livelihood' by Chiang Kai-shek and the KMT as official dogma has certainly influenced the latter's policy in this respect (Fields 1995). Fearing the negative consequences of a concentration of capital,[12] Sun preferred a mixed economy in which the state would own and manage companies that produced goods or services that are monopolistic in nature; are beyond the financial and technological means of private investors; or yield low returns or share other risks that would deter private investors. In addition to ideology, military arguments of course also played a role in reserving some sectors to state control. Moreover, historical events, such as the hyperinflation that occurred under KMT rule in mainland China (and which is seen as an important reason for the Nationalists' defeat) have also attributed to the ideological support for a strong control of (private) capital (Gold 1986, Whitley 1994).

Another important motivation seems to be the subethnic division of labour in Taiwan, where the political power is controlled primarily by the KMT and the private sector by the local, Taiwanese majority (Amsden 1992). As an external occupying force, the KMT had few ties to existing elites; indeed it had largely destroyed the old intellectual political elites and had no capitalist or financial class to contend with. As a result, the KMT could act rather autonomously with respect to native dominant groups (Wade 1990). Instead of supporting a small group of privileged capitalists, their attention was more on the smaller and medium sized enterprises that employ most people. The authorities have always feared that the dominant position which the ethnic Taiwanese have built up in the private sector may result in a demand for more political power.

State-owned enterprises are also important business actors in the Taiwanese economy. For a long time, the public sector in Taiwan has been one of the largest in the non-communist world. After Japanese colonisation, the new ruling KMT authorities inherited a substantial public sector. In some fields, the authorities in the 1950s and 1960s expanded even further their control and eventually included among others the financial system, and a large part of the transportation and communication sectors. In addition, production activities in such sectors as shipbuilding, heavy machinery and petroleum were monopolised by the state. In spite of recent privatisation programmes, state-owned enterprises in 1995 still accounted for 14.4 per cent of gross fixed capital formation.[13]

Summary

This overview of the most important business actors in Korea and Taiwan revealed important differences between both economies. In Korea, the tight relationship between the authorities and the leading companies has resulted in a limited number of very dominant conglomerates. In Taiwan, on the other hand, the relationship between the authorities and most private companies is much more distant. Historical and political considerations have led to a business structure in which small and medium sized enterprises are dominant in many outward oriented sectors of the economy and where conglomerates, such as those in Korea, are almost non-existent.[14]

The forerunners: international trading companies

Inspired by the successes of the Japanese trading companies ('*sogo shoshas*') Korea and Taiwan commenced in the 1970s to promote the establishment of their own trading companies that were specialised in penetrating foreign markets for local producers.[15] Since such trading companies potentially can play a pivotal role in the further internationalisation of indigenous companies, this phenomenon is addressed at greater length here. In this section, we discuss how Korea and Taiwan have attempted to boost their exports by the establishment of trading companies and why they failed in the case of Taiwan.

Korea

The development of the general trading companies is a fine example of the direct impact of state intervention in Korea (Whitley 1994). Why did the Korean authorities decide to introduce these trading companies? As was pointed out in Chapter 3, the adoption of an export-oriented industrialisation strategy in the 1960s resulted in a rapid increase of exports. Because of this success, the incentive package that had been introduced to stimulate export activities (such as cheap finance facilities and tax exemptions) had become exceedingly costly (Fields 1995). In addition, these incentives attracted many small-scale companies without any experience or expertise in overseas marketing which led to overcompetition among exporters. Furthermore, many exporters had become too dependent on these support programmes and often did not base their investment decisions on market conditions and opportunities. And finally, Korea's export performance was threatened by the oil crisis of 1973 which ushered in a more protectionist attitude in industrialised economies *vis-à-vis* Korean products.

In the mid-1970s the government therefore introduced a new instrument to expand Korean exports by penetrating new markets and by realising scale advantages (Fields 1995). At the same time, the financial burden on the government had to be partly shifted to the private sector. To be designated as a general trading company, initially the following minimum requirements were set:

- Paid-up capital of US$2.5 million;
- Annual exports of US$50 million;
- Handling of seven products with an export value of more than US$500,000 each;
- Ten branches overseas;
- Ten receiving countries with export value of goods purchased of over US$1 million each;
- Public offering of company stock.

These requirements clearly show that the authorities aimed at large-scale companies operating in a relative large number of countries and selling a diversified range of products.

The companies that were able to obtain the GTC status received various subsidies and incentives (Fields 1995). Examples are the privileged access to export financing and the financing of imports of raw materials; the right to import raw materials for their own use and for resale to domestic producers; and tax reductions and exemptions for GTC trade commissions. Most of these subsidies were also gradually provided to other large-scale exporters that had not been designated as general trading companies. Nevertheless, companies were very eager to obtain the GTC status since it improved the company's credibility in and outside Korea (Cho 1987). In addition, it could help in winning government-initiated projects. Furthermore, the high esteem attached to the GTC status enabled these companies to attract the best university graduates.

In 1975, Samsung, Ssangyong and Daewoo were the first GTCs to be designated. As can be seen in Table 4.4, in total thirteen companies have received GTC status. Since then, however, only eight have maintained their status.

The government continuously pressured the GTCs to expand exports by raising the export value requirement (Song 1994). The minimum export value that had to be realised went up from US$50 million in 1975, to US$100 million one year later and US$150 million in 1977. From 1978 onwards companies had to realise an export value of 2 per cent of Korean GNP which for some GTCs was too much to handle. The trading companies of Hyundai, Samsung, Daewoo and LG have become most dominant realising no less than 80 per cent of all exports of Korean GTCs in 1992 and almost 38 per cent of total Korean exports.[16] As opposed to their Japanese examples, the Korean GTCs were never stimulated to diversify to other functions such as finance (which for a long time was state

Table 4.4 Korea's general trading companies

GTC	*Designation date*	*Status*
Samsung	May 1975	In operation (1998)
Ssangyong	May 1975	In operation (1998)
Daewoo	May 1975	In operation (1998)
Kukje-ICC	Nov. 1975	Bankrupted in 1985
Hanil Synthetic	Dec. 1975	Disqualified in 1980
Koryo	April 1976	Government owned and operated – in operation
Hyosung	Aug. 1976	In operation (1998)
Bando	Nov. 1976	Became Lucky Goldstar (LG) – in operation
Sunkyong	Nov. 1976	In operation (1998)
Samhwa	Dec. 1976	Disqualified in 1980
Kumho	Dec. 1976	Disqualified in 1984
Hyundai	Feb. 1978	In operation (1998)
Yulsan	Feb. 1978	Bankrupted in 1979

Source: Fields (1995)

controlled), insurance and transportation. In addition, and different from the *sogo shoshas*, the GTCs have developed primarily as export companies; import activities, especially in the past, were of minor importance.

At the end of the 1970s, the survival of the GTCs was challenged because of their rapidly worsened financial situation. During the first years of their existence, the GTCs, who paid less than half of the regular interest rate to finance their exports, could ignore profitability to increase their shares of the export pie.[17] Large-scale malpractice soon emerged, however (Fields 1995). Export figures were inflated to meet export requirements and concessionary loans were used to diversify into sectors in which the companies had no expertise. Real estate speculation and other questionable ventures meant to raise capital led to costly failures and higher debts. The substitution of a fixed for a floating exchange rate system in January 1980 and the concurrent devaluation of the Korean won imposed a heavy burden for the GTCs who had to pay off their long-term loans abroad (Cho 1987). In 1982, the average debt ratio of GTCs amounted to 1,000 per cent; in the case of Ssangyong this figure even amounted to 1,900 per cent. The second oil crisis in the beginning of the 1980s led to a lowered attention of the authorities for the GTCs and a subsequent gradual abolition of their privileges.

To improve their financial situation, the companies turned to their chaebol sisters. Among others, by merging with more profitable members and charging higher commission fees to sister companies, the GTCs were able to improve their financial position. In recent years, an extension of tasks is observed. LG International, the trading company of the LG Group, at present, for instance, not only exports the entire range of goods produced by the LG Group, but is also active in the manufacturing of ready-to-wear apparel, overseas resource development, construction, and the import of a broad range of products. Samsung Co. has expanded into natural resource development, manufacturing an own brand of clothing, construction, the exploitation of a hypermarket and department store and the import of such products as industrial electronics, machinery, and raw materials.[18]

Notwithstanding the criticism that arose in the past, it appears beyond doubt that the trading companies can be labelled a success. They certainly helped to expand Korean exports to new markets introducing new products. The extensive support of the authorities, of course, has contributed to this achievement. In addition, being part of the chaebol provided a guaranteed source of goods to be sold. An additional explanation for their success is that export channels have been very limited in Korea (Fields 1995). Only since the 1980s have foreign trading companies been officially allowed to handle foreign transactions. Before 1992, Japanese companies, for instance, were not even allowed to establish wholly-owned subsidiaries in Korea.

For the future, competition from other export channels will remain a major challenge for the further expansion of the GTCs. The Japanese *sogo shoshas*, which are in a position to combine their financial, technological and information assets with a geographical, cultural and linguistic closeness to Korea are among the greatest threats. Other important competitors are local producers who increasingly tend to sell their products abroad themselves. As will be illustrated in Chapter 7, other members of the chaebol are rapidly increasing their sales and production channels abroad, thus making less use of the services of the GTCs. Foreign (primarily American) retailers are also competitors already possessing an established market abroad and offering SME producers attractive prices. Perhaps the most important obstacle for the future growth of the GTCs, however, are the Korean authorities who do not allow them to undertake business in major, profitable, import sectors (Fields 1995).[19]

Taiwan

As we saw in Chapter 3, Taiwan is even more dependent on international trade than Korea. By the end of the 1970s, however, about 70 per cent of Taiwan's cross-border trade was undertaken by foreign (especially Japanese and American) companies. In addition, in geographic terms, trade activities were too much confined to Asia and the USA. Furthermore, growing protectionism in target markets and rising wage levels in Taiwan and high oil prices meant a considerable deterioration of the trade regime (Fields 1995). It was therefore no wonder, also in view of the successes of the Japanese *sogo shoshas* and the Korean GTCs, that the Taiwanese authorities decided to promote the establishment of their own trading companies.

These so-called 'large trading companies' (LTCs) had to realise a wide range of goals (Fields 1995). The companies were supposed to establish international trade and information networks and indigenous brand names. In addition, the target markets and products sold by Taiwanese companies had to be diversified. Furthermore, the dependence on foreign trading companies and the over-competition among local firms had to be diminished. Likewise, the scale of trade and production activities were supposed to be expanded. These goals had also been set in the case of the GTCs and the *sogo shoshas*. On top of these targets, however, the Taiwanese LTCs had to support the development of small and medium sized enterprises. Furthermore, a concentration of welfare had to be avoided. The ambiguity regarding these objectives is also reflected in the minimum requirements formulated to receive a LTC status:

- Paid-in capital of US$ 5.6 million;
- Annual exports of US$ 10 million;
- Three branches overseas;
- Composed of the following units, with the first, and either the second or the third, as necessary shareholders:
  - (a) a bank dealing predominantly in foreign exchange or one with a wide reputation abroad;
  - (b) standard trading company engaged primarily in exporting;
  - (c) small or medium sized enterprise or merger between such enterprises;
  - (d) a manufacturer producing predominantly export products.

If these requirements are compared with those for Korean GTCs, we note that, although the Taiwanese requirements with regard to the amount of paid-in capital were more strict than in Korea, the criteria concerning the export amount as well as the geographical and product diversity were much milder. Another important dissimilarity with the Korean GTCs is that in Taiwan new companies were established with a SME, trading company or commercial bank as partners. The state made no effort to encourage the participation of existing business groups and in fact saw to it that the LTCs remained independent of existing groups.

The privileges granted to LTCs were rather modest and related to preferential finance possibilities; the duty-free import of raw materials used for the production of export goods; the right to provide import financing guarantees to SMEs and a maximum of 25 per cent on taxable income as corporate income tax and value-added tax. The most important of these privileges were also available to other reputable companies (Cho 1987).

In September 1978, the first LTC (namely, the Pan Overseas Corporation) was selected. Over the years, six more LTCs were formed: Collins Company, Nanlien International and Great International in 1979; E-Hsin International and Taipoly in 1980; and Peacock in 1981. As a result of increased designation requirements and the failure of some companies, as of 1987 only two companies remained (namely, Collins Company and E-Hsin International). In fact, the LTCs have never been able to play a significant role in Taiwan's export activities. In the peak year 1985, the LTCs realised a very modest 1.46 per cent of total exports. In 1991, all LTC incentives were terminated.

How can their failure be explained (Cho 1987, Fields 1995)? An important obstacle has been the enormous competition from other trading companies, the most important one being the Japanese *sogo shoshas*. Unlike Korea, Taiwan had in the 1950s already actively attracted Japanese companies to invest in its economy. Since then, the *sogo shoshas* have acquired a strong position as the commercial link for the Japanese subsidiaries with the outside world. In addition, long-term relations were built up with many indigenous SME producers whose products are exported and for whom raw materials and intermediates are imported – primarily from Japan. Furthermore, local SME producers who feared to lose their export licence have kept trading themselves.[20] An additional complication was that the LTCs also were not able to retain experienced staff. Building up such skills as foreign language ability takes time and many valuable staff were lured away by other companies or started their own business.

Perhaps most crucial was the unwillingness of the authorities to really support the development of the LTCs (Fields 1995). Their limited debt capacity (caused by the unwillingness of the conservative financial system in Taiwan to support the LTCs and the fact that they were not associated with the established business groups) made it difficult to finance the working capital needed and the expansion of overseas marketing networks. In addition, until recently, the import of bulk commodities and export of such profitable products as canned pineapples, soy beans and grain were almost monopolised by state-sanctioned institutions.[21] However, the lack of substantive government support cannot be merely explained by the fear of the Taiwanese authorities of losing income sources. In Taiwan, as we saw earlier, the policy to create large-scale private companies has always been controversial.

Summary

The Korean GTCs are among the best known 'imitations' of the Japanese *sogo shoshas*. Although not yet as advanced as their Japanese examples, these trading companies have contributed substantially to the country's export performance. In addition, they have served as important instruments for the accumulation of experience in operating in foreign markets. As was already pointed out by the Scandinavian School (Chapter 2), this is an important prerequisite for their later internationalisation through FDI. It is no surprise that the chaebol – to which the most successful GTCs belong – are also among the most important present-day outward investors. Analysing the reasons behind the failure to establish trading companies in Taiwan revealed that not only economic explanations played a role in this. The historical and political considerations that have influenced the policy *vis-à-vis* the LTCs have hampered the rise of large-scale trading companies, resulting in a continued domination of international trading activities by foreign companies.

Sources of expansion: finance in Korea and Taiwan

Despite the similarities in Korea's and Taiwan's political economies, it was illustrated in the foregoing that companies from both economies are vastly different in terms of organisational structure, scale, dominance in their respective economies, relations with the state and therefore their development paths. Because financial capital can be considered one of the most important environmental factors for any enterprise, an investigation of the financing of Korean and Taiwanese companies is needed to understand this variance (Fields 1995).

Korea

Since President Park nationalised all private banks in 1961, the tight state control over the formal financial system has been a key to dominating the private sector. One of the most important instruments became the differentiated access to cheap credits of commercial and specialised banks under government control which guaranteed a stable flow of bank loans regardless of short-term performance. In addition, foreign loans required government authorisation which meant that the authorities could use this instrument to support selected companies or industries. According to Kang (1996), the system of credit rationing in the 1970s strongly favoured large (HCI) enterprises at the expense of SMEs. Since government controlled credits primarily went to the chaebol, SMEs often had to turn to the informal credit market where interest rates were often twice as high as those at the government controlled banks (Business Korea Yearbook 1996).

In the 1980s the state-owned commercial banks were gradually liberalised, resulting in 1982 in the abolition of preferential lending rates for export loans. Moreover, in an attempt to curb the growing power of the chaebol, the size of total loans provided to these companies was limited. However, to maintain a high employment and investment level, the restrictions for an important part were lifted again in 1986 (Fields 1995). Even after the privatisation programmes of the 1980s the government maintained a strong influence on the formal bank sector by determining the interest rates, rationalising policy loans, allocating operating funds and appointing bank executives.

What are the most important sources of financial capital in Korea? The relative importance of the banks has gone down from 33.5 per cent for the period 1963–65 to 17.0 per cent for 1987–91 (Fields 1995). For most chaebol, however, the relative importance of the banks as a finance source is greater. Many of the largest chaebol are important shareholders of the commercial nation-wide 'city' banks where they borrow.[22] Since individual companies are not allowed to own more than 8 per cent of the national banks, some chaebol have effectively extended their control by holding bank stock through a variety of associated companies.

In an attempt to curtail the informal curb market,[23] since the 1970s the government allowed the establishment of so-called Non-Bank Financial Institutions (OECD 1994). With the exception of the development institutions (e.g. Korea Development Bank, Export-Import Bank) these institutions are able to offer high interest rates, face lower entry barriers and have more leeway in managing their assets than 'regular' banks. These characteristics, combined with the rapid economic growth, have made these institutions important financial sources providing 19.0 per cent of total finance to the corporate sector in the period 1987–91. The most important subcategory are the 'investment and finance companies' (IFCs) that primarily offer short-term credits. (On 1 July 1996, these were converted into merchant banks.) The annual interest rates are only (on average) 1 per cent higher than that of normal banks and 3–5 per cent lower than curb rates. In addition, no collateral is required. It is estimated that – through interlocking directorships and shareholding – no less than three-quarters of the IFCs are owned by the chaebol (Fields 1995).

Although 22.9 per cent (in 1987–91) of all financial funds of the corporate sector consisted of equity shares, a large number of member companies of the chaebol are not listed on the Korean Stock Exchange. The stock market is further constrained by extensive, stable cross-holdings among chaebol members. Moreover, a substantial part of the listed shares are owned by investment trust companies and security houses, so that the supply of 'floating stock' is only about 20 per cent (Fields 1995). Bonds, commonly perceived as another form of debt, accounted for 14.5 per cent of total funds in the period 1987–91.

Although their relative importance has gone down since the peak by the end of the 1960s, foreign loans still play an important role in financing the activities of Korean companies.[24] The recent internationalisation of Korean firms has also increased the interest of the companies in borrowing abroad. Since Korean interest rates are 5–6 percentage points higher than in overseas markets, companies have found it very attractive to finance their activities abroad through overseas loans.

Although a wide variety of actors are active in the informal sector, often professional brokers or moneylenders and relatives or friends of the business owners are the primary lenders. Despite efforts of the authorities to suppress the market, it is still quite substantial, estimated at 20.2 per cent for the period 1987–92.

Even more important than informal sources in the case of the chaebol, are intra-group financial sources and institutional arrangements accounting for 26.4 per cent of total sources in the same period. Especially important are cross-lending and cross-guaranteeing of loans and cross-holdings of equity among members. Because the chaebol are not legally seen as single units, there are no consolidated company accounts. This allows a highly indebted company which is unable to obtain bank loans on its own to receive financing by relying on an affiliate with an apparently healthy balance sheet as guarantor of the loans (Fields 1995).[25] Although cross-lending and cross-guarantees have generated valuable working capital, group expansion has been realised mainly through various forms of cross-investment and equity sharing arrangements among member companies. This allows the family of the founding owners of the chaebol to maintain control of the group firms with relatively small personal holdings. Notwithstanding attempts to reduce the control of the families, in 1991 the average of total shareholdings of the founding families of the largest 30 chaebol was still no less than 44.9 per cent. According to Kang (1996), in the case of the biggest conglomerates, this share is even higher.

Although further liberalisation plans were carried out in the 1990s, the OECD (1996: 80) recently concluded that 'the banks still have to respect broad government guidelines on lending with the result that some sectors still have little access to bank loans.' The collapse of Hanbo Steel (see Box 4.1), the country's second

Box 4.1 The collapse of Hanbo Steel

Hanbo borrowed nearly US$6 billion (more than half of the company's capitalisation) from 61 banks and financial institutions to build one of the largest steel mills in the world but had to declare bankruptcy in January 1997 after banks cut off lending due to the project's escalating costs. The banks decided to stop lending to Hanbo when they calculated that projected sales from the steel mill would be unable to cover annual interest payments. According to the Office of Bank Supervision leading banks lent the money to Hanbo without adequate credit analysis and feasibility studies of the project. According to the Office, little consideration had been given to Hanbo's financial state and concerns that its ambition was unrealistic. Interestingly, the banks believed the alleged participation of the state industrial bank in financing the steel project indicated that it enjoyed government support. Incessant rumours about the involvement of top politicians in the financial scandal and questions posed about the blocked entry of the Hyundai Group into the steel business to protect Hanbo subsequently led to the fall of the government.

Sources: FT (28.01.97, 30.01.97, 10.02.97, 13.02.97, 21.02.97, 22.02.97)

largest steel company, in the beginning of 1997 turned out to be a prelude to the economic and financial crisis that hit Korea so severely by the end of that year. After that first Thailand, the Philippines, Malaysia, Singapore, and Indonesia had run into the most serious economic problems in years, then in the period November 1997–January 1998 the Korean Stock Exchange collapsed. At the same time, the Korean won rapidly lost half of its value. The foreign exchange reserves of the Bank of Korea were insufficient to support the exchange rates.

Although some have suggested otherwise, these disastrous events did not come as a complete surprise, however. In fact, the crisis can to a large extent be ascribed to the very unhealthy financial sector which for a long time has unconditionally supported the chaebol. Notwithstanding the recent changes in the structure of the financial sector just described, the extraordinarily rapid expansion of large companies was still financially supported by the government. Stagnating sales growth led to increasing demand for money which could only be satisfied by concluding short term loans abroad. During the crisis, an acute shortage of foreign exchange made it impossible to pay off these foreign loans. The fact that the crisis coincided with presidential elections thwarted an adequate response by the Korean authorities.

In December 1997 an agreement was reached with the IMF about a financial support package amounting to some US$58 billion. The IMF, however, demanded rigorous reforms of the Korean economy. The chaebol are forced to scale down their activities, the economy (especially the financial sector) has to be liberalised further and the monetary situation has to become much more transparent. During the first months of 1998 many companies had already gone bankrupt as a result of the crisis, with the prediction that one million people would soon lose their jobs.

Taiwan

The high savings rate in Taiwan is often mentioned as an important determinant of its economic success. At the same time, however, the financial system is frequently criticised for being inefficient and is even considered a serious impediment for the future growth of the economy (Shea and Yang 1994). Central in the criticism is the excessive government intervention in setting interest rates and financial intermediation.

Until recently, almost all banks were wholly or partly state-owned. Moreover, all foreign exchange transactions were subject to state control (Whitley 1994). Despite significant financial liberalisations since 1987,[26] government control is, according to Fields (1995) still tighter than in any other (non-communist) Asian economy. Nevertheless, the nature of government–business relations has meant that their influence is neither as direct nor as predictable as one might expect. State control is intended partly to prevent a concentration of private (financial) power, and partly to maintain price stability. The goal of controlling inflation is not only based on economic arguments, but is also a result of the fear for the potential social and political consequences. The relative high interest rates held in Taiwan has indeed kept inflation at relatively low levels and resulted in one of the highest saving rates in the world. At the same time, however, this policy had significant consequences for Taiwan's companies in terms of costs of funds, ability to expand and freedom to participate in financial activities.

Although exporters (who received a subsidised rate) and large enterprises (which could obtain bank loans) have in fact had favourable treatment relative to other private sector borrowers,[27] as compared to Korea, the Taiwanese authorities have always been much more reluctant to use such instruments as concessionary credit allocations and tariffs and quotas for selective protection. Not only would this lead to an (ideologically disapproved of) increased concentration of capital, but the special position of the KMT in Taiwanese society does not permit accusations of favouritism. Although subsidised loan rates were offered to exporters, interest rates were also not as heavily subsidised as in Korea and have always remained positive in real terms. Furthermore, loans have never taken up such a large share of total capital requirements as in Korea.[28]

Notwithstanding the strict government control of the formal financial system, a large spectrum of financial actors can be distinguished in Taiwan. Unfortunately, weighing the relative importance of the various sources of funds is extremely difficult since financial information is usually treated very secretly by the familial owners of Taiwan's companies (Fields 1995). The domestic banks are still considered the backbone of the formal financial system (Shea and Yang 1994). Before 1991, only twenty-four banks were allowed of which twenty-one were majority government-owned. Since then, new private banks have been allowed to open but it is still too early to assess their impact on the total financial system. Next to the domestic banks, co-operatives, which are also used by non-related private companies, are allowed to create money. Of the institutions that are not allowed to create money but do mobilise money to finance investments, the postal savings system plays a substantial role (but it is *not* allowed to extend loans to the public). The so-called investment trust companies (ITCs), some of which are closely related to the *guanxiqiye*, are the most controversial institutes of Taiwan's formal financial system. Despite government intentions to limit their scope, private ambition and the lack of a clear public mandate has led the ITCs to gradually take on functions more similar to merchant or investment banks in other countries.

According to Fields (1995), although the money market is still relatively small, its importance as a source of short-term finance has grown. The capital (namely bond and stock) markets are still rather underdeveloped. Corporate bond issues are not popular since they are fully taxed and also the Taiwan Stock Exchange is not an important source of corporate finance (Orrù *et al.* 1991). A majority of the shares listed on the market are controlled by the KMT, state-owned enterprises or family companies that are all reluctant to sell their shares (Fields 1995) (only 10 per cent of the shares have to be sold to the public). The requirement to publish annual reports and disclose financial records has hindered the stock market's ability to become much of a vehicle for investors, but it is very attractive for speculators (Shea and Yang 1994).

Next to the formal financial market, the informal 'curb' market in Taiwan is a very important source of finance, also for larger companies. Their interest rates are much higher than on the regular market and vary between 4–9 per cent per month (Shea and Yang 1994). Given the conservative lending policies of the formal banks and the substantial collateral required, the curb markets are very popular – especially among private companies.[29] It was estimated that 40 per cent of Taiwan's capital needs are being met by this source (Fields 1995).

What internal sources are used by Taiwanese companies to finance their activities? Most large companies accept deposits from their employees and extend loans, like a bank. Although officially it is illegal, business groups also lend money to other companies within the group – an activity that in practice is not restricted by the authorities. Inter-company loan guarantees are also very common – often the parent or flagship company will act as a guarantor for newer, smaller or financially less sound companies. Unlike Korean companies (where state-controlled debt-financing is more crucial), in Taiwan, companies turn especially to family and friends as a major source of financing (Fields 1995).

Although the Asian crisis did not leave Taiwan untouched, the damage caused has been much less severe than in Korea. The limited size of the financial crisis in Taiwan cannot merely be explained by the enormous foreign exchange reserves available. The fact that, as we have already discussed, the private sector has to a large extent had to look after its own financial resources with little support of the authorities, no doubt plays an important role in this. Nevertheless, the events drew attention to the importance of modernising the financial system not only in Korea but also in Taiwan.

Summary

Although both in Korea and Taiwan the state – at least until recently – controlled the formal financial system, important differences exist that can be understood by the different motivations for their strict financial policy. In Korea, where the authorities intentionally channelled highly subsidised funds to chaebol, the financial system has been the most important instrument for implementing its economic policy. According to C.H. Lee (1992), the government and large private enterprises can be characterised as constituting a 'quasi-internal' organisation. The financial system should therefore be seen as an internal capital market. This set-up was an important reason for the financial and economic crisis which hit Korea by the end of 1997. In Taiwan, on the other hand, the authorities have primarily controlled the flow of funds to curb inflation and counter a concentration of private (financial) power. The latter also implies that for the larger private enterprises in Taiwan the financial funds available from the formal sector were very limited. This no doubt helped in keeping the effects of the Asian crisis limited in the case of Taiwan.

The organisation of markets

In this section, we will briefly discuss how the relations between the business actors in Korea and Taiwan are organised. More specifically, the co-ordinating role that the authorities play in this respect is addressed. In addition, the most prevalent forms of co-operation between companies are discussed.

Korea

The foregoing sections clearly illustrated that the Korean authorities have strongly directed the development of selected sectors of the economy. Although the control over an important part of the financial sources was their most important instrument, other devices (such as the tax system and the control of licences) were also employed to realise economic goals. It should not come as a surprise that the 'strategic choices' of the chaebol – as well as those of other companies – were strongly influenced by the often very restricted manoeuvring space granted by the Korean authorities (Whitley 1994). In the oligopolistic market structure that was frequently created by the state, at least two chaebol were allowed to develop an industry; the total number of companies was limited so that sufficient profit and expansion possibilities were offered. Companies in a given market do not only compete in a direct sense, but also in an indirect manner in building up or maintaining strong links with political leaders and bureaucrats at the centre. As was also illustrated in Box 4.1, the business–government nexus is still crucial in present-day Korea.

Because of the diversity of their activities and the fact that many economic activities are incorporated within their authority structure, the chaebol are quite independent of each other. In other words, there is little evidence of collusive behaviour between the chaebol (OECD 1996). Moreover, the government generally distrusts inter-firm alliances since this may undermine its own dominant position. The cumbersome relation among chaebol is exemplified by the failure of the co-operative R&D programme in computer technology in the mid-1980s when all leading companies involved wanted to develop their own prototype (Wade 1990). Similar problems appeared in the 1990s with the joint development programme of high definition TV technology.

As compared with Taiwan or Japan, subcontracting in Korea takes place on a more modest scale – certainly outside the group. Hyundai Motors, for example, only subcontracts about 40 per cent of value added in a standard car and a large share of this is manufactured by other firms within the Hyundai group. In addition, it usually concerns relatively standardised components. Furthermore, the relation between the subcontractor and subcontractee is much less reciprocal (i.e. both acting as partners) and more short term than, for instance, in Japan.

Taiwan

Unlike Korea, the authorities have not played a dominant role in directing the development of the private sector in Taiwan (Whitley 1994). We have already noted that informal circuits have been a more important source of finance for most private companies and that the relation between government and these companies has been rather distant. Although tax and other incentives have been in place to stimulate the development of priority sectors, the strategic direction of companies in general was not decisively influenced by these measures. In fact, perhaps the most direct attempt by the authorities to set up a private industry, i.e. the automobile sector, has turned into a failure (Chu 1994). Through the sizeable public sector, the KMT authorities – certainly in the past – nevertheless controlled certain lines of business (Fields 1995).

Given the relatively small size of most companies and the limited external funding available to expand, many have pursued a highly specialised (niche) strategy. As a result, most individual companies show a high degree of interdependency. Or, as Whitley (1994: 56) put it, 'each firm is connected to a large number of other ones through a complex web of deals, obligations, personal ties and joint activities'. The importance attached to doing business on the basis of personal networks is also observed in the overseas investments of Taiwanese firms. As will be illustrated in the next chapter, in Asia most FDI has gone to countries with a major Chinese community. Ethnicity is often considered an important determinant in building up a business relationship overseas.

According to Redding (1990), producers of finished goods strongly depend on orders from trading companies whereas components suppliers heavily depend on local factories. Since flexibility is considered to be of utmost importance, unlike for instance, in Japan, non-group subcontracting relations are usually neither exclusive nor long-lasting (Orrù *et al.* 1991). The most important criteria for concluding a contract are speed of delivery, costs and reputation of a producer.

Summary

In this section it has been shown that also in terms of the way in which the relations between the various business actors are organised, substantial differences exist between Korea and Taiwan. The much more direct interference with the private sector in Korea has resulted in more or less independent conglomerates with relatively little mutual interaction. In Taiwan, on the other hand, the lack of a co-ordinating institution has led to a widespread network of companies splitting up value-added tasks among themselves.

Government policy with regard to outward FDI

During the last two decades, government policies with regard to outward investment from late industrialising economies has gone through various stages. During the first wave of FDI from LIEs, few home governments had developed a specific FDI policy. Cross-border investments were subdued under general capital-control policies which were usually quite restrictive (UNCTAD 1995). Most countries faced capital scarcity and considered foreign exchange as a serious constraint to their economic development. In more recent years, a growing number of LIEs have introduced a more liberal outward FDI policy (Dunning *et al.* 1997). In this section, the development of government policy in Korea and Taiwan with regard to FDI is discussed.

Korea

Although the first Korean overseas investment project was registered in 1968, the authorities maintained an extensive capital control system until the mid-1980s (UNCTAD 1995). In view of a lack of domestic capital, technology and managerial capabilities, the policy with regard to outward investment was until the mid-1970s extremely restrictive (Kim and Wang 1996). The regulations were somewhat eased in the second half of the 1970s when applications for overseas projects would in principle, be approved with some restrictions.[30] However, applicants still had to satisfy very restrictive qualifications and, in addition, the Korean government continuously monitored each project by analysing the performance of the overseas subsidiaries (Kim and Wang 1996).

Although in 1981 the restrictions on the qualification requirements were somewhat eased and the application procedures simplified, only in 1987 were the foreign exchange restrictions genuinely liberalised. In that year, Korea experienced a current account surplus. In addition, as will be illustrated in Chapter 5, the rise of industrial wages and the appreciation of the Korean won were not compensated for by productivity gains which began to deteriorate Korea's international competitiveness (UNCTAD 1995). Furthermore, protectionist measures in major target markets made it increasingly difficult to sell products solely through exports. As a result of the liberalisation, minimum project values were specified, requiring only certification from foreign exchange banks for small investments, a notification procedure for mid-range projects and screening and approval for large overseas projects. In addition, many restricted categories were opened for foreign investment leaving only fourteen business areas closed for overseas investment.

Notwithstanding current account deficits re-emerging in the 1990s, the Korean authorities further liberalised their outward FDI policy. To achieve a more advanced economy and '*Segyehwa*' ('globalization'), the Korean government has recognised the need to support business activities abroad in order to enhance the competitiveness of domestic economies (*Business Korea Yearbook* 1996). In addition, Korea has joined the OECD (in 1996) which made them subject to the Organization's liberalisation codes. Next to investments in developing regions to maintain competitiveness in more labour-intensive industries ('natural labour resource seeking'), FDI in industrialised economies is also stimulated, not only to avoid trade barriers ('market seeking/retaining') but also to gain access to new technology ('strategic asset seeking').

As indicated in Appendix 4.1, at present investment projects amounting to less than US$10 million only have to be reported to one of the foreign exchange banks, whereas those projects smaller than US$50 million have to be reported to the Bank of Korea or a designated foreign exchange bank if the investor had a loan from that bank extending to more than 50 per cent of the investment amount. Only if an application exceeds an investment amount of US$50 million or relates to a restricted business category, is permission needed from the Bank of Korea and will only be granted after approval of the so-called 'Overseas Investment Advisory Committee'. No more than three business areas are in principle barred from overseas investment (namely leasing as well as selling in lots of real estate and the construction and operation of golf courses abroad). However, if carried out by the right actor, even investment in these areas is possible.[31] The Korean government has plans to pursue its liberalisation efforts with regard to outward FDI even further.[32]

At present, a wide range of financial, fiscal and insurance benefits are in place for overseas investors. The Export-Import Bank of Korea provides FDI credits with preferential rates up to 90 per cent of the investment amount. Foreign exchange banks offer foreign exchange loans up to 70 per cent of the investment amount (80 per cent for SMEs and leading export enterprises). Resource access is promoted by the Korea Petroleum Development Corporation and the Mining Promotion Corporation, which support surveys to develop essential foreign sources and may provide initial and operating capital for FDI projects.[33] Tax incentives include loss-reserve provisions, a reduction of corporate taxes paid abroad, an imputed foreign tax scheme and dividend income tax exemptions for overseas resource development projects. Furthermore, the government grants insurance benefits against non-commercial risks.

Apart from the above incentives, the Korean authorities also grant various other forms of services, such as providing information, consulting and advice about overseas investment projects through various institutions (see Appendix 4.1). In addition, a large number of bilateral agreements on FDI have been concluded with prospective host countries facilitating overseas business operations for Korean companies.

Taiwan

Taiwan's first outward investment dates back to 1959; in 1962 the government started to pay serious attention to it. In that year for the first time the 'Regulations Governing the Screening and Handling of Outward Investment and Outward Technical Co-operation Projects' were proclaimed. Basically, it was considered unreasonable, or a legal loophole for capital flight in some cases, that a local firm should decide to invest abroad when the country had to improve its own economic conditions (IDIC 1993: 10). As a result, very strict financial requirements were formulated stifling capital export from Taiwan.

In view of the energy crisis and global recession, by the end of the 1970s, incentives were launched to stimulate investments abroad to secure the supply of raw materials and natural resources for industrial use. The financial requirements for such overseas projects were relaxed and the restriction on reinvestment was loosened. In addition, tax incentives were offered for outward FDI solely in extractive industries involved in shipping natural resources back to Taiwan (UNCTAD 1995).

To take advantage of the US Caribbean Basin Initiative[34] in 1984 the government expanded the possibilities of outward investment by further relaxing the restrictions on the financial criteria for FDI in that region.[35] The most far-reaching change, however, took place in mid-1987 when the Taiwanese authorities decided to considerably relax their long-standing controls on foreign exchange and further revise regulations regarding outward investment. Minimum paid-in capital requirements were abolished. Since 1987 individuals have been allowed to remit up to US$ 5 million annually, out of the country. The approval procedures for corporations have been reviewed several times since then. Since 1996, a company that intends to invest an amount below US$ 20 million only has to register with the Investment Commission (Ministry of Economic Affairs); the proposal will be automatically approved. Investment projects amounting to more than US$ 20 million have to be approved by the Investment Commission. In principle, all projects have to meet one of the conditions listed in Appendix 4.2 that are considered to meet national interest criteria. Given the very broad nature of the criteria, in most cases *de facto* this is not an obstacle to approval.[36]

As in Korea, financial, fiscal and insurance benefits are provided for investment projects from Taiwan (Appendix 4.2). Firms that have received official approval from the Ministry of Economic Affairs to invest abroad are eligible to apply for an outward investment loan provided by the Export-Import Bank of the Republic of China up to 80 per cent of the investment amount. In addition, a sum up to 20 per cent of the total investment amount may be set aside as a reserve to cover possible losses on the project. Furthermore, a tax deduction possibility is granted for income generated abroad and for which taxes have been paid in the country where it was generated. Through the Export-Import Bank the government also supports an overseas investment insurance covering non-commercial risks such as expropriation, nationalisation, war and inconvertibility of or restriction on the remittance of capital and profit from the host country.[37] In order to encourage FDI in countries in the Caribbean and Latin America area that have formal diplomatic ties with Taiwan, subsidies are available to small and medium sized companies that invest a total of less than NT$ 75 million.[38]

Next to the aforementioned incentives, the Industrial Development and Investment Centre (IDIC) of the Ministry of Economic Affairs supplies companies with information on overseas investment environments, sponsors investment seminars and provides potential investors with overseas investment opportunities. In addition, the Taiwanese government has concluded a growing number of double taxation avoidance agreements.

The policy as described in the foregoing holds for Taiwanese FDI in all countries except the PRC. As is illustrated in Box 4.2, the regulations with regard to

Box 4.2 Formosa Plastics Group and the KMT: a fragile relationship

A striking example of the often cumbersome relationship between the Nationalist authorities and the Taiwanese business community is provided by the activities of the Formosa Plastics Group (FPG), Taiwan's largest private industrial group. FPG's chairman Y.C. Wang, who is not identified with the KMT, has frequently clashed with the authorities. The biggest dispute concerned the establishment of a huge industrial complex (costs: US$10.9 billion) at a well-known beauty spot on the island. The plans for the project were first announced in 1985 and quarrels over it have extended well into the 1990s. Frustrated with the authorities' red tape, Wang frequently threatened to shift new investments to the PRC. In 1992, for instance, FPG eventually bowed to government pressure and backed away from building a planned US$6 billion petrochemical complex in Fujian province. In August 1996, a project to build a US$3 billion power plant in Fujian was suspended after President Lee Teng-hui called for new guidelines on FDI in China, which according to him hurt investments at home and damages Taiwan's international competitiveness. The fact that FPG threatens to invest in the 'occupied territory' on the mainland makes these disputes especially sensitive. Lee's call to slow down FDI in the PRC should be seen in the light of China's push for an agreement to establish direct shipping, air and postal links; Taiwan, however, is very reluctant to sign such an agreement fearing the mainland's 'suffocating' embrace.

Sources: FT (29.12.95, 30.05.96, 17.08.96, 06.12.96), IDIC (July 1996)

investments in the PRC appear to be much less clear-cut. First of all, only 'indirect' investments (e.g. through a shell company in Hong Kong) in the PRC are allowed; direct investments made from Taiwan are prohibited. Another important difference with 'regular' FDI projects is that not only the Ministry of Economics Affairs (i.e. the Investment Commission), but also the Mainland Affairs Council is involved in the approval procedure, underlining the sensitive political nature of these investments. Experience has shown that, for those enterprises that formally seek approval for their investment in the PRC (as will be illustrated in Chapter 5, many companies do not), the chances that SMEs operating in labour-intensive industries gain approval are much higher than large, technology or capital-intensive companies that consider shifting production across the Taiwan Strait. At the same time, however, the possibility for individuals to take US$ 5 million out of the country enables them to invest directly in the PRC without formal approval. The authorities appear to fear that, if large, high-profile companies invest in the PRC, the flood of production investments (and hence, the de-industrialisation of Taiwan) can no longer be stopped.

Summary

In sum, we noted that both economies have since 1987 drastically liberalised outward FDI policy. Changing locational conditions certainly influenced the decision to give more freedom to their companies to operate abroad through FDI. In Korea, the recent membership of the OECD has been an important stimulus for the country's more recent liberalisation measures. Especially in the case of Taiwan, however, FDI policy is still not always transparent. This holds especially for FDI in the PRC for which considerable space for discretionary intervention still exists. Next to fears for de-industrialisation, political considerations no doubt also play a role. The latter is not confined to the relation with the PRC but was also illustrated by the special policy with regard to FDI in Latin American countries where investment projects by Taiwanese companies are closely related to diplomatic ties.

Summary and conclusions

In this chapter the way in which economic activities are organised in Korea and Taiwan has been described. The various sections revealed the substantial variations that exist between the economies in this respect which – for an important part – are related to their different institutional context. It was shown that in Korea the authorities deliberately created large conglomerates to lead the economy's growth. In Taiwan, on the contrary, the government–business relation has always been much more distant resulting in a dominance of SMEs in many (certainly the export-oriented) industries. The lack of unconditional government support combined with contradictory policy goals also explains the failure in Taiwan to create international trading companies. In Korea, in imitation of their Japanese examples, the general trading companies have become important forerunners of their present-day MNEs.

Although the Korean and the Taiwanese authorities both strongly influenced the expansion possibilities of their private sector through a strict financial policy, the considerations and outcomes have been quite different. In Korea, the formal financial system was utilised to support a limited number of selected companies. Although the former policy of privileged credit supply is not – at least formally – in place anymore, the recent severe crisis has revealed that banking practices in Korea are still very unhealthy and need serious restructuring. As a result of historical events and the political goal of limiting the accumulation of private wealth and power, in Taiwan the formal financial sources available to private companies have always been very restricted. These differences in institutional context have resulted in local markets that in Korea are often characterised by an oligopolistic structure in which chaebol operate rather independently. In Taiwan, on the other hand, complex networks of specialised companies that are highly interdependent are, in most cases, the rule.

These characteristics are no doubt significant determinants of the strategies of the 'average' Korean and Taiwanese company. Levy and Kuo (1991) in this regard distinguished between a so-called 'high volume assembly' strategy and a 'bootstrap' strategy. In case of a high volume assembly strategy, which is frequently observed in Korea, companies are prepared to invest in realising high volume production and to push their export performance – even if unit costs may exceed prevailing market prices. The large volume is supposed to result in rapid travel along the learning curve, leading to higher productivity. The price is considered the most important competitive edge in attempting to gain big market shares competing with existing market leaders. The large size of the chaebol also enables them to engage into long-term (e.g. R&D) projects. The interwovenness of these companies with the financial system and the authorities until very recently enabled these companies to take relatively high risks. In cases where a well connected chaebol ran into problems, it was backed up by the state. Given the large number of recent bankruptcies and in view of the requirements of the IMF and pressure of the international (financial) community, it is expected that this situation will change in the coming years. The pressure to reform the business system in Korea is also expected to lead to the down-sizing of the chaebol, forcing them to confine their activities to a limited number of sectors.

At the core of a bootstrap strategy, as that is often observed in Taiwan, is the continuous exploration of new niche markets for non-standardised products. In Taiwan, where, as was noted, government support for the private sector is very limited, the profit orientation is usually short term and projects requiring a long-term commitment are less frequent. A bootstrap strategy is particularly successful with products having a short life cycle, where the flexibility to follow innovations is of crucial importance. This characteristic implies a rather opportunistic and flexible approach emphasising short pay-back periods on investments. Since most companies internalise only a part of the production process (the rest is subcontracted to others) the required investments in production facilities are minimal.

Extending this line of thinking, the organisation of economic activities as discussed in this chapter no doubt also influences the OLI constellation that determines outward investment. Examples of factors that – according to MNE theory – *ceteris paribus* influence the size and nature of FDI are company size and the strong state support. In the remainder of this study, many more relationships that are relevant in the case of outward FDI from Korea and Taiwan are expected to come to the fore, underlining that the influence of the government in this respect is not confined to its FDI policy, as has been discussed in this chapter.

Appendices

Appendix 4.1 Outward FDI policy, Korea (as of end 1996)

Approval procedure	*Restrictions*	*Incentives*	*Government services*
* > US$50 m or restricted category: permission Bank of Korea only after approval Overseas Investment Advisory Committee * > US$10 m: report to Bank of Korea or designated FOREX bank if invested been loaned > 50% investment amount * < US$ 10 m: report to FOREX bank	* Leasing in lots of real estate * Selling in lots of real estate * Construction and operation of golf courses	* Financial: • Export-Import Bank Korea: overseas investment funds • FOREX banks: foreign exchange loans • Korean Mining Promotion Co.: overseas resource development fund • Korean Petroleum Development Co.: petroleum business fund loans * Fiscal: • Reserve against loss from FDI • Reduction of corporate taxes paid abroad • Impact foreign tax scheme • Dividend income tax exempt for overseas resource development projects * Insurance benefits	* Information, consulting and advice about FDI opportunities: Bank of Korea, Export-Import Bank of Korea, Small and Medium Industry Promotion Corporation * Bilateral agreements on FDI: • Investment guarantee agreements • Double taxation avoidance agreements

Source: *Business Korea Yearbook* (1997)

Appendix 4.2 Outward FDI policy, Taiwan (excluding FDI in the PRC) (as of end 1996)

Approval procedure	*Restrictions*	*Incentives*	*Government services*
* Individuals: < US$5 m per year abroad without restrictions * Corporations: • < US$20 m: automatically approved • > US$20 m: needs approval by Investment Commission	* Project not only beneficial to domestic economic development, but also meet one of the following conditions: • Provide raw materials required by domestic industries • Expansion of export domestic products or secure markets • Assist acquisition of pivotal technical know-how • Promotion export of technical know-how and therefore boosting FOREX earnings • Promotion international economic co-operation • Support adjustment domestic industrial structure • Introduction of technical know-how indirectly through investment in foreign company * If project provided with cash in form of FOREX: additional requirements • paid-in capital > NT$20 m • liability ratio (i.e. liability/net worth) <30%	* Financial: Export-Import Bank of ROC: outward investment loans * Fiscal: • Reserve against losses from FDI • Tax deduction * Insurance benefits: insurance against commercial risks * Government support for FDI into Latin America and Caribbean Area: • Travel expenses • Preferential loans • Preferential insurance	* Through IDIC: • Information on investment environments in other countries • Information on investment opportunities • Sponsoring investment sessions * Double taxation avoidance agreements

Source: Information provided by Investment Commission (Ministry of Economic Affairs)

5 Korea and Taiwan as outward investors

An aggregate empirical exploration

Introduction

In this chapter, we empirically explore the position of Korea and Taiwan as home countries of MNEs. The emphasis is on a presentation and analysis of aggregate data. The investigation at a more detailed (meso and firm) level is conducted in Chapters 6 and 7. Although Korea and Taiwan will be frequently compared, we have to underline that foreign direct investment is not defined in the same manner in both economies. With all its limitations (Cantwell 1992), FDI data are nevertheless used here as a proxy of MNE activities.

For Korea, data are available on both approval (see Table 5.1; column 1) and arrival (or realised) basis (column 2). In addition, net realised investment flows are published (column 3). The Bank of Korea (BOK) defines outward direct investments as 'any transactions related to the acquisition of shares or lending funds needed for conducting business in a foreign country, with a repayment period of more than one year' (UNCTC 1992a: 233).[1] *Realised* outward FDI differs from *approved* outward FDI due to time lags involved in the actual implementation of Korean investments abroad and to cancellations after obtaining authorisation from the government. *Net realised* FDI flows refer to the value of actual total investments minus the reduction of the value of investment, due to the liquidation of companies abroad or redemption of lending. FDI stock data are estimated as cumulative net realised FDI outflows since 1968.

For Taiwan, total outward FDI flows are published by two distinct sources: the Investment Commission (IC) of the Ministry of Economic Affairs (MOEA; Table 5.1, columns 4–6) and the Central Bank of China (CBC; column 7). Outward investment is defined as 'the situation where a domestic corporation invests capital independently or in association with another domestic corporation or corporations, or jointly with a foreign government, juristic persons, or individuals to establish a new enterprise abroad, increase capital to expand an existing overseas enterprise, or purchase the stocks of existing foreign companies' (IDIC 1993).[2] Although Taiwan is not a member of the IMF, the Central Bank of China separately publishes balance-of-payments statistics in a format similar to that of the Balance of Payments Statistics Yearbook of the IMF. A comparison of columns 6 and 7 in Table 5.1 shows a wide divergence between MOEA/IC and CBC

Table 5.1 Korean and Taiwanese outward direct investment flows and the number of investment projects (between brackets)

	Korea			Taiwan			
	(1) *Approved FDI (BOK)*	(2) *Realised FDI (BOK)*	(3) *Net realised FDI (BOK)*[a]	(4) *Approved FDI (MOEA/IC)*	(5) *Indirect approved FDI in PRC (MOEA/IC)*	(6) *Total approved FDI (4) + (5) (MOEAC/IC)*	(7) *FDI (CBC)*
1975	58[b] (117)	45[b] (102)	42[b] (91)	26[c] (94)	na	see (4)	0[d]
1976	18 (39)	7 (31)	6 (15)	4 (8)	na	see (4)	3
1977	18 (61)	21 (52)	19 (45)	14 (13)	na	see (4)	7
1978	38 (87)	34 (82)	31 (70)	5 (8)	na	see (4)	4
1979	98 (50)	21 (45)	16 (42)	9 (12)	na	see (4)	4
1980	21 (45)	17 (40)	12 (16)	42 (17)	na	see (4)	42
1981	109 (64)	28 (49)	27 (36)	11 (10)	na	see (4)	60
1982	121 (54)	101 (49)	98 (32)	12 (4)	na	see (4)	32
1983	83 (67)	109 (56)	103 (44)	11 (7)	na	see (4)	19
1984	67 (49)	50 (46)	48 (33)	39 (22)	na	see (4)	70
1985	219 (43)	113 (38)	64 (9)	41 (23)	na	see (4)	80
1986	364 (73)	183 (50)	158 (31)	57 (32)	na	see (4)	66

Table 5.1 Korean and Taiwanese outward direct investment flows and the number of investment projects (between brackets) (*cont.*)

	Korea			*Taiwan*			
	(1) Approved FDI (BOK)	*(2) Realised FDI (BOK)*	*(3) Net realised FDI (BOK)*[a]	*(4) Approved FDI (MOEA/IC)*	*(5) Indirect approved FDI in PRC (MOEA/IC)*	*(6) Total approved FDI (4) + (5) (MOEAC/IC)*	*(7) FDI (CBC)*
1987	367	410	320	103	na	see (4)	703
	(109)	(91)	(59)	(45)			
1988	474	216	157	219	na	see (4)	4,120
	(250)	(173)	(141)	(110)			
1989	943	567	392	931	na	see (4)	6,591
	(368)	(269)	(246)	(153)			
1990	1,611	959	812	1,552	na	see (4)	5,243
	(515)	(338)	(316)	(315)			
1991	1,511	1,115	1,027	1,656	174	1,830	1,854
	(527)	(443)	(420)	(364)	(237)	(601)	
1992	1,207	1,218	1,098	887	247	1,134	1,869
	(631)	(497)	(463)	(300)	(264)	(564)	
1993	1,888	1,260	1,015	1,661	3,168	4,829	2,451
	(1,049)	(682)	(617)	(326)	(9,329)	(9,655)	
1994	3,587	2,305	2,036	1,612	962	2,574	2,460
	(1,947)	(1,475)	(1,407)	(324)	(934)	(1,258)	
1995	4,914	3,059	2,748	1,357	1,093	2,450	2,678
	(1,560)	(1,285)	(1,193)	(339)	(490)	(829)	

Sources: BOK (various years), CBC (various years), MOEA/IC (various issues)

Notes: na = not applicable
a Excluding liquidations, etc.
b Accumulated FDI 1962–75
c Accumulated FDI 1952–75
d Accumulated FDI 1974–75

sources, especially since the second half of the 1980s. While one would expect that approved outflows as published by the MOEA/IC would exceed the actual foreign investments appearing in the balance-of-payments (UNCTC 1992a), until 1992 and again in 1995 the opposite phenomenon is observed. An important explanation is that, as pointed out in Chapter 4, due to the relaxation of foreign exchange controls after the mid-1980s, investment projects representing an amount smaller than US$ 5 million need not be registered. (In 1996 this amount was increased to US$20 million.) Many cases of outward investment – often involving smaller scale companies – therefore have not gone through the approval procedures of the Taiwanese authorities. Convincing evidence that actual FDI flows are indeed much more substantial than the MOEA/IC figures suggest is rendered by publications of host countries reflecting inward direct investments from Taiwan. This holds especially for investments in Asia, where many smaller Taiwanese companies have invested. An investigation conducted by Kuo and Chen (1996) using the latter sources for the period 1986–95, for instance, put Taiwanese outward investment in Asia at US$ 54.8 billion, which is more than six times the total *officially* approved outward investment (MOEA/IC) figure from Taiwan! An additional complication is constituted by Taiwan's investments into the People's Republic of China (PRC). Only in early 1991 did the Investment Commission state that all enterprises that have indirectly (i.e. via a third country, often Hong Kong) invested in the PRC, have to register with the MOEA/IC – these figures are shown in column 5 of Table 5.1. According to Kuo (1993), however, by the end of 1992 more than 10,000 investment projects had already been approved by the PRC authorities instead of the 501 listed in the table. It is generally acknowledged that the total FDI outflow figures published by the Central Bank approximate the actual investment flows better than the MOEA/IC data (UNCTC 1992a: 297). The most serious flaw may concern the (officially prohibited) investments by private persons which are carried out directly, i.e. not via a third country, in the People's Republic of China; since individuals are allowed to take US$ 5 million out of the country, it is very difficult to prevent them from investing in the PRC. These investments are also not registered in the balance-of-payments statistics. Outward stock data are not available as such, but are estimated as cumulative approved outflows since 1959 (UNCTC 1992a: 292).

Although the first investment abroad by a Korean firm was already made in 1968 (in a forestry project in Indonesia) and a Taiwanese company invested overseas as early as 1959 (in a cement factory in Malaysia), FDI from Korea and Taiwan has only recently gained momentum. In the case of Korea, we note that – with the exception of 1983 – only since 1986 have net realised FDI flows exceeded US$ 100 million. Since then, the size of annual FDI flows has increased substantially. Since 1991 annual net outward direct investments has amounted to more than US$ 1 billion. For Taiwan, the rapid rise of FDI is even more pronounced. As in Korea, the emergence of Taiwanese investors abroad really took off in the second half of the 1980s. The approved FDI (according to MOEA/IC sources) skyrocketed from US$ 103 million in 1987 to US$ 4,829 million in 1993. According to the balance-of-payments, outward direct investments jumped from US$ 66 million in 1986 to US$ 703 million in the following year and reached a peak of US$ 6,951 in 1991. Although since then the amounts have stabilised at a lower level, ever since 1988 annual outflows (according to the Central Bank of China) have exceeded US$ 1 billion.

In the following sections of this chapter, the characteristics of outward direct investments from Korea and Taiwan are further investigated. We will first study the emergence of both economies as outward investors in an international context. Subsequently, the geographical and industrial distribution of Korean and Taiwanese FDI is explored. Next, the investments by Korea and Taiwan in Europe are discussed in more detail. Locational changes that stimulated the emergence of FDI from Korea and Taiwan are then discussed. The question whether Korea and Taiwan are not only 'late industrialisers' but also 'late internationalisers' is thereupon addressed. Finally, the findings of this chapter are summarised and conclusions are drawn.

Korea and Taiwan as outward investors: an international comparison

In the introduction we noted a strong growth of the size of outward direct investments from Korea and Taiwan in recent years. How important are both economies as home countries of MNEs as compared to other late as well as early industrialised economies?[3] FDI stocks or cumulative flows represent the book value of assets (valued at their historic costs) and, consequently, underestimate the FDI *stock* of countries with a strong tradition of MNE activity (Cantwell 1992). Therefore, in Table 5.2 the most important regions and countries in terms of FDI *flows* are depicted for the period 1981–95.[4] Both first and second wave LIE investors are included.[5]

From the table, a number of interesting trends can be observed. Although the share of early industrialised economies (EIEs) has decreased from 94.43 per cent in 1981–85 to 87.48 per cent ten years later, these countries are still a dominant source of outward FDI flows. Within the advanced world, however, a limited number of countries account for the majority of cross-border direct investment activities. In the first half of the 1990s, the United States were the single biggest investor in the world. As compared to the first half of the 1980s, their share even went up slightly from 21.68 per cent to 23.83 per cent of global FDI flows. The second and third biggest outward investors in the world were the UK (10.47 per cent in 1991–95) and Japan (10.11 per cent). However, the share of global FDI coming from the UK has become much more modest than it used to be in the past. In the first half of the 1980s 18.20 per cent of global outward FDI still originated in the UK. Other important EIE home countries are France (9.78 per cent in 1991–95), Germany (8.99 per cent), the Netherlands (5.29 per cent) and Switzerland (3.46 per cent). The table illustrates that the European Union is still by far the most important home region of MNEs. Although their share has gone down, the four EU members listed in the table (namely France, Germany, the

Netherlands and the UK) still accounted for 34.53 per cent of global FDI flows in the first half of the 1990s. Although the EU countries account for the largest share of outward (as well as inward) FDI, it is fair to note that a very substantial proportion of these investments is made in the European Union. By the end of 1993, for instance, intra-EU outward FDI stock amounted to US$ 339 billion,

Table 5.2 Home regions and countries of outward direct investment flows (annual averages in US$ million/%)

	1981–85		*1986–89*		*1991–95*	
	(US$m)	*(%)*	*(US$)*	*(%)*	*(US$)*	*(%)*
Total	50,625	100.00	173,402	100.00	237,112	100.00
EIEs	47,807	94.43	162,709	93.83	207,416	87.48
USA	10,975	21.68	21,820	12.58	56,512	23.83
Japan	5,092	10.06	32,084	18.50	23,963	10.11
France	2,695	5.32	16,687	9.62	23,192	9.78
Germany	4,474	8.84	15,882	9.16	21,310	8.99
Netherlands	3,835	7.58	10,021	5.78	12,547	5.29
Switzerland	2,068	4.08	5,130	2.96	8,210	3.46
UK	9,215	18.20	28,090	16.20	24,826	10.47
Other countries	9,453	18.67	32,995	19.03	36,856	15.54
LIEs	2,818	5.57	10,693	6.17	29,696	12.52
Africa	1,167	2.31	1,008	0.58	638	0.27
Asia	1,044	2.06	7,429	4.28	25,607	10.80
Hong Kong	285	0.56	1,845	1.06	15,046	6.35
Korea	82	0.16	469	0.27	1,791	0.76
Malaysia	246	0.49	296	0.17	1,314	0.55
Philippines	9	0.02	2	0.00	2	0.00
PRC	225	0.44	711	0.41	2,956	1.25
Singapore	134	0.26	591	0.34	1,820	0.77
Taiwan	52	0.10	3,417	1.97	2,262	0.95
Thailand	1	0.00	77	0.04	386	0.16
Other countries	10	0.02	21	0.01	30	0.01
Central and Eastern Europe	4	0.01	22	0.01	237	0.10
Middle East	131	0.26	772	0.45	789	0.33
Latin America and Caribbean	468	0.92	1,453	0.84	2,422	1.02
Argentina	10	0.02	30	0.02	7	0.00
Brazil	179	0.35	329	0.19	813	0.34
Chile	7	0.01	9	0.01	492	0.21
Colombia	40	0.08	29	0.02	123	0.05
Mexico	74	0.15	142	0.08	511	0.22
Venezuela	12	0.02	211	0.12	455	0.19
Other countries	166	0.33	703	0.41	21	0.01
Other countries	4	0.01	9	0.01	3	0.01

Source: BOK (various years); CBC (various years); DTCI/UNCTAD Database on FDI; Plummer and Montes (1995)

Notes:
Missing data concern 1981 and 1982 for Switzerland; 1981 for the PRC; 1981–83 for Chile and 1981 and 1983 for Venezuela. For these countries, the annual average for the 1981–85 period is based on the remaining years for which FDI flow data were available. The minus value for Argentina reflects a net disinvestment by Argentinian companies: more investments were withdrawn than invested overseas.

whereas the FDI stock in the US which originated in EU countries amounted to US$ 249 billion. The total amounted invested in Japan did not exceed a modest US$ 10 billion (UNCTAD 1995).

Not only within the group of early industrialised economies but also among late industrialised economies (LIEs) substantial shifts in relative importance during the last decade-and-a-half can be observed. Most striking is the increased importance of developing Asia (excluding Japan) as a home region for MNEs. Its share in global FDI flows went from 2.06 per cent in the first half of the 1980s to 10.8 per cent ten years later. The other third world home region of some importance is Latin America whose share only slightly increased from 0.92 per cent to 1.02 per cent. The share of Africa in the same period dropped from 2.31 per cent to 0.27 per cent.

As noted earlier, the importance of developing Asia as a source of foreign direct investments has increased drastically. Whereas in the first half of the 1980s a modest annual outflow of only US$ 1.0 billion for the region was registered, one decade later this amount had gone up to US$ 25.6 billion. Taiwan has become one of the biggest outward investors outside the early industrialised world. Its share went from 0.10 per cent in the first half of the 1980s to 0.95 per cent of global FDI ten years later. Although less spectacular, Korea's global role also has become more considerable, having increased its FDI share from 0.16 per cent to 0.76 per cent in the same period. Perhaps surprising is the substantial share of global FDI flows originating in the PRC in recent years; in the first half of the 1990s no less than 1.25 per cent came from China making it one of the largest outward investors among developing countries (Zhang and van den Bulcke 1996).[6] The share of Singapore, an important outward investor described in the first wave literature (cf. Aggarwal 1985, Lim and Teoh 1986), has not grown as rapidly as the previous cases (from 0.26 per cent to 0.77 per cent). Moreover, a considerable part of FDI from Singapore is undertaken by foreign companies which have an important base in the city state (Lim and Teoh 1986).[7] The figures for Hong Kong have to be interpreted with great caution. According to Table 5.2, no less than 6.35 per cent of global FDI flows in the first half of the 1990s came from the city state, making it one of the biggest outward investors in the world. It is known, however, that a (very) substantial part of outward investments from Hong Kong is undertaken by non-indigenous firms. As will be discussed more elaborately in the next section, a very important example of the latter are the Taiwanese firms that invest from their Hong Kong base in the People's Republic of China (van Hoesel 1996). The role of the Philippines has become negligible since the first wave of studies on Third World multinationals. After a sharp drop in the second half of the 1980s, Malaysia's share in global FDI has gone up again to 0.55 per cent in 1991–95. A newcomer is Thailand which in 1981–85 hardly registered any FDI and ten years later had captured 0.16 per cent of global FDI.

In Latin America, Brazil is still the most important home country of MNEs although its share has decreased slightly from 0.35 per cent in 1981–85 to 0.34 per cent ten years later. Remarkable growers in Latin America are Venezuela and

Chile. In the period 1981–85 only 0.02 and 0.01 per cent of global FDI came from these countries, respectively; ten years later their shares had gone up to 0.19 and 0.21 per cent. Mexico increased its share from 0.15 per cent to 0.22 per cent. Colombia's share, on the other hand, went down from 0.08 per cent to 0.05 per cent. Argentina, which is considered the oldest home of 'Third World multinationals' (Lall 1983), has become an insignificant player in terms of outward investment.

Although the foregoing provides an overview of the global position of individual countries as outward investors, it does not render any insight into how important outward investment is in proportion to their domestic economies. For this purpose two ratios (namely, FDI stock/GDP and FDI stock/Export) are listed for a number of years in Table 5.3. In the following, the discussion is confined to Korea and Taiwan; for comparison purposes the same countries that were listed in Table 5.2, are also included in Table 5.3.

The FDI stock/GDP ratio is listed to normalise for the size of the home economies. How important is outward investment for Korea and Taiwan if related to the size of their domestic economies? For Korea, we observe that its FDI stock/GDP ratio has gone up 10 times from 0.002 in 1980 to 0.02 by the end of 1994.[8] In the case of Taiwan, the rapidly increased importance of overseas direct

Table 5.3 FDI stock/GDP and FDI stock/export ratio for various years

	FDIs/GDP			*FDIs/EXP*		
	1980	*1990*	*1994*	*1980*	*1990*	*1994*
USA	0.081	0.079	0.091	0.982	1.118	1.214
Japan	0.019	0.069	0.059	0.155	0.719	0.723
France	0.038	0.087	0.124	0.220	0.500	0.829
Germany	0.053	0.101	0.112	0.235	0.387	0.478
Netherlands	0.271	0.424	0.416	0.624	0.894	1.061
Switzerland	0.223	0.271	0.355	0.515	0.848	0.732
UK	0.149	0.235	0.274	0.734	1.270	1.364
Hong Kong	0.005	0.177	0.455	0.008	0.161	0.389
Korea	**0.002**	**0.009**	**0.020**	**0.008**	**0.037**	**0.082**
Malaysia	0.018	0.053	0.087	0.032	0.080	0.111
Philippines	0.005	0.004	0.002	0.030	0.019	0.011
PRC	0.000	0.007	0.027	0.002	0.048	0.135
Singapore	0.056	0.110	0.096	0.036	0.084	0.071
Taiwan	**0.002**	**0.080**	**0.086**	**0.005**	**0.192**	**0.225**
Thailand	0.000	0.005	0.008	0.002	0.017	0.025
Argentina	0.000	0.003	0.001	0.009	0.034	0.025
Brazil	0.003	0.014	0.013	0.032	0.076	0.124
Chile	0.000	0.006	0.037	0.008	0.021	0.173
Colombia	0.004	0.011	0.012	0.034	0.057	0.070
Mexico	0.001	0.002	0.005	0.009	0.021	0.035
Venezuela	0.000	0.027	0.046	0.001	0.057	0.146

Sources: IMF (various years), UNCTAD (1995)

investments for the domestic economy is even more remarkable. The figures suggest a growth of the FDI stock/GDP ratio between 1980 and 1994 from 0.002 to 0.086 or a forty-fold increase. Taiwan's ratio in 1994 is not only higher than that of most LIEs but also exceeds that of Japan and is almost as high as the FDI stock/GDP ratio of the USA.

The second ratio listed in Table 5.3, FDI stock/EXP(orts), could serve as a proxy for the relative importance of international business by means of FDI as compared with exports from the domestic economy. In the case of Korea, the FDIs/EXP ratio also has gone up about ten times between 1980 and 1994 from 0.008 to 0.082. For Taiwan we note an even sharper increase between 1980 and 1994 from 0.005 to 0.225. With the exception of Hong Kong, this is the highest FDIs/EXP ratio in the late industrialised world.

Summing up, we conclude that Asia's share in global outward investment has increased substantially in the last decade-and-a-half. Moreover, we observed that within Asia, Korea and Taiwan are among the most important home economies of MNEs leaving important early (first wave) outward investors such as Singapore, Brazil and Argentina behind. In addition, we can conclude that, although not yet as important as in most leading outward investors in the world, both in the case of Korea and (especially) Taiwan the rapidly grown role of outward investment is also very apparent if related to the size of their domestic economies. Moreover, although exports remain the dominant mode of doing business with other countries, a clear shift in importance towards FDI for both Korea and Taiwan was observed.[9]

The geographical and industrial distribution of Korean and Taiwanese outward FDI

In this section, the geographical and industrial distribution of outward direct investment from Korea and Taiwan is examined. First, FDI patterns from both economies are discussed separately. Subsequently, a comparison of the destinations of Korean and Taiwanese outward investments is made.

Korean FDI

In Figure 5.1 total Korean FDI stock in the various regions (namely, Asia, North America, Europe, and 'other areas') is depicted against time (excluding investment in real estate). The figure again illustrates the rapid expansion of FDI since the second half of the 1980s. Since 1994, Asia has received most Korean FDI. By the end of 1995 44.7 per cent of Korean FDI (or US$ 4,574 million) had gone to that region. Although not indicated in the figure, we note that within the region, no less than 41.2 per cent of this amount has gone into China – a position that has materialised since 1992 when investments in China started to grow rapidly. The other major host countries are Indonesia (20.7 per cent), followed by Vietnam (6.7 per cent) and Malaysia (6.3 per cent).

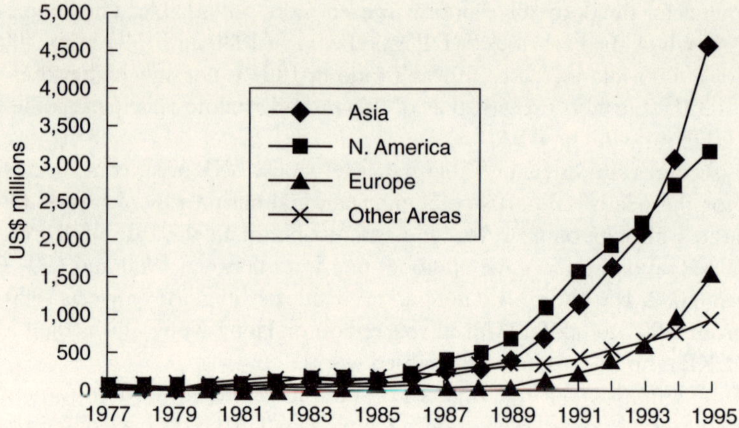

Figure 5.1 Regional FDI stock – Korea
Source: BOK (various issues)

Until 1993, North America had been the most popular destination for Korean investors; since 1987 substantial annual FDI flows exceeding US$ 100 million into the region have been recorded. By the end of 1995, North America was the second most popular region for Korean investors having hosted 30.1 per cent of total investments (or US$ 3,162 million). Within North America, the United States received the biggest share (85.9 per cent), Canada attracting the remaining 14.1 per cent.

Until recently, Europe was relatively unknown territory for most Korean investors. Only since 1992, have annual Korean FDI flows into Europe exceeded US$ 100 million. As of the end of 1995, Korean companies had invested US$ 1,550 million in Europe, representing 14.8 per cent of total outstanding FDI. The distribution of Korean FDI over the various countries in Europe will be discussed in more detail in a separate section in this chapter.

By the end of 1995 the other regions in the world had in total received US$ 957 million or 9.1 per cent of total Korean FDI. The Bank of Korea sources show that, within this category, Australia is the most important host country having received 17.5 per cent followed by the Sudan (12.5 per cent) and Algeria (7.9 per cent).

In Table 5.4, an overview of the sectoral distribution of outstanding Korean FDI stock for the various regions is provided. In terms of amounts invested, we observe that more than half of all Korean FDI (57.3 per cent) had gone into manufacturing activities, whereas 19.0 per cent of FDI has gone into trade activities. This distribution reflects the situation as of the end of 1995. In the past, however, the sectoral distribution looked quite different. In the early 1970s, investors in timber projects and construction were more dominant, whereas in the late 1970s a shift was observed to trading, manufacturing and construction (Han and Brewer 1987). In the beginning of the 1980s, the relative importance of investments in the (coal) mining industry was more substantial than nowadays.

Although Korean companies still invest in mining, as a result of the fast rise of investments in the manufacturing industry, its share has decreased. As a resource-poor country, securing access to natural resources remains an important motivation to invest abroad, however.

Table 5.4 shows that Asia is the most important host region for Korean FDI in the mining, manufacturing, transportation and real estate sectors. North America attracts most investors in construction, trade and 'miscellaneous' sectors. With regard to Korean investments in forestry, Papua New Guinea in particular, is a popular destination. Latin America (especially Argentina and Panama) – next to North America – attracts substantial Korean investments in fisheries. These countries are included under 'other areas'.

Substantial variations exist with regard to the sectoral distribution between the various regions. In Asia, manufacturing is more important than in the other regions, representing 72.5 per cent of all Korean FDI in that region. As compared to North America and Europe, trade investments in the region are relatively unimportant (5.8 per cent), underlining the importance of Asia as a production location instead of as a target market. Although not specified in the table, major differences can also be discerned within the region. In China, the biggest recipient, by the end of 1995 no less than 88.5 per cent of Korean FDI had gone into manufacturing. The rapidly increased popularity of China as a production location was triggered by improved diplomatic relations between the two countries in 1992 (Kim 1995). In addition, production costs are lower than in most of the other countries in the region where Korean FDI initially primarily went to. Mounting infrastructural problems in the latter countries, and abundant natural resources have also contributed to the importance of China as a host country. As a result of the prohibition on selling products in China, trade investments represent a scant 0.3 per cent in China. The sectoral distribution of Korean FDI in the PRC differs importantly from that, for instance, in Indonesia. Although manufacturing in Indonesia had also attracted 60.4 per cent of total Korean investments, 33.9 per cent of Korean FDI went into mining activities. In Vietnam, a great majority of FDI (86.4 per cent) has gone into manufacturing.

The sectoral distribution of Korean FDI in North America and Europe is rather similar. The most important difference concerns the share of FDI in manufacturing that is higher in Europe than in North America (54.4 per cent versus 42.4 per cent). The distribution within North America is, however, different. In Canada, 73.9 per cent of Korean capital has been invested in manufacturing activities and 19.0 per cent in mining. In the United States, on the other hand, only 37.3 per cent of Korean FDI was aimed at manufacturing. The biggest sector was trade, representing 41.4 per cent of all investments. Also in Europe, investments in trade activities received a very substantial share of total Korean FDI, namely 32.1 per cent.

In the 'other areas', manufacturing represents 34.3 per cent of Korean FDI, followed by mining (25.7 per cent) and other 'miscellaneous' activities (16.0 per cent). As will be shown on the basis of the three biggest recipients in this category, important inter-country variations exist. The Bank of Korea sources show that in

Table 5.4 Industrial distribution of Korean FDI per region (as of end 1995) (US$m; %)

	Asia		*N. America*		*Europe*		*Other Areas*		*Total*	
Mining	328 6.8%	46.8%	113 3.6%	16.1%	14 0.9%	2.0%	246 25.7%	35.1%	701 6.7%	100.0%
Forestry	30 0.6%	38.5%	0 0.0%	0.0%	0 0.0%	0.0%	48 5.0%	61.5%	78 0.7%	100.0%
Fisheries	16 0.3%	17.8%	22 0.7%	24.4%	3 0.2%	3.3%	49 5.1%	54.4%	90 0.9%	100.0%
Manufacturing	3,502 72.5%	58.2%	1,340 42.4%	22.3%	843 54.4%	14.0%	328 34.3%	5.5%	6,013 57.3%	100.0%
Construction	72 1.5%	34.1%	82 2.6%	38.9%	5 0.3%	2.4%	52 5.4%	24.6%	211 2.0%	100.0%
Transportation & storage	55 1.1%	71.4%	12 0.4%	15.6%	2 0.1%	2.6%	7 0.7%	9.1%	77 0.7%	100.0%
Trade	278 5.8%	14.0%	1,144 36.2%	57.4%	497 32.1%	24.9%	73 7.6%	3.7%	1,992 19.0%	100.0%
Miscellaneous	292 6.1%	27.5	439 13.9%	41.3%	180 11.6%	16.9%	153 16.0%	14.4%	1,063 10.1%	100.0%
Real estate	256 5.3%	94.1%	10 0.3%	3.7%	6 0.4%	2.2%	1 0.1%	0.4%	272 2.6%	100.0%
Total	4,829 100.0%	46.0%	3,162 100.0%	30.1%	1,550 100.0%	14.8%	957 100.0%	9.1%	10,497 100.0%	100.0%

Source: BOK (1996)

Note:
Reading from the left to the right, the percentages indicate the relative importance of the various regions as hosts for Korean FDI in the various sectors. The percentages in one column reflect the relative importance of the various sectors for the respective regions.

the case of Australia, the biggest host in this category, no less than 57.5 per cent of Korean FDI has gone into mining; manufacturing represents only 9.8 per cent. For the Sudan, on the other hand, 62.4 per cent of Korean FDI has been invested in the manufacturing sector, followed by construction (27.6 per cent). Finally, in Algeria, a great majority (93.3 per cent) has flown into the non-specified 'miscellaneous' category; the remaining 6.7 per cent of Korean investments have been used to initiate trade activities.

In Table 5.5, a further distinction into the various branches of the manufacturing industry is made. Before going into geographical details, we observe that no less than 40.2 per cent of all Korean investment abroad in the manufacturing sector can be accounted for by companies active in fabricated metals (including electronics). Other important outward investors operate in textiles and clothing (12.1 per cent), basic metals (9.1 per cent) and petroleum products (8.8 per cent).

The table further underlines the dominant position of Asia as a production region for Korean investors. With the exception of basic metals, for all branches Asia is the most important host region. More than 80 per cent of Korean investments in the leather and footwear, non-metals and 'miscellaneous' manufacturing industries have gone into Asia. The same holds for more than 70 per cent of the food, petroleum, and machinery and equipment industries. In terms of amounts invested in the region, fabricated metals (32.3 per cent) is the most important branch of industry, followed by textiles and clothing (13.9 per cent), petroleum (11.6 per cent) and 'miscellaneous' branches of industry (10.5 per cent).[10] Investments in the fabricated metals industry primarily refer to the assembly of electronics goods – a relatively labour-intensive operation. This also holds for the textiles and clothing industry, which underlines the attraction of Asia as a cheap production region.[11]

Comparing the distribution of Korean FDI over the branches of industry for the three biggest Asian recipients of investments in the manufacturing sector (namely, China, Indonesia, and Malaysia; these country figures are not specified in the table) reveals substantial variations between these countries. In China, the fabricated metals industry has received 29.0 per cent, the textiles and clothing industry 15.1 per cent and the 'miscellaneous' category (e.g. toys) 11.4 per cent of total Korean investments. In Indonesia, fabricated metals has also received most Korean FDI (23.5 per cent), followed, however, by the petroleum (21.4 per cent) and food (12.8 per cent) industries. Finally, in Malaysia, fabricated metals is by far the most important target for Korean investors (48.6 per cent), followed by non-metals (18.1 per cent) and 'miscellaneous' industries (12.4 per cent).

As noted earlier, the basic metals industry is the only exception for which Asia is not the most important recipient of Korean FDI. In this line of business, no less than 82.4 per cent of total Korean invested capital went to North America. It should be noted, however, that two take-overs account for an important part of the total amount invested in that sector. In 1984, Pohang Steel Co. (POSCO) acquired US Steel for US$ 186 million.[12] Sammi Special Steel Co. in 1989 bought Rio Algoon Ltd., one of the largest steel manufacturers

Table 5.5 Manufacturing Korean FDI per region (as of end 1995) (US$m; %)

	Asia		*N. America*		*Europe*		*Other Areas*		*Total*	
Food	181 5.2%	78.4%	25 1.9%	10.8%	16 1.9%	6.9%	9 2.8%	3.9%	231 3.8%	100.0%
Textile & clothing	488 13.9%	66.8%	44 3.3%	6.0%	33 3.9%	4.5%	165 50.5%	22.6%	730 12.1%	100.0%
Leather & footwear	233 6.7%	85.7%	7 0.5%	2.6%	8 0.9%	2.9%	24 7.3%	8.8%	272 4.5%	100.0%
Wood & furniture	90 2.6%	60.0%	40 3.0%	26.7%	16 1.9%	10.7%	4 1.2%	2.7%	150 2.5%	100.0%
Paper & printing	55 1.6%	57.3%	39 2.9%	40.6%	2 0.2%	2.1%	0 0.0%	0.0%	96 1.6%	100.0%
Petroleum	408 11.6%	76.7%	51 3.8%	9.6%	49 5.8%	9.2%	24 7.3%	4.5%	532 8.8%	100.0%
Non-metals	294 8.4%	80.5%	54 4.0%	14.8%	10 1.2%	2.7%	7 2.1%	1.9%	365 6.1%	100.0%
Basic metals	73 2.1%	13.4%	450 33.6%	82.4%	19 2.3%	3.5%	4 1.2%	0.7%	546 9.1%	100.0%
Fabricated metals	1,132 32.3%	46.8%	584 43.6%	24.1%	628 74.4%	26.0%	76 23.2%	3.1%	2,420 40.2%	100.0%
Machinery & equipment	181 5.2%	75.4%	20 1.5%	8.3%	39 4.6%	16.3%	0 0.0%	0.0%	240 4.0%	100.0%
Miscellaneous	368 10.5%	85.4%	24 1.9%	5.8%	24 2.8%	5.6%	14 4.3%	3.2%	431 7.2%	100.0%
Total	3,503 100.0%	58.3%	1,339 100.0%	22.3%	844 100.0%	14.0%	327 100.0%	5.4%	6,013 100.0%	100.0%

Source: BOK (1996)

Note:
Reading from the left to the right, the percentages indicate the relative importance of the various regions as hosts for Korean FDI in the various sectors. The percentages in one column reflect the relative importance of the various sectors for the respective regions.

in Canada, for US$ 112 million. For the North American region, fabricated metals industry nevertheless is the most important target for Korean investors (43.6 per cent), followed by basic metals (33.6 per cent). As can be read from the table, the remaining 22.8 per cent of Korean FDI in that region went to branches of industry all representing shares smaller than 5 per cent of the total amount invested. In Canada, the relative share of the fabricated metals industry is higher than in the US (57.9 versus 39.0 per cent). In the US, the basic metals industry has almost received as much Korean capital as fabricated metals (namely, 33.4 per cent). Also in Canada the basic metals industry is the only other highly popular manufacturing industry for Korean companies, having received 34.3 per cent of total investments in manufacturing.

As far as the 'other areas' are concerned, it is interesting to note that a substantial share of all Korean investments has gone into the textiles and clothing industry (22.6 per cent). For the regions themselves, this branch of industry represents no less than 50.5 per cent of all Korean FDI in manufacturing, followed by fabricated metals (23.2 per cent), and the petroleum industry (7.3 per cent). For these branches of industry, a remarkable geographical concentration of Korean FDI can be noticed (BOK 1996). Of all Korean investments in the textiles and clothing industry in these countries, no less than 61.5 per cent has gone to Latin American countries (especially Panama, Honduras and Guatemala). Mexico has received 67.0 per cent of the investments in the fabricated metals industry, whereas 53.8 per cent of the investments in the petroleum industry by the end of 1995 had gone to Saudi Arabia.

As already indicated, Korean FDI in Europe will be studied in more detail later on in this chapter.

Taiwanese FDI

As was pointed out earlier, the FDI statistics published by the Investment Commission of the Ministry of Economic Affairs (MOEA/IC) are on an approval basis and have to be interpreted with great caution. Since this is the only source publishing more detailed information on outward investments (including industrial and regional breakdowns) these data nevertheless are used here. In Figure 5.2, total approved cumulative FDI to the various regions is shown. We note that, as in the case of Korea, the great majority of direct investment flows from Taiwan are of a recent nature.

Because of the special relationship between the two economies, indirect FDI flows from Taiwan into the People's Republic of China (PRC) are depicted separately. Until the end of the 1970s, the PRC and Taiwan were in a state of serious hostility (Yu 1993). Since then, the relationship has shifted from a military orientation to peaceful coexistence and from economic blockade to trade exchange and (unilateral) investments. In 1979, when Beijing initiated its economic reform programme, a prudent start of indirect trade activities (via Hong Kong) was made. This trend gained momentum especially after the abolition of martial law in Taiwan after 1987. Although officially approved indirect FDI in the PRC is only registered since 1991, the figure shows that since 1993 the PRC has become the single most important host country for Taiwanese companies, having received US$ 5,642 million or 35.5 per cent of total Taiwanese outward investment by the end of 1995 from the island. (FDI fom the PRC to Taiwan has not yet been registered.) Within the Asian region, the PRC has received 58.1 per cent of FDI from Taiwan. Other studies have shown that in reality, the amounts invested in the PRC are more substantial, which would make their position as a recipient even more predominant.[13] The PRC is not only considered an attractive production location providing cheap labour at a short physical and psychological distance (sharing Chinese culture and history) and abundant raw materials, it also represents a fast growing market. In addition, as was discussed in Chapter 4, political considerations sometimes also play a role in the case of investments in the PRC.

The dominant position of Asia as a host region is further underscored by the position of the 'rest of Asia' which by the end of 1995 had attracted another 25.6 per cent of total Taiwanese FDI. As was discussed in the introduction to this chapter, since many smaller Taiwanese companies in Asia have invested relatively small amounts, a substantial share of these investment projects will not have been registered, suggesting that in reality, FDI amounts in Asia are substantially higher (Chen and Chen 1995). Other important host countries – next to the PRC – in the regions that are not indicated separately in the figure are Malaysia (12.3 per cent of all FDI in Asia, including the PRC), Hong Kong (7.2 per cent) and Thailand (6.4 per cent). All three countries have sizeable Chinese communities which illustrates the importance of ethnic (business) links for Taiwanese investors in selecting a host country. Nevertheless, lately these countries have seen their share of Taiwanese FDI decline in favour of investments in the PRC.

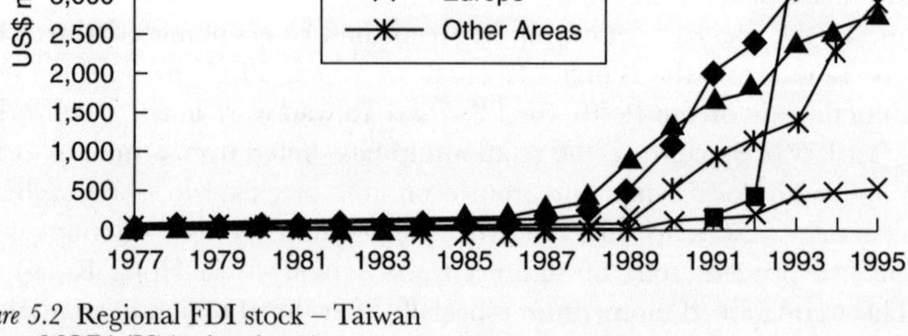

Figure 5.2 Regional FDI stock – Taiwan
Source: MOEA/IC (various issues)

During the late 1970s, when FDI from Taiwan was still very low, North America received more than half of total investments. Its position has since become more modest. By the end of 1995, US$ 2,765 million or 17.4 per cent of total Taiwanese investments abroad had gone to North America. Of that amount, no less than 97.8 per cent went to the US and only 2.2 per cent to Canada. At the end of 1995 Europe had received a modest US$ 559 million or 3.5 per cent of total approved Taiwanese FDI.

By the end of 1995, the 'other areas' (namely, Latin America, Africa, Oceania) in the world had received US$ 2,862 million (or 18.0 per cent of Taiwanese investments). MOEA/IC sources show that the most important recipients in this category are located in Latin America; this region had attracted no less than 94.4 per cent of Taiwanese FDI going to these other areas. As was pointed out in Chapter 4, special incentives exist for investment projects in that region. Only two countries are specified (namely, Panama and Mexico); since the investment amounts which have flowed to these two countries are rather modest (US$ 122 million and US$ 44 million respectively), the most important host countries unfortunately remain unknown.

Table 5.6 gives a sectoral distribution of Taiwanese FDI over the various regions. On approval basis, we note that by the end of 1995 no less than 65.9 per cent of all direct investments abroad were made in setting up or expanding manufacturing activities overseas. Other major targets have been banking and insurance (17.8 per cent), trade (7.7 per cent) and 'miscellaneous' non-specified sectors (5.6 per cent). FDI in banking and insurance can be considered a response to the world-wide liberalisation of financial markets and the deregulation of financial markets in Taiwan (J. Lee 1992). Moreover, these investments reflect a trend of banks and other financial service companies to follow Taiwanese manufacturing and trading companies overseas in order to support their international business activities. Although in the second half of the 1970s 5.0 per cent of Taiwanese investments abroad were still aimed at the mining sector, it only represented a modest amount (namely, US$ 3.8 million). For the remainder, FDI intended to secure access to natural resources has always been of limited importance for Taiwan – unlike many other countries, among which is Korea (J. Lee 1992).

With the exceptions of construction, transportation, and banking and insurance, Asia is the most important host region for all sectors. Taking FDI in Asia and the PRC together, no less than 81.3 per cent of FDI in the mining industry has flowed to the region, 100.0 per cent in the case of agriculture and forestry, 81.4 per cent in the fishery and animal husbandry industry, 79.0 per cent of manufacturing investments, 47.0 per cent in the case of trade and 40.4 per cent of Taiwanese FDI in 'miscellaneous' sectors. With regard to the construction industry, North America (33.6 per cent) and 'other areas' (39.0 per cent) are the biggest recipients of Taiwanese investments.[14] Taiwanese FDI in banking and insurance especially has found its way to 'other areas' (64.0 per cent) and North America (22.5 per cent). No less than 98.2 per cent of FDI in the 'other areas' category has gone to non-specified countries in the Latin American region, probably – at least partly – to the tax havens in the Caribbean region.

Table 5.6 Industrial distribution of Taiwanese FDI per region (as of end 1995) (US$m; %)

	Asia		*PRC*		*N. America*		*Europe*		*Other Areas*		*Total*	
Mining	3 0.1%	18.8%	10 0.2%	62.5%	0 0.0%	0.0%	0 0.0%	0.0%	3 0.1%	18.8%	16 0.1%	100.0%
Agriculture & forestry	15 0.4%	42.9%	20 0.4%	57.1%	0 0.0%	0.0%	0 0.0%	0.0%	0 0.0%	0.0%	35 0.2%	100.0%
Fishery & animal husbandry	1 0.0%	3.7%	21 0.4%	77.7%	0 0.0%	0.0%	0 0.0%	0.0%	5 0.2%	18.5%	27 0.2%	100.0%
Manufacturing	3,024 74.3%	28.9%	5,246 93.0%	50.1%	1,543 55.8%	14.7%	317 56.7%	3.0%	338 11.8%	3.2%	10,468 65.9%	100.0%
Construction	23 0.6%	15.8%	17 0.3%	11.6%	49 1.8%	33.6%	0 0.0%	0.0%	57 2.0%	39.0%	146 0.9%	100.0%
Transportation	55 1.4%	21.6%	37 0.7%	14.5%	20 0.7%	7.8%	0 0.0%	0.0%	143 5.0%	56.1%	255 1.6%	100.0%
Trade	427 10.5%	34.9%	148 2.6%	12.1%	256 9.3%	20.9%	85 15.2%	7.0%	307 10.7%	25.1%	1,223 7.7%	100.0%
Banking & insurance	300 7.4%	10.6%	2 0.0%	0.1%	638 23.1%	22.5%	79 14.1%	2.8%	1,813 63.3%	64.0%	2,832 17.8%	100.0%
Miscellaneous	220 5.4%	24.6%	141 2.5%	15.8%	259 9.4%	29.0%	78 14.0%	8.7%	196 6.8%	21.9%	894 5.6%	100.0%
Total	4,068 100.0%	25.6%	5,642 100.0%	35.5%	2,765 100.0%	17.4%	559 100.0%	3.5%	2,862 100.0%	18.0%	15,896 100.0%	100.0%

Source: MOEA/IC (1996)

Note:
Reading from the left to the right, the percentages indicate the relative importance of the various regions as hosts for Korean FDI in the various sectors. The percentages in one column reflect the relative importance of the various sectors for the respective regions.

As with Korean investments abroad, in the case of Taiwan the sectoral distribution also differs substantially between the various regions. In Asia (excluding the PRC) almost three-quarters of total approved Taiwanese investments has gone to the manufacturing sector. For the PRC, this share was as high as 93.0 per cent. Other substantial targets in Asia are trade (10.5 per cent) and banking and insurance (7.4 per cent). If we compare the three biggest recipients in the region (excluding the PRC) (these countries are not specified separately in the table), we note that in two cases (namely, Malaysia and Thailand), manufacturing is by far the most important sector having received 97.3 per cent and 92.5 per cent, respectively. In the case of Hong Kong, however, the sectoral distribution is very different. The share of Taiwanese FDI in manufacturing is a modest 8.9 per cent. The most important sectors in Hong Kong are trade (41.1 per cent), followed by services (25.3 per cent; in Table 5.6 included in the 'miscellaneous' category) and banking and insurance (19.3 per cent). As was indicated earlier, Hong Kong plays a special intermediary role in Taiwan–PRC business activities. Taiwanese businessmen cannot travel directly to China and transactions have to be made through an intermediary or a neutral territory (Kwok 1995). In most cases, Hong Kong is the necessary middleman for Taiwanese companies in providing a transfer point for personnel visits and financial transactions, and transit point for business operations between headquarters in Taipei and production facilities in the PRC. This explains the high share of investments in the trade and service sectors.

As can be read from Table 5.6, in North America by the end of 1995 55.8 per cent of approved Taiwanese investments had been directed at manufacturing activities. A substantial share (namely, 23.1 per cent) had gone to banking and insurance, whereas trade activities attracted 9.3 per cent.[15] The relative importance of Taiwanese FDI in the manufacturing sector in Europe is similar to that in the US (56.7 per cent). Other sizeable receiving sectors are trade (15.2 per cent), banking and insurance (14.1 per cent) and miscellaneous activities (14.0 per cent).

Interestingly, for the 'other areas' in the world that are not specified further in the table, the distribution over the sectors is quite different from that in the regions mentioned earlier. Here, banking and insurance is the most important sector having attracted 63.3 per cent of total approved inward investments, followed by manufacturing (11.8 per cent) and trade (10.7 per cent). In Latin America (including Panama and Mexico), the biggest recipient in this category, banking and insurance had received as much as 65.9 per cent of Taiwanese FDI. Other substantial sectors in Latin America are trade (10.9 per cent), and manufacturing (10.6 per cent).

Table 5.7 provides further details of Taiwanese FDI in the various branches of manufacturing industry. The electronic and electrical appliances (or 'electronics') industry is the most important source of FDI (by the end of 1995: 24.2 per cent), followed by chemicals (14.4 per cent), basic metals and metal products (9.7 per cent), rubber and plastic products (9.4 per cent) and food and beverages (9.2 per cent).

Table 5.7 Manufacturing Taiwanese FDI per region (as of end 1995) (US$m; %)

	Asia		*PRC*		*N. America*		*Europe*		*Other Areas*		*Total*	
Food & beverages	106	11.0%	654	67.8%	160	16.6%	0	0.0%	44	4.6%	964	100.0%
	3.5%		12.5%		10.4%		0.0%		13.1%		9.2%	
Textiles	411	51.2%	318	39.7%	0	0.0%	55	6.9%	18	2.2%	802	100.0%
	13.6%		6.1%		0.0%		17.4%		5.4%		7.7%	
Leather & fur	4	5.2%	73	94.8%	0	0.0%	0	0.0%	0	0.0%	77	100.0%
	0.1%		1.4%		0.0%		0.0%		0.0%		0.7%	
Garments &	37	15.5%	171	71.5%	1	0.4%	0	0.0%	30	12.6%	239	100.0%
footwear	1.2%		3.3%		0.1%		0.0%		9.0%		2.3%	
Lumber & bamboo	66	22.6%	207	70.9%	18	6.2%	0	0.0%	1	0.3%	292	100.0%
	2.2%		3.9%		1.2%		0.0%		0.3%		2.8%	
Paper products &	225	61.6%	136	37.3%	4	1.1%	0	0.0%	0	0.0%	365	100.0%
printing	7.4%		2.6%		0.3%		0.0%		0.0%		3.5%	
Chemicals	354	23.6%	386	25.7%	734	48.8%	1	0.1%	28	1.9%	1,503	100.0%
	11.7%		7.4%		47.6%		0.3%		8.4%		14.4%	
Rubber & plastic	118	12.0%	805	81.6%	50	5.1%	1	0.1%	12	1.2%	986	100.0%
products	3.9%		15.3%		3.2%		0.3%		3.6%		9.4%	
Non-metallic minerals	293	46.6%	325	51.7%	6	1.0%	0	0.0%	5	0.8%	629	100.0%
	9.7%		6.2%		0.4%		0.0%		1.5%		6.0%	
Basic metals &	503	49.4%	484	47.5%	19	1.9%	0	0.0%	13	1.3%	1,019	100.0%
metal products	16.7%		9.2%		1.2%		0.0%		3.9%		9.7%	
Machinery equipment	27	15.0%	144	80.0%	8	4.4%	0	0.0%	1	0.6%	180	100.0%
	0.9%		2.7%		0.5%		0.0%		0.3%		1.7%	

Table 5.7 Manufacturing Taiwanese FDI per region (as of end 1995) (US$m; %) (*cont.*)

	Asia		*PRC*		*N. America*		*Europe*		*Other Areas*		*Total*	
Electronic & electric appliances	846 28.0%	33.5%	883 16.8%	34.9%	535 34.7%	21.2%	88 27.8%	3.5%	177 52.8%	7.0%	2,529 24.2%	100.0%
Transport equipment	21 0.7%	4.4%	279 5.3%	59.0%	3 0.2%	0.6%	170 53.6%	35.9%	0 0.0%	0.0%	473 4.5%	100.0%
Precision equipment	10 0.3%	2.5%	382 7.3%	94.8%	3 0.2%	0.7%	2 0.6%	0.5%	6 1.8%	1.5%	403 3.9%	100.0%
Total	3,021 100.0%	28.9%	5,247 100.0%	50.2%	1,543 100.0%	14.7%	317 100.0%	3.0%	335 100.0%	3.2%	10,461 100.0%	100.0%

Source: MOEA/IC (1996)

Note:
Reading from the left to the right, the percentages indicate the relative importance of the various regions as hosts for Korean FDI in the various manufacturing industries. The percentages in one column reflect the relative importance of the various manufacturing industries for the respective regions.

As in the case of Korea, the table underlines the dominance of the Asian region as a production location for Taiwanese companies. With only one exception (namely, chemicals), Asia is by far the most important host region for Taiwanese investments in the various branches of manufacturing. Taking Asia and the PRC together, by the end of 1995 they had attracted no less than 78.8 per cent of Taiwanese FDI in the food and beverages industry, 90.9 per cent in textiles, 100.00 per cent in leather and fur products, 87.0 per cent in garments and footwear, 93.5 per cent in lumber and bamboo products, 98.9 per cent in paper products and printing, 93.6 per cent in rubber and plastic products, 98.3 per cent in non-metallic minerals, 96.9 per cent in basic metals and metal products, 95.0 per cent in machinery equipment and 68.4 per cent in the electronic and electric appliances industry. Only in the case of chemicals is the share of Taiwanese FDI attracted by North America almost as large as Asia (48.8 per cent versus 49.3 per cent). As can be read from the table, in a number of cases the lion's share of Taiwanese investments is hosted by the PRC.

A comparison of the three biggest recipients (excluding the PRC) of FDI in manufacturing branches (namely, Malaysia, Thailand and Indonesia) reveals substantial differences in the relative shares of the various lines of industry. In Malaysia, basic metals and metal products (37.8 per cent) and electronics (32.2 per cent) are by far the most attractive branches for Taiwanese investors, followed – at a distance – by textiles (13.0 per cent) and chemicals (9.1 per cent). In Thailand, the electronics industry is the biggest recipient of Taiwanese FDI in the manufacturing sector (47.1 per cent). Other branches of industry that have attracted Taiwanese investors are chemicals (17.3 per cent), food and beverages (9.6 per cent) and textiles (6.4 per cent). Indonesia shows a very different picture. The sector which attracted most Taiwanese investments is paper products and printing (25.4 per cent), followed by non-metallic minerals (18.5 per cent), textiles (17.9 per cent), rubber and plastic products (13.0 per cent) and electronics (9.9 per cent).

An interesting recent phenomenon is the active involvement of Taiwan in the development of a special economic zone in Subic Bay, the Philippines (van Hoesel 1996). A joint venture between the Philippine government, the Subic Bay Metropolitan Authority and Taiwan's United Development Co., commenced building in 1994, a 300 hectare industrial park on the former US naval base. A satisfactory infrastructure – initially constructed by the US army and improved by the joint venture – the availability of cheap labour and various (tax) facilities offered by the Philippine government made it fit for export-oriented light industries, such as computer assembly. By the end of 1995, accumulated Taiwanese FDI in the Philippines amounted to US$ 293 million. In Vietnam, Taiwan is also involved in establishing industrial zones to house Taiwanese companies. At present, Taiwan is among the biggest investors in that country with FDI stock amounting to US$ 414 million by the end of 1995.

In North America, almost half (namely, 47.6 per cent or US$ 734 million) of Taiwanese approved FDI in the manufacturing sector has gone into the chemicals industry. This also makes North America overall the most attractive host region for Taiwanese FDI in this branch of industry. Other popular lines of business in which Taiwanese companies have invested are electronics (34.7 per cent), and the food and beverages industry (10.4 per cent).

With regard to the 'other areas', we note that more than half (52.8 per cent) has been invested in the electronics industry, followed by food and beverages (13.1 per cent), garments and footwear (9.0 per cent) and the chemicals industry (8.4 per cent). For all these branches of industry Taiwanese companies have invested predominantly in Latin American countries.

Taiwanese investment activities in Europe are discussed in a separate section.

Korean and Taiwanese FDI compared

The outward direct investment patterns from Korea and Taiwan share important characteristics. In Korea and Taiwan, outgoing FDI is of a (very) recent nature. In both economies, a major part of FDI is directed towards the Asian region. In Taiwan (61.1 per cent), however, the share of total FDI that has gone to Asian countries is substantially higher than in Korea (46.0 per cent). This is clearly a result of the popularity of the PRC as a potential host country. For both economies, manufacturing is the most important target sector, although its relative importance for Taiwan (65.9 per cent) is higher than for Korea (57.3 per cent).

Important differences between the Korean and Taiwanese FDI pattern were also revealed. By the end of 1995, Taiwan had invested a considerably larger amount abroad (especially on a per capita basis) than Korea. At the same time, however, Korean investments abroad show a wider geographical divergence than Taiwanese FDI. Next to the Asian region, substantial amounts have been invested in other parts of the world, especially in North America and, to a lesser degree, Europe. In the case of Taiwan, the shares that have been invested in these regions – certainly in Europe – are substantially smaller. In addition, the distribution of FDI over the various sectors shows some interesting discrepancies. In the case of Korea, next to manufacturing (57.3 per cent), trade (19.0 per cent) and mining (6.7 per cent) are important target sectors. In Taiwan, on the other hand, the second most popular sector next to the manufacturing sector (65.9 per cent), is banking and finance (17.8 per cent). FDI in trade has captured only a comparatively small share (7.7 per cent).

Unfortunately, the fact that the subcategories distinguished for the Korean and Taiwanese manufacturing industries do not match exactly makes a comparison more difficult. In the case of Korea, 'fabricated metals' (including electronics) is by far the most important source of outward investment (40.2 per cent); though less dominant, in Taiwan electronics firms are also the most investment prone (24.2 per cent). Basic metals in both economies represent a similar share of FDI (9.7 per cent for Taiwan and 9.1 per cent for Korea). For Taiwan, other important sectors are chemicals (14.4 per cent), rubber and plastic products (9.4 per cent) and food and beverages (9.2 per cent). In Korea, the picture looks quite different. Textiles and clothing are the second most important industry (12.1 per cent), followed by basic metals (9.1 per cent) and petroleum products (8.8 per cent).

We conclude this section by comparing the size of Korean and Taiwanese investments abroad. In Appendix 5.1, an overview of the average size of investment projects in the various regions and sectors is provided. Overall, investment projects of Korean firms are larger in size than those initiated by Taiwanese companies. Although the sectoral categories are not equivalent, it seems to hold for investments in most kinds of economic activities. The major exception concerns FDI in Asia (excluding the PRC), where Taiwanese firms have invested in considerably larger projects than Korean firms; this also holds for FDI in manufacturing in that region. Interestingly, the size of Taiwanese investment projects (including those in manufacturing) in the PRC is very small. Although not reflected separately in the table, the size of Korean FDI in the PRC is also smaller than in other countries. How can this be explained? Many of these investment projects from Korca and Taiwan are made by small and medium sized enterprises facing worsening L advantages (such as rising wage levels) at home. In addition, the small average size of investment projects in the PRC can be explained by the fact that Korean and Taiwanese entrepreneurs are very cautious about these investments due to uncertainties of China's investment climate (Chen 1995, Kim 1995).

Korean and Taiwanese investments in Europe

Until recently, the size of Korean and Taiwanese FDI in Europe was very modest. It was not until the end of 1990, that FDI stock from both economies exceeded US$ 100 million. Since then, the Korean presence in Europe in particular, has expanded considerably, amounting to more than US$ 1.5 billion by the end of 1995. The small discrepancies between the figures in Table 5.8 and those mentioned in Tables 5.4 and 5.5 can be ascribed to the rounding off effect. Taiwanese approved FDI by that time had reached US$ 559 million. Although the Taiwanese economy is much smaller than Korea's, we have seen that *overall* global outward investment from the island economy is much larger than Korean FDI underlining the modest role Europe plays as a host country for Taiwanese firms. From a European perspective, it is fair to note that firms from both economies have until now contributed only marginally to FDI stock in the region. By the end of 1995, Korean FDI represented 0.14 per cent of total FDI in Europe and in Taiwan a scant 0.06 per cent.

In which countries in Europe have Korean firms invested? From Table 5.8 we note that by the end of 1995, 68.1 per cent of all FDI had been aimed at EU countries. Within the EU, Germany used to be the biggest recipient of Korean FDI (18.6 per cent of total FDI in Europe), followed by the UK (15.5 per cent), the Netherlands (11.3 per cent) and France (11.0 per cent).[16] It should be stressed that this 'static' overview is due to change considerably. As will be pointed out in Chapter 7, large-scale new investments in the UK will lead to a major increase in the share of Korean FDI going to the UK. Striking is the rapidly increased importance of Central and Eastern Europe as a host region for Korean companies. From virtually no investment five years ago, their share by the end of 1995 had gone up to no less than 30.4 per cent of all Korean FDI in Europe. The privatisation programmes in the former East bloc offer relatively cheap opportunities for Korean firms to enhance their technological capabilities by acquiring former state-owned companies (van Hoesel 1998a). As will be discussed further in Chapter 7, Samsung's picture tube factory (previously named 'FGT') was acquired in former East Germany. Most active in this respect, however, has been Daewoo which has invested in former state companies in Poland, the Czech Republic, and Rumania.[17] In the case of the Polish FS Osobowych factory, for instance, the production of the outdated Polonez cars is gradually replaced by Daewoo models which, by the way, are based on products from other car manufacturers. Although Daewoo expects to sell cars successfully in Central and Eastern Europe, the company exports an important share of its production to Western Europe. An additional advantage is that many of the components of Daewoo cars in the future will be made locally, so that cars from these factories will be sold almost tariff-free in Western Europe. By acquiring majority shares in Czech Avia and Polish FS Lublin factories Daewoo hopes to learn quickly how to build light trucks and vans. By the end of 1995, the most important host countries in the former East bloc were Rumania (10.2 per cent of total FDI in Europe), Uzbekistan (7.1 per cent) and Russia (4.3 per cent).

How does the industrial distribution of Korean FDI in Europe look? For Europe as a whole, more than half (54.6 per cent or US$ 843.5 million) has been invested in manufacturing activities. In the EU, however, the relative share of Korean FDI in the manufacturing sector (38.8 per cent) is much lower than in Central and Eastern Europe where no less than 91.1 per cent of all Korean FDI had gone into manufacturing activities. The latter clearly illustrates the importance of Central and Eastern Europe as a production region. Within the manufacturing sector, fabricated metals was by far the most popular target having received 40.7 per cent of all FDI in Europe and 74.4 per cent of all FDI in the manufacturing sector. The dominance of fabricated metals as a manufacturing target holds both for FDI in the EU (68.0 per cent) and Central and Eastern Europe (80.7 per cent). The only other important sector for Korean firms in Europe is the trading sector which attracted 32.3 per cent of FDI stock (namely, US$ 497.5 million). In the EU, the share of FDI in trading activities is even higher (namely, 45.9 per cent). In Central and Eastern Europe trade activities represent a modest 3.0 per cent of outstanding Korean FDI illustrating the still limited relative importance of the region as a target market.

Unfortunately, in Taiwan MOEA/IC sources specify only five individual countries and a (sizeable) 'other countries' category. In Table 5.9 we can read that by the end of 1995 53.2 per cent of approved FDI in Europe (or US$ 297.5 million) had gone to the UK. It is fair to note, however, that one very large investment project in the UK has not yet materialised (in transportation equipment, amounting to more than US$ 170 million) and it therefore distorts the picture. On the other hand, recent individual project announcements (among others by Acer Peripherals) underline the attractiveness of the UK as a host for Taiwanese companies. Other important host countries are the Netherlands (9.8 per cent) and

Table 5.8 Korean FDI in Europe (as of the end of 1995) (US$ 1,000; %)

		Textile & clothing		*Petroleum*		*Fabricated metals*		*Manufacturing (total)*		*Trade*		*Total*
European Union		1,678	0.2%	33,536	3.2%	277,833	26.4%	408,482	38.8%	482,744	45.9%	1,051,608
		5.0%		68.9%		44.3%		48.4%		97.0%		68.1%
	France	1,500	0.9%	4,197	2.5%	91,192	53.8%	97,290	57.4%	69,806	41.2%	169,393
		4.5%		8.6%		14.5%		11.5%		14.0%		11.0%
	UK	78	0.0%	2,624	1.1%	60,606	25.2%	104,274	43.4%	130,457	54.3%	240,087
		0.2%		5.4%		9.7%		12.4%		26.2%		15.5%
	Germany	100	0.0%	0	0.0%	46,178	16.1%	60,273	21.0%	222,275	77.4%	287,249
		0.3%		0.0%		7.4%		7.1%		44.7%		18.6%
	Netherlands	0	0.0%	0	0.0%	11,991	6.9%	20,211	11.6%	6,685	3.8%	174,097
		0.0%		0.0%		1.9%		2.4%		1.3%		11.3%
	Italy	0	0.0%	0	0.0%	14,520	46.3%	14,520	46.3%	16,549	52.8%	31,345
		0.0%		0.0%		2.3%		1.7%		3.3%		2.0%
Central & Eastern Europe		31,511	6.7%	3,000	0.6%	344,743	73.5%	427,321	91.1%	13,945	3.0%	469,086
		94.9%		6.2%		54.9%		50.7%		2.8%		30.4%
	Uzbekistan	1,670	1.5%	2,500	2.3%	105,000	95.6%	109,170	99.4%	650	0.6%	109,849
		5.0%		5.1%		16.7%		12.9%		0.1%		7.1%
	Rumania	0	0.0%	0	0.0%	157,710	100.0%	157,710	100.0%	0	0.0%	157,760
		0.0%		0.0%		25.1%		18.7%		0.0%		10.2%

Table 5.8 Korean FDI in Europe (as of the end of 1995) (US$ 1,000; %) (*cont.*)

	Textile & clothing		*Petroleum*		*Fabricated metals*		*Manufacturing (total)*		*Trade*		*Total*
Poland	0	0.0%	0	0.0%	41,192	93.3%	41,192	93.3%	2,960	6.7%	44,152
	0.0%		0.0%		6.6%		4.9%		0.6%		2.9%
Russia	163	0.2%	400	0.6%	4,320	6.5%	35,823	54.1%	3,550	5.4%	66,196
	0.5%		0.8%		0.7%		4.2%		0.7%		4.3%
Hungary	0	0.0%	9,228	23.1%	21,741	54.4%	35,678	89.3%	4,282	10.7%	39,960
	19.0%		0.0%		3.5%		4.2%		0.9%		2.6%
Total	33,189	2.1%	48,666	3.2%	627,840	40.7%	843,450	54.6%	497,465	32.2%	1,544,383
	100.0%		100.0%		100.0%		100.0%		100.0%		100.0%

Source: BOK (1996)

Note:
The percentages in one column reflect the relative importance of the countries for a particular sector. Since some countries which are not a member of the EU and are not located in Central and Eastern Europe (e.g. Switzerland) are included, the percentages may not add up to 100.0%. Reading from the left to the right, the percentages indicate the relative importance of the various sectors for a particular host country.

Table 5.9 Taiwanese FDI in Europe (as of the end of 1995) (US$ 1,000; %)

	Textiles		*Electronic/electric appliances*		*Transport equipment*		*Manufacturing (total)*		*Trade*		*Banking & insurance*		*Total*
UK	0	0.0%	64,531	21.7%	170,370	57.3%	235,338	79.1%	10,791	3.6%	4,800	1.6%	297,512
	0.0%		73.4%		100.0%		74.2%		12.7%		6.1%		53.2%
Germany	0	0.0%	11,861	26.6%	0	0.0%	14,408	32.3%	13,236	29.7%	0	0.0%	44,625
	0.0%		13.5%		0.0%		4.5%		15.6%		0.0%		8.0%
France	0	0.0%	2,309	18.2%	0	0.0%	2,309	18.2%	3,818	30.0%	5,593	44.0%	12,720
	0.0%		2.6%		0.0%		0.7%		4.5%		7.1%		2.3%
Netherlands	0	0.0%	5,073	9.2%	0	0.0%	5,073	9.2%	39,293	71.5%	10,400	18.9%	54,985
	0.0%		5.8%		0.0%		1.6%		46.2%		13.2%		9.8%
Switzerland	0	0.0%	718	92.5%	0	0.0%	718	92.5%	58	7.5%	0	0.0%	776
	0.0%		0.8%		0.0%		0.2%		0.1%		0.0%		0.1%
Other	54,800	36.8%	3,404	2.3%	0	0.0%	59,204	39.8%	17,902	12.0%	58,193	39.1%	148,765
countries	100.0%		3.9%		0.0%		18.7%		21.0%		73.7%		26.6%
Total	54,800	9.8%	87,896	15.7%	170,370	30.5%	317,050	56.7%	85,098	15.2%	78,986	14.1%	559,383
	100.0%		100.0%		100.0%		100.0%		100.0%		100.0%		100.0%

Source: MOEA/IC (1996)

Note:
The percentages in one column reflect the relative importance of the countries for a particular sector. Reading from the left to the right, the percentages indicate the relative importance of the various sectors for a particular host country.

Germany (8.0 per cent). The non-specified countries have attracted 26.6 per cent of total FDI in Europe. Until recently, investments in Eastern Europe were forbidden, but it is expected that their importance as a host region for Taiwanese firms will increase in the future.[18]

In which industries did these firms invest? By the end of 1995, 56.7 per cent (or US$ 317.1 million) had been invested in manufacturing activities, especially in electronic and electrical appliances and textiles.[19] Other important sectors are trade (15.2 per cent)[20] and banking and insurance (14.1 per cent).[21] In addition, we observe that the sectoral distribution varies substantially between countries. The production of electronic and electrical appliances, for instance, for a major part (73.4 per cent) takes place in the UK whereas almost half of FDI in trading (including distribution) activities was made in the Netherlands.

In sum, we conclude here that for Korean firms in particular, Europe in recent years has become an increasingly important host region. For Taiwanese companies, Europe remains rather more unknown territory. Moreover, not only in absolute but also in relative terms Korean companies are more heavily engaged in manufacturing activities in the region than Taiwanese firms. It appears, however, that the severe crisis in Korea which occurred at the end of 1997 has serious effects for the pace of internationalisation of the companies. As will also be illustrated in Chapter 7, the actual operations of some of the major projects that were in announced in 1996 and 1997 are seriously delayed or perhaps even abandoned.

Changing L advantages

Although the focus in the remainder of the study will be on the firm-level analysis of cross-border investment activities, changes in L advantages have unmistakably also played a substantial role in the recent growth in outward FDI. Although some locational aspects were touched upon in the foregoing, it makes sense to briefly address the most important ones more systematically.

As discussed in Chapter 3, the rapid increase of wage levels has made Korea and Taiwan less suited for labour-intensive production activities. In Korea, the basis for this new situation was laid in the 1970s. The second half of that decade saw the end of the period of excess labour supply and labour shortages set in (Cho 1994). This tendency was stimulated by the spurt of investment in heavy and chemical industries which was accompanied by the competitive bidding up of wages by firms in priority industries. In addition, the construction boom in the Middle East created severe bottlenecks on the labour market (Sakong 1993).[22]

This trend was further reinforced by the democratisation movement that emerged in the second half of the 1980s which gave more room for wage demands by labour unions.[23] In Figure 5.3 the annual growth rates of industrial wages and labour productivity since 1980 are depicted. The figure clearly shows that for most years, the growth in wage levels was not matched by an increase in labour productivity, implying an erosion of Korea's traditional comparative advantage in labour-intensive production.

A similar trend was observed in Taiwan. The rapid growth of the Taiwanese economy in the 1960s and 1970s had led to a situation of nearly full employment. As a result, in the 1970s industrial wages started to increase very rapidly (Fields 1992). Since then, the labour market – especially for the low-skilled workforce – has remained very tight. The Labour Standards Law (LSL), promulgated in 1984, also had an important impact on the continued rise in wages. Regulations with regard to minimum wage levels, pension and severance payments, overtime premiums, etc., increased the costs of labour substantially (van Hoesel 1996).

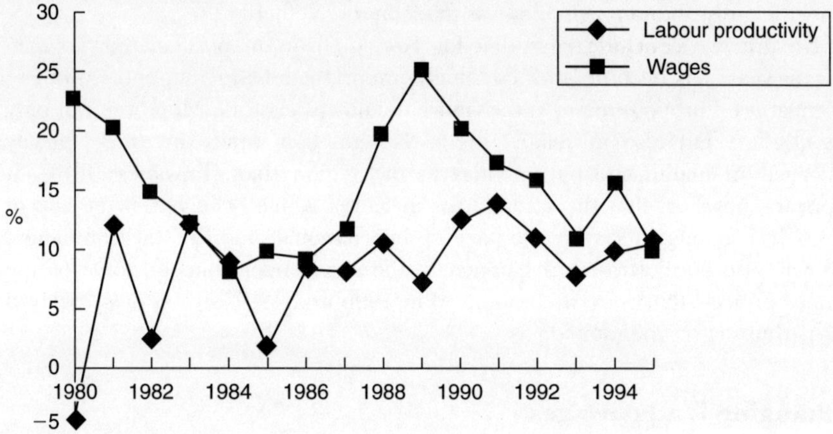

Figure 5.3 Growth rates in labour productivity and industrial wages – Korea
Source: NSO (various issues)

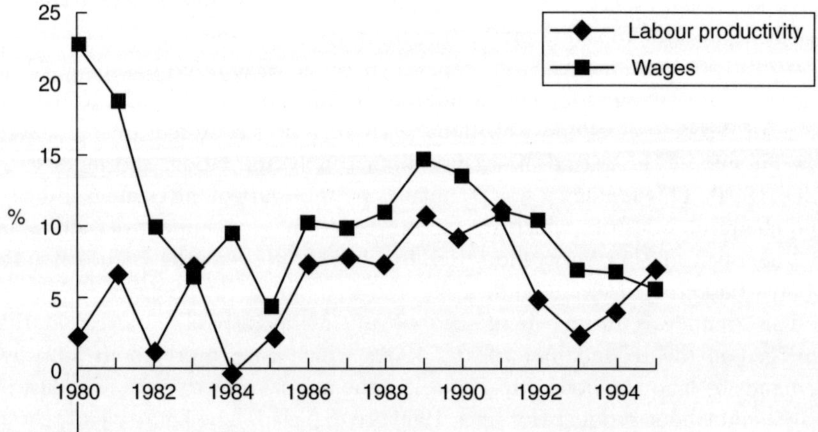

Figure 5.4 Growth rates in labour productivity and industrial wages – Taiwan
Source: CEPD (various issues)

As illustrated in Figure 5.4, wage rates in Taiwan also grew considerably faster in most years than labour productivity, weakening the opportunities of Taiwan as a cheap production location.

The upgrading of the Korean and Taiwanese economy as discussed in Chapter 3 has also led to more intense direct competition with producers from major trading partners. To curb the success of Korean and Taiwanese products abroad, authorities – especially in advanced economies – increasingly erected various kinds of trade and non-trade barriers against imports from both economies.

Another response to their export successes was that the US authorities (for both the most important market), in the second half of the 1980s, forced the Korean won and Taiwanese dollar to appreciate *vis-à-vis* the US dollar. The exchange rates against the US dollar are depicted in Figure 5.5. This appreciation *during those years* clearly made exporting from the home economies less attractive, stimulating FDI in production facilities overseas.

The financial and economic crisis that hit Asia by the end of 1997 has changed this picture quite drastically. At the beginning of January 1998 the value of the Korean won had plunged to about 1,750 won per US dollar, which was about half of its value one year earlier. Although the exchange rate had gone up again to 1,400 won per US dollar by mid-April, it still implied that the 'expensive' won was no longer a push factor to investment abroad. In the case of Taiwan, the effects have been less extreme. At the beginning of January 1998, the exchange rate had gone down to 34 NT dollars per US dollar implying a drop of about 25 per cent in comparison with one year earlier. By mid-April, the value had improved again somewhat to 33 NT dollars per US dollar.

Other L conditions that have stimulated the rapid increase of FDI since the second half of the 1980s are escalating real estate prices at home (especially in

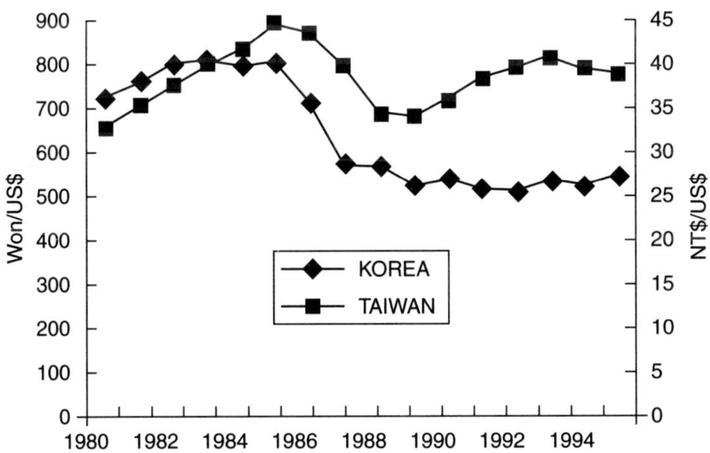

Figure 5.5 Exchange rates vis-à-vis US dollar for Korea and Taiwan
Source: CEPD (various issues), NSO (various issues)

Taiwan; Kuo 1994) and growing environmental concerns in Korea and Taiwan, making it less acceptable to utilise polluting production techniques at home.

Notwithstanding these changing L conditions, the rapid increase in Korean and Taiwanese FDI would not have been possible without the major change in government policy with regard to outward investment. As was discussed at length in Chapter 4, since 1987, the regulations with regard to capital outflows in Korea and Taiwan have been liberalised substantially.

Late industrialisers, late outward investors?

According to the Late Industrialisation paradigm, LIE firms are relatively late in multinationalising their manufacturing operations (Hikino and Amsden 1994: 307). The main reason for this delay is that the O advantages of these firms (primarily based on incremental process and product improvements) are often too intangible to be easily transferred overseas. Because individual LIE firms do not possess a technological edge over leading firms from EIEs, they, *by and large*, typically rely on exports to serve foreign markets and postpone manufacturing overseas in these economies. This at least holds for the markets which are intended to consume their outputs instead of supply their inputs. Only significant trade barriers can force them to start overseas greenfield operations as a defensive measure, especially in foreign markets in which technologically advanced domestic firms occupy a dominant position (i.e. in DCs). According to the LI paradigm, the establishment of R&D facilities in advanced economies and the acquisition of ailing companies abroad with a high knowledge content can be seen as exceptions to the rule in this respect.

The foregoing suggests a distinct pattern of outward direct investment from Korea and Taiwan *in terms of timing and direction* as compared to FDI from early industrialised economies. Although the possible exceptions to the rule as indicated by the LI paradigm (e.g. defensive investment, mergers and acquisitions) make it difficult to formulate clear-cut hypotheses, we will attempt to measure some aspects of this deviant FDI pattern on an aggregate level. More specifically, the following questions are addressed:

1 Are there clear indications that outward investment from Korea and Taiwan is delayed?
2 Does this delay relate more specifically to investments in economically advanced countries? In other words: do LIEs (namely, Korea and Taiwan) first primarily invest in LDCs and only (much later) invest in advanced economies?

(1) *Timing of outward investment*

Unfortunately, the LI paradigm does not define 'delay'. In principle, one way to measure this delay is to take the level of economic development of Korea/Taiwan as a starting-point and compare the size of outward FDI with that from EIEs when they reached similar levels of economic development. Such an approach would, however, render serious data problems. To gain a better understanding of the moment of outward FDI we therefore developed a 'Relative Investment Development' (RID) ratio. For this purpose, first the relative economic development and relative outward investment positions of Korea and Taiwan *vis-à-vis* the major early outward investors listed earlier in this chapter are calculated. The relative economic development position is measured as the ratio of the per capita real GDP of Korea/Taiwan and that of the respective EIEs (j), or:

$$RDratio_{kor/twn,j} = \frac{p.c.RGDP_{kor/twn}}{p.c.RGDP_j} \ .$$

If RD<1, Korea's/Taiwan's level of economic development is lower than that of country j, whereas RD=1 indicates a similar development level and RD>1 a higher level of economic development for Korea/Taiwan than that for country j. The relative outward direct investment position is measured as the ratio between the per capita outward investment stock of Korea and Taiwan and that of the respective EIEs (j), or:

$$RIratio_{kor/twn,j} = \frac{p.c.FDIs_{kor/twn}}{p.c.FDIs_j} \ .$$

If RI<1, Korea/Taiwan – *per head* – by the end of the year has less outstanding investment abroad than country j, whereas in the case of RI = 1 per capita outward investment stock between both economies is equal; RI>1 suggests a higher per capita outward investment level for Korea/Taiwan than for country j. For the individual values of the RD and RI ratios we refer to Appendix 5.2.

Subsequently, the Relative Investment Development (RID) ratio is simply defined as the ratio of ($RI_{kor/twn,\,j}/RD_{kor/twn,\,j}$). As long as RID<1, the relative outward investment position of Korea/Taiwan (*vis-à-vis* country j) lags behind its relative economic development position (*vis-à-vis* country j). If RID>1, the reverse holds true whereas RID = 1 suggests that a 'balance' between the relative development and investment positions *vis-à-vis* country j exists. In Table 5.10, the RID values for 1971, 1975, 1980, 1985, 1990 and 1994 are depicted.[24]

What conclusions can be drawn from the table? First of all we observe that, with only one exception, in all cells, RID ratios < 1. In other words, for almost all years and as compared with all (early industrialised) economies listed, relative outward investment lagged behind Korea's and Taiwan's relative levels of economic development. If we take a closer look at the situation in 1971 (see Appendix 5.2), we observe that, although Korea and Taiwan also lagged substantially behind in terms of economic development, the gap behind Korea and Taiwan and the other countries as outward investors was much bigger than in recent years. Consequently, the RID ratios were also very low. This clearly illustrates the fact that, although the economic development (based primarily on export-oriented labour-intensive industrialisation) by that time had taken off,

Table 5.10 RID ratios for Korea and Taiwan vis-à-vis other countries

KOREA	*1971*	*1975*	*1980*	*1985*	*1990*	*1994*
USA	0.00751608	0.01441618	0.01878660	0.04827385	0.08458152	0.23027271
Japan	0.04085760	0.03561800	0.07139074	0.09880060	0.07174814	0.28280815
France	0.19270490	0.03096021	0.03186374	0.05551206	0.05843714	0.16633462
Germany	0.01822805	0.02631477	0.02019667	0.03904852	0.04880722	0.19721601
Netherlands	0.00205056	0.00425862	0.00450161	0.01069741	0.01456325	0.05208938
Switzerland	0.00200908	0.00224558	0.00499370	0.01376758	0.01374656	0.05911411
UK	0.00462566	0.00856280	0.00845450	0.01938557	0.02689039	0.09063311
TAIWAN	*1971*	*1975*	*1980*	*1985*	*1990*	*1994*
USA	0.00860224	0.01284874	0.01932517	0.00495810	0.81427983	0.99016527
Japan	0.04676197	0.03174532	0.07343735	0.01014758	0.69073083	1.21606598
France	0.22055287	0.02759397	0.03277720	0.00570152	0.56258368	0.71523353
Germany	0.02086221	0.02345362	0.02077566	0.00401058	0.46987496	0.84802252
Netherlands	0.00234689	0.00379559	0.00463066	0.00109871	0.14020272	0.22398267
Switzerland	0.00229941	0.00200143	0.00513685	0.00141404	0.13234035	0.25418878
UK	0.00529412	0.00763179	0.00869688	0.00199105	0.25887808	0.38971945

Sources: Dunning and Cantwell (1987), IMF (1995), Penn World tables, UN (various issues), UNCTAD (1995), UNCTC (1978)

both economies had not yet accumulated sufficient L and O advantages to realise substantial outward FDI. More striking is that also in 1990, RID ratios for Korea were still very low, suggesting that outward investments still lagged behind its economic position. Even in 1994, although Korean RID ratios had improved, they are nowhere near 1. In the case of Taiwan, however, we see that the convergence of the relative development and investment positions had progressed much further. RID ratios have gone up steeply since 1990. In 1994, the RID ratio *vis-à-vis* Japan even exceeded 1. Also if related to the US, France and Germany, the RID ratios approach 1. When interpreting these results, one has to take into account that the RID ratios merely render a rough indication of the relative investment position of an economy in comparison with its relative welfare position. As is argued by the IDP, the idiosyncratic investment paths of individual countries are influenced to a large extent by their economic structure. Since in terms of, for instance, size and dependence on international business Taiwan bears a closer resemblance to a country like the Netherlands, a comparison with such economies is in fact more appropriate. From the table it shows that, although also *vis-à-vis* the Netherlands Taiwan's position has improved considerably, the RID ratio is still relatively low. The comparison of Taiwan with the Netherlands also has, however, to be interpreted with some caution. Although the economies of the Netherlands and Taiwan are much more alike than those of Taiwan and, for instance, the United States, the history of the Netherlands as a former colonial power has no doubt influenced the country's outward investment position.

Nevertheless we can conclude that for Taiwan, the gap between relative outward investment positions and relative economic positions *vis-à-vis* major early industrialised economies that has existed in the past has rapidly become smaller. In the case of Korea, a similar trend was observed although here the RID ratios are still substantially lower.

(2) *Developed versus developing host regions*

In Figures 5.6 and 5.7, the shares of outward direct investment stock for Korea and Taiwan that have gone to developing countries (LDCs) and to advanced economies (DCs) since 1977 are depicted.[25] From the figures, a number of interesting observations can be made. Although by the end of the 1970s a clear majority of FDI from both economies had gone to LDCs, by the beginning of the 1980s the share of Korean and Taiwanese investments going to advanced economies had started to increase rapidly. This is rather remarkable, since, as we saw in Chapter 2, the (first wave) literature that was published in that same period almost exclusively dealt with activities by LIE firms in developing countries. It also contradicts the notion as put forward by the Scandinavian school which argued that companies first invest in nearby countries with similar characteristics and only gradually extend their overseas presence to locations with a larger 'psychic distance'. By the end of 1980, however, 27.4 per cent of Korean outward investment and no less than 46.8 per cent of investments originating in Taiwan had gone to industrialised countries – especially the United States. The early popularity of the USA as a host country for Taiwanese companies should be seen

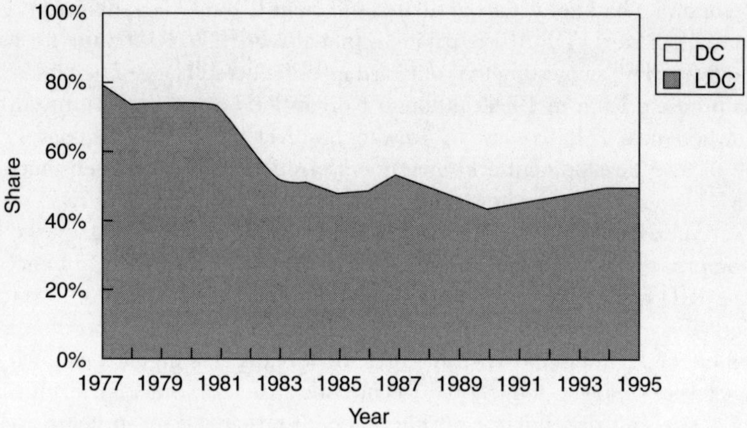

Figure 5.6 Destination of Korean FDI stock
Source: BOK (various issues)

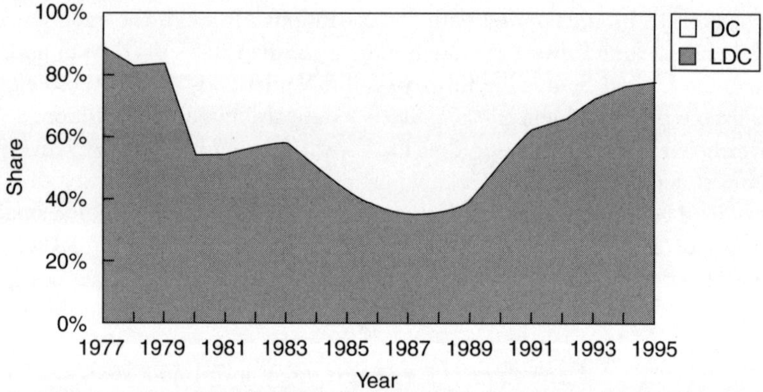

Figure 5.7 Destination of Taiwanese FDI stock
Source: MOEA/IC (various issues)

in the context of the special ties the country has built up with Korea and Taiwan since the post-war period. Although by that time it concerned only limited amounts (US$ 98.1 million total FDI from Taiwan and US$ 138.3 million from Korea), the stock of direct investments (in absolute amounts) in advanced economies since then has grown considerably. In the case of Korea, the share of total FDI going to DCs has continued to increase from 27.4 per cent of total FDI stock by the end of 1980 to 51.8 per cent by the end of 1995. For Taiwan, the pattern looks quite different. After a substantial rise in the relative importance of DCs as a host of Taiwanese FDI – reaching a peak of 64.8 per cent of FDI stock in 1987 – a reorientation towards LDCs can be observed. This was primarily a result of the 'PRC effect'. As was shown earlier, since the end of the 1980s the

PRC has rapidly become the most popular destination for Taiwanese FDI; as a result, by the end of 1995 the PRC had received no less than 35.5 per cent of total outstanding investment from Taiwan. The share of Taiwan's FDI stock that can be found in DCs had gone down to 21.7 per cent. We can nevertheless conclude that this graph does not unconditionally support the suggestion that LIE firms first primarily invest in LDCs and only in a (much) later stage in DCs.

Summary and conclusions

In this chapter, we empirically explored the nature of Korean and Taiwanese FDI at an aggregate level. This exercise has rendered some interesting insights. As was shown earlier, when compared with the first wave of outward FDI from late industrialised economies, the most important home countries are no longer to be found in Latin America, but instead in Asia. In the Asian region, due to a recent steep increase in outward FDI, Korea and Taiwan no doubt belong to the leading investors. The importance for the Korean and Taiwanese economies is, however, still relatively modest in comparison with most early industrialised economies. The latter observation is confirmed if outward FDI is normalised for the size of the home economies or if related to their export volumes.

The conclusion was drawn that FDI from both economies was still, for an important part, a regional phenomenon (especially for Taiwan) and that in particular, the recent rise of the PRC as a popular destination has considerably influenced the geographical distribution of outward investment. For both economies, North America is the second most popular destination outside Asia, whereas Korean companies recently have also discovered Europe. In sectoral terms, manufacturing in both cases represents more than half of outward FDI (for Taiwan as much as 65.9 per cent), although important regional differences can be discerned. Of the various branches of industry, both for Korea and Taiwan, fabricated metals and electronics are the most important contributors. The internationalisation of this industry is examined in more detail in Chapter 7.

A closer look at the investment activities in Europe revealed that Korean FDI in manufacturing in absolute as well as in relative terms is much larger than that of Taiwan. In both cases, the UK and Germany are the most popular host economies, although Eastern Europe (at least for Korea) has also attracted a considerable share of FDI.

Of course, the recent increase in outward investment from Korea and Taiwan can be ascribed to a multitude of reasons. A number of locational explanations are repeatedly mentioned in the literature and therefore were discussed separately. Although not determined empirically, it appears that especially the disproportionately fast rise in industrial wages, the deterioration of the exchange rates *vis-à-vis* major currencies and the change in government policy are changes in L conditions that triggered off FDI from both economies.

Finally, the question was raised to what extent Korea and Taiwan are not only late industrialisers but also late outward investors as is suggested by the Late Industrialisation paradigm. The so-called Relative Investment Development (RID) ratios showed that *vis-à-vis* most EIEs, Korea's and Taiwan's outward investment position indeed for a long time lagged behind their relative economic welfare position – although during the last few years Taiwan has partially caught up in this respect. Since the LI paradigm suggested that this delay in outward investment would especially manifest itself with regard to FDI in advanced economies, we also looked at the division between investments in both categories of countries. In the case of Korea, in a relative short time span the share of FDI flowing to advanced economies has gradually increased and at present represents about half of total outstanding FDI. For Taiwan we observed that at present the majority of outstanding FDI stock can be found in developing countries, although a considerable share has also been invested in advanced economies. These outcomes therefore at least do not fully support the notion voiced by the LI paradigm in this respect.

Appendices

Appendix 5.1 Average size of Korean and Taiwanese investments abroad as of end 1995 (US$ million; number of cases)

Korea

	Asia	*N. America*	*Europe*	*Other Areas*	*Total*
Mining	17.27	11.30	7.13	11.16	13.23
Forestry	3.74	0.00	0.13	5.34	4.34
Fishery	0.36	2.20	0.38	0.78	0.72
Manufacturing	1.18	5.96	7.67	1.58	1.71
Construction	0.95	3.18	1.68	1.91	1.60
Transportation	1.13	0.73	0.51	1.05	1.01
Trade	0.78	3.52	3.19	1.21	2.22
Others	1.08	3.11	5.80	2.46	2.11
Real estate	12.79	1.06	1.55	0.31	19.47
Overall average	1.26	4.14	4.89	2.08	1.96
No. of cases	3,822	763	317	460	5,362

Source: BOK (1996)

Taiwan

	Asia	*PRC*	*N. America*	*Europe*	*Other Areas*	*Total*
Mining	1.50	0.56	0.00	0.00	3.00	0.76
Agriculture & forestry	1.67	0.15	0.00	0.00	1.00	0.25
Fisheries & animal husbandry	0.33	0.54	0.00	0.00	1.00	0.57
Manufacturing	4.02	0.52	4.13	7.04	5.62	0.92
Construction	1.21	0.18	9.80	0.00	5.70	1.15
Transportation	2.89	0.44	2.22	0.00	20.43	2.14
Trade	1.21	0.63	1.13	0.80	6.82	1.27
Banking & insurance	3.57	0.50	11.60	6.08	16.33	10.61
Others	4.07	0.27	2.23	5.57	7.00	1.22
Overall average	3.14	0.50	3.52	3.14	10.68	1.15
No. of cases	1,295	11,254	785	178	268	13,780

Source: MOEA/IC (1996)

Appendix 5.2 RI and RD ratios for Korea and Taiwan vis-à-vis EIEs

Korea

	RI ratio 71	*RI ratio 75*	*RI ratio 80*	*RI ratio 85*	*RI ratio 90*	*RI ratio 94*
USA	0.00103088	0.00244765	0.00379908	0.01228550	0.03126246	0.07360480
Japan	0.00987437	0.00989724	0.02192331	0.03539564	0.03340837	0.07755476
France	0.03667010	0.00698461	0.00838334	0.01917863	0.02804596	0.05438595
Germany	0.00345163	0.00605599	0.00524063	0.01313663	0.02271045	0.06821179
Netherlands	0.00039317	0.00096468	0.00123391	0.00390944	0.00745879	0.01814350
Switzerland	0.00027345	0.00040155	0.00108003	0.00390594	0.00555776	0.01263218
UK	0.00097003	0.00213610	0.00257203	0.00727498	0.01357642	0.03559214
	RD ratio 71	*RD ratio 75*	*RD ratio 80*	*RD ratio 85*	*RD ratio 90*	*RD ratio 94*
USA	0.13715711	0.16978512	0.20222295	0.25449608	0.36961338	0.31964189
Japan	0.24167776	0.27717456	0.30708896	0.35825333	0.46563394	0.27423098
France	0.19029147	0.22559969	0.26309969	0.34548583	0.47993383	0.32696711
Germany	0.18935837	0.23013671	0.25947987	0.33641803	0.46530925	0.34587349
Netherlands	0.19173885	0.22652365	0.27410493	0.36545628	0.51216517	0.34831476
Switzerland	0.13610799	0.17881610	0.21627858	0.28370560	0.40430173	0.21369149
UK	0.20970537	0.24946306	0.30421953	0.37527810	0.50488008	0.39270572

Appendix 5.2 RI and RD ratios for Korea and Taiwan vis-à-vis EIEs (*cont.*)

Taiwan

	RI ratio 71	*RI ratio 75*	*RI ratio 80*	*RI ratio 85*	*RI ratio 90*	*RI ratio 94*
USA	0.00155494	0.00285674	0.00563393	0.00163046	0.36366114	0.42144917
Japan	0.01489409	0.01152240	0.03251163	0.00469749	0.38862345	0.44406595
France	0.05531164	0.00815197	0.01243225	0.00254527	0.32624513	0.31140513
Germany	0.00520630	0.00706815	0.00777170	0.00174341	0.26417975	0.39056964
Netherlands	0.00059304	0.00112591	0.00182986	0.00051884	0.08676449	0.10388674
Switzerland	0.00041246	0.00046866	0.00160165	0.00051837	0.06465072	0.07232983
UK	0.00146315	0.00249312	0.00381424	0.00096549	0.15792797	0.20379481
	RD ratio 71	*RD ratio 75*	*RD ratio 80*	*RD ratio 85*	*RD ratio 90*	*RD ratio 94*
USA	0.18076022	0.22233592	0.29153318	0.32884731	0.44660463	0.42563517
Japan	0.31850866	0.36296385	0.44271247	0.46291734	0.56262647	0.36516600
France	0.25078633	0.29542585	0.37929568	0.44641979	0.57990506	0.43538944
Germany	0.24955660	0.30136715	0.37407718	0.43470283	0.56223415	0.46056517
Netherlands	0.25269385	0.29663579	0.39516129	0.47222463	0.61885026	0.46381597
Switzerland	0.17937758	0.23416211	0.31179638	0.36659042	0.48851863	0.28455161
UK	0.27637204	0.32667526	0.43857578	0.48491590	0.61004767	0.52292698

Sources: Dunning and Cantwell (1987), IMF (1995), Penn World tables, UN (various issues), UNCTAD (1995), UNCTC (1978)

6 Firm-level determinants of outward FDI from the electronics industry

An econometric analysis

Introduction

During the past decades, many empirical studies about the various aspects of foreign direct investment have been carried out. Dunning (1993) distinguishes three categories of empirical research in this field. One type of research is mainly interested in the choices which MNEs have to make (such as whether to export or engage in foreign production and where to locate export platforms). A second group of investigations attempts to explain the sectoral composition of international production (why is FDI concentrated in certain sectors?). A third category of empirical research focuses on testing the theory of the MNE *per se* (such as the explanatory power of the internalisation theory with regard to the licensing versus investment decision).

These approaches are not, however, mutually exclusive. The empirical analysis discussed in this chapter simultaneously relates to two 'types' of research. Our primary goal is to investigate which firm-level factors determine whether a Korean or Taiwanese firm invests abroad or exports instead. This question is typically related to the first category of research. However, as Chen (1992) noted, most of the indicators that are commonly tested in such research are typically ascribed to MNEs from early industrialised economies (EIEs). It therefore follows that a second question almost automatically has to be raised, namely whether the determinants that are derived from conventional MNE theory also apply in the case of LIE MNEs. In other words, are the characteristics of Korean and Taiwanese MNEs similar to those of MNEs from EIEs?[1] This second question is of a different nature and perhaps fits better into the third category of research questions.

The numerous empirical studies carried out to test firm-level determinants of outward investment were in most cases based on samples of American, European or, especially in more recent years, Japanese companies.[2] In view of the fact that most FDI from Korea and Taiwan is still very recent, it is not surprising that the evidence from these late industrialising economies is scarce.

An early attempt to test variables that were expected to determine FDI by Korean manufacturing companies was made by Jeon (1992). This investigation concerned companies that had invested abroad before 1990, so that the sample was restricted to the very early outward investors. In addition, the equations estimated were certainly not limited to firm-level variables. 'National' variables (such as the growth of the Korean economy and geographic proximity) were also tested. Similarly, a typical industry characteristic (i.e. market concentration) was included.[3] In addition, the sample did not distinguish between firms from different sectors, ignoring industry differences as far as the tendency to invest abroad is concerned (cf. Dunning 1993).[4]

Chen (1992) tested a number of possible determinants of outward FDI by manufacturing companies in Taiwan. For this purpose, the information about investment intentions of individual companies was collected through a survey carried out at the beginning of 1988. Unfortunately, the value of the dependent variable (0 or 1) was based on *intentions* instead of the actual decision to invest abroad or export. Since, in 1988, investments from Taiwan were still very limited, it was not possible to use realised FDI data. In addition, no distinction was made between investments which took place in industrialised and those in developing economies. Since the motivations to invest in either category of countries can be very different, it is likely that – at least in part – the characteristics of the companies investing in one of the categories will also differ. Moreover, although the author admitted that important industrial differences may exist with regard to the propensity to invest abroad, firms from all manufacturing subsectors were put together in one sample.[5]

In a more recent study, Chen, now with Ku and Liu (1995) obviated important objections to the 1992 paper just mentioned. In this study, a distinction was indeed made between FDI in industrialised and in developing economies. Also, *intentional* investment projects were replaced by actual (albeit approved) investments. Although a dummy was included to take into account differences among industries, this did not work out very well. The industry dummy was only inserted to distinguish between industries that were the most active in FDI. The results even underlined the dominant effect of industrial origin on the propensity to invest abroad, which suggests that a sample across industries is highly inappropriate.

The above remarks clearly call for a test that uses firm-level data within a specific industry. For these reasons the analysis in this chapter is confined to electronics firms in Korea and Taiwan. As was already mentioned in Chapter 5 and will be elaborated further in Chapter 7, the electronics industry in both economies is among the most important sources of outward investment. The remainder of this chapter is set up as follows. First, the model used for the econometric estimates and the data collection are elucidated. Subsequently, hypotheses are formulated and variables defined. Next, the results are discussed, followed in the final section by the conclusions.

The model and data collection

It is assumed here that, in principle, each firm is a potential outward investor. In practice, however, not all electronics companies will have invested overseas.

Hence, for a relative large number of observations the dependent variable has a value of zero. The dependent variable, FDI_i, cannot have negative values, but is either zero (case I) or positive (case II). The econometric equation should 'explain' not only whether case I occurs or not, but also, if case II occurs, what the size of FDI_i is. For such a task tobit models were developed that are specifically equipped to deal with dependent variables that in part are discrete and in part continuous.[6] The tobit model looks as follows:

$$X_i\beta + u_i > 0 \Rightarrow FDI_i = X_i\beta + u_i$$
$$X_i\beta + u_i \leq 0 \Rightarrow FDI_i = 0$$

FDI_i expresses both the investment or non-investment status of the i_{th} company and the extent of foreign involvement (measured by the size of FDI by individual companies). X_i represents a vector of explanatory variables. In the tobit model, the β's are estimated through a maximum likelihood procedure.[7] To suppress the scale of variation, all independent variables (with the exception of the dummy variables) were expressed in natural logarithms ('linear-log' or 'semi-log').[8]

Another assumption made is that the explanatory variables discussed in the foregoing are not relevant in every case. Since we assume that the motivations to invest in industrialised economies to some extent differ from those applicable to investments in developing economies, we argue that part of the explanatory variables relevant in explaining FDI are different. We therefore estimate three equations for both economies separately:

1 Firm level determinants of FDI in all economies;
2 Firm level determinants of FDI in industrialised economies;
3 Firm level determinants of FDI in developing economies.

For both Korea and Taiwan, investment projects up to the end of 1993 are included. The data needed for this analysis were gathered from various sources. The data collection was rather cumbersome, partly because some of the data was not available from published sources, in which case the co-operation of the local authorities was required. Moreover, some sources were published in Korean and Chinese. The variables used for the Korean sample refer to their values in 1993. The information for most explanatory variables (namely, firm size, technological capabilities, commercial capabilities, human capital, productivity, labour intensity and foreign ownership) was extracted from *Maekyung: Annual Corporation Reports*. Data on antidumping measures against electronics companies were derived from *Import Restrictions against Korean products by major industrialised countries*. Export figures were taken from the *Annual Directory of Electronic & Electrical Manufacturers in Korea*. Information on Korean conglomerates came from the *Business Korea Yearbook* (1994/95). Data on individual investment projects (namely, size and destination of investments) were provided by the Bank of Korea. In principle only those companies for which information could be gathered for all variables were included in the sample, yielding 141 observations.[9] A major exception concerns technological capabilities, for which, unfortunately, information was only available for a very limited number of companies.

Starting-point for the Taiwanese sample were the electronics companies (defined as 'electronics and electrical' (CCIS 22) and 'computers and peripherals' (CCIS 25)) belonging to the top 1,000 manufacturing companies as published in *The Largest Corporations in the Republic of China 1992* (which records 1991 data). From this list, companies with total sales exceeding NT$ 400 million in 1991 were selected. In total, this concerned 173 companies in CCIS 22 and 75 companies belonging to CCIS 25. The company information listed in this source (namely, firm size, labour intensity and foreign ownership) was complemented with variables (namely, technological capabilities, human capabilities, export orientation and productivity) included in (or calculated from) the 1991 census data (*Industrial and Commercial Census; Taiwan-Fukien Area, the Republic of China*).[10] Since the most recent census data referred to 1991 data, information on the other explanatory variables were also collected for that year. Because the companies listed in the Census are anonymous, the values of the variables had to be matched with those published in *The Largest Companies*. Eventually, 64.5 per cent of the 248 companies (n=160) were identified, forming the total sample for the estimates. Moreover, information on the membership of the Brand international promotion association (Bipa) was provided by the association itself. Unpublished data on individual investment projects were provided by the Investment Commission of the Ministry of Economic Affairs. The most serious limitation of the sample is that no data on investment projects in the People's Republic of China are available. Given the political sensitivity of these investments, the Investment Commission does not provide information on these projects. Even if these data would be available, however, they reflect a serious bias since a great majority of the projects are small-scale and go unregistered.

Hypotheses and variables

In this section, hypotheses are formulated about the relationships between firm characteristics (namely, explanatory variables) and outward investment. The proxy variables used to measure these characteristics are also defined. As noted above, outward investment refers to the likelihood that a company actually invests abroad and – if investment has already taken place – the consequent size of outward investment. For matters of convenience, in the text we mostly refer to 'FDI' or 'outward investment'. If a hypothesis is assumed to be relevant only in case of investments in either industrialised or developing economies, this is mentioned explicitly.

Firm size

There are many reasons why the size of a firm is often regarded as an important determinant of outward investment.[11] Horst (1972b), for instance, refers to the fixed costs incurred in investing in foreign production for which the larger firms often can get financing more easily than smaller firms. Moreover, a certain minimum size ('critical mass') of tangible and intangible assets is needed to invest abroad (cf. Jeon 1992). In addition, due to their sheer size, larger firms reach the limits of domestic markets and involvement in overseas markets more rapidly than smaller firms (UNCTC 1992b). Moreover, large firms are often associated with concentrated (or even oligopolistic) market structures which imply certain O advantages that encourage FDI (see for instance UNCTC 1992b). *It is therefore argued here that a positive relationship exists between the size of a firm and outward investment.* Firm size will be measured by total sales of companies (SIZE).

Intangible assets

According to mainstream MNE theory, the possession of intangible assets is an essential prerequisite for FDI (Caves 1971a, Dunning 1993, Helpman 1984). In order to undertake FDI, a firm must possess intangible assets that can be transferred internationally at low opportunity costs for use within the firm (Pugel *et al.* 1996). In this study, we distinguish between technological capabilities, marketing strengths and the quality of human capital.

Technological capabilities

Some form of technological superiority is mostly seen as indispensable to compete successfully in foreign markets (cf. Caves 1971b, 1993, Dunning 1993, UNCTC 1992b). It usually is assumed that this superiority is based on R&D efforts. Investment projects in advanced economies in particular, are often market seeking and therefore have to be competitive in the host country. *We will test the hypothesis that technological capabilities are positively correlated with overseas investment in industrialised economies.* Technological capabilities (RDINT) are proxied by the ratio of R&D expenditures and total sales. For 69.4 per cent of the firms in the Taiwanese sample, data on RDINT were available. We therefore estimated two equations: one for the total sample excluding RDINT and one for the sample for those companies for which data on RDINT were available. In the Korean sample, unfortunately only for 23.9 per cent of the firms could data on RDINT be collected. Since the degrees of freedom would be too few, an estimation of the full equation was not even attempted. Instead, a simple Pearson's correlation coefficient will be calculated to get an impression of the relationship between RDINT and FDI.

Marketing capabilities

According to Kumar (1994), the goodwill associated with a firm's brand names or trade marks in the form of consumer loyalty is considered to be one of the most important intangible assets possessed by EIE MNEs. Other researchers (cf. Caves 1971b, UNCTC 1992b) have also argued that skills in marketing and product differentiation can develop into key O advantages that may lead to overseas production. A firm's ability to differentiate products is supposed to be more easily transferable than the introduction of a new technology. Close co-ordination with, for instance, research and development functions is not needed as much in the case of product differentiation. On the contrary, Lall (1980) argued that a successful marketing policy requires to be carried out in proximity to the final markets. If such expertise is to be exploited effectively, this, *inter alia*, implies feedback to the production units, for instance with regard to the recommended product adaptations or developments which again may require implementation by local production facilities. Since most goods produced in developing economies are not marketed in the host country but are sold elsewhere, marketing capabilities are assumed to be especially relevant in the case of FDI in industrialised economies. *It will therefore be tested whether a positive relationship exists between a company's marketing capabilities and its foreign involvement through FDI in industrialised economies.* In the case of Korea, the ratio of (direct and indirect) selling expenditures and total sales is used as a proxy for marketing capabilities (MARKINT1). Unfortunately, such data are not available for Taiwanese firms. As a result, we had to look at the membership of the Brand international promotion association (Bipa) as a proxy. Bipa in Taiwan acts as a platform for indigenous companies that intend to build up or improve international brand awareness for their products. A dummy variable (MARKINT2) taking a value of 1 if a company is a member of Bipa and 0 otherwise, will be used.

Human capital

Next to the more specific technological and commercial determinants of FDI, a more general, complementary, indicator for the possession of intangible assets is tested that reflects the quality of human capital. It has been argued by Chen *et al.* (1995), Lall (1980), and UNCTC (1992b), that many firms develop important advantages in a form that is not specifically related to their technological or commercial capabilities. Such skills cannot by definition be transferred outside of a company but can contribute to successful operations overseas. In addition, the complicated co-ordination of cross-border production operations requires a high quality of the firm's human capital – both in industrialised and developing economies. Caves (1971b) refers in this respect to 'entrepreneurial resources' as a determinant of outward investment. According to Caves, MNEs expand abroad in order to give full employment to co-ordinating activities of their fixed stock of entrepreneurial talent. *It is hypothesised that the general quality of human capital embodied in a firm is positively related to the amount of FDI.* Following Lall's approach (1980), average annual earnings per employee are used as a proxy variable to measure the general quality of human capital (HUMCAP) for both Korean and Taiwanese firms.

Productivity

As was argued in Chapter 2, the strength of (leading) firms in such LI economies as Korea and Taiwan are typically based on their shopfloor operations. The question then arises whether manufacturing capabilities are a relevant determinant in explaining outward investment from Korea and Taiwan. We therefore included labour productivity in the analysis. It is difficult, however, to determine a priori the sign of this relationship. On the one hand, one might argue that high productivity levels make overseas production redundant, since such companies are less sensitive to rising industrial wage levels (Jeon 1992). On the other hand, high productivity might be perceived as a source of O advantage typically needed to overcome the disadvantages related to being alien in a foreign market *vis-à-vis* indigenous firms. In any case, *we hypothesise that a relationship exists between (labour) productivity and outward investment.* For Taiwan, labour productivity (LABPROD2) was proxied by the total value of production per person employed. Since this information was not available in the case of Korea – following Pilat (1993) – value added per employee was used to measure labour productivity (LABPROD1).

Export orientation

A common assumption of this type of analysis is that firms which for a substantial part depend on sales outside the home economy will – at some stage, and if other characteristics are present – move production to important target countries (Agarwal 1980). This is often referred to as the 'export-market defence' motive (Chen 1992, Chen *et al.* 1995). With regard to investments in developing economies we can add that companies selling on the international market face continuous price pressure from global competitors. In case of strong domestic factor price (e.g. wage level) increases this may lead to a decision to shift production to other countries with more favourable factor prices (namely, wages). *We therefore suspect that the export orientation is positively related with FDI.*[12] Export orientation (EXP) is measured by the export value divided by total sales.

Labour intensity

Similarly, it is argued that the effect of wage increases (and consequently, the need to shift production abroad) is more strongly negative for labour-intensive companies than for those firms for which labour costs account only for a modest part of their total production costs. This is likely to be especially relevant to economies such as Korea and Taiwan that traditionally possessed a comparative advantage in labour-intensive industries but in recent years have experienced extremely fast wage rises in the industrial sector. To make use of low wage levels abroad, companies are assumed to invest in developing economies. Following Chen (1992) and Jeon (1992), *we therefore expect that FDI in developing economies is positively related to the labour intensity of individual firms.* Labour intensity (LABINT) is proxied by the ratio of the number of employees to the value of fixed assets.

Foreign ownership

Chen *et al.* (1995) drew attention to the relationship between the origin of the equity ownership and the decision to invest abroad. Referring to the so-called 'corporate decision' approach to FDI (Aharoni 1966), it is argued that competitive pressure and opportunities to make profits are not sufficient conditions for FDI. Other aspects such as the objectives of individual managers, their limited decision-making horizons, and the existence of information costs may still prevent companies to invest overseas. Accordingly, foreign managers of MNE subsidiaries in Taiwan show limited initiative to invest abroad. In more general terms, it has to be admitted that, in general, investment decisions by most MNEs are still taken and implemented at the level of corporate headquarters in the home economy rather than at the subsidiary level in a host country.[13] The hypothesis therefore reads that the likelihood of investing abroad is related to the nationality of the ownership. In other words: *firms that for an important part are owned by foreign companies are less likely to invest abroad.* To capture the influence of foreign ownership on outward investment we include a dummy variable (FOR) taking 1 in case foreign ownership of a firm in the sample exceeds 50 per cent and 0 otherwise.

Protectionism

Horst (1972b) has already argued that exporters would rather establish production facilities behind a country's trade wall than continue to serve markets through exports. In other, more recent, studies it is also argued that barriers to imports tend to increase FDI as local production is used as a means of supplying those markets (cf. Pugel *et al.* 1996, UNCTC 1992b). Belderbos (1994) showed that the surge in Japanese FDI in the electronics industry in North America and Europe since the second half of the 1980s was, for a substantial part, a response to export barriers, especially antidumping measures. Dumping occurs when a good is sold abroad for a lower price than the seller charges for the same good in his home market (Hindley 1988). GATT(WTO) authorised the imposition of antidumping duties on dumped exports.[14] To avoid paying antidumping levies the Japanese companies set up production activities in North America and the EU.[15] The exports of Korean electronics companies are also known to have been frequently hit by antidumping duties in industrialised economies (Bark 1993).[16] Since there are no a priori reasons to assume that Korean electronics companies respond differently to such measures, *we expect that protectionist (*namely, *antidumping) measures against individual (Korean) companies are positively related to the decision to invest in industrialised economies.* To measure this effect a dummy variable (DUMP) is included with a value of 1 in case a company has been hit by antidumping measures and 0 otherwise. Since no such measures have been observed against Taiwanese firms operating in the electronics industry, this hypothesis will only be tested for Korean firms.

Company structure

A frequently hypothesised determinant of FDI is the possibility of raising capital at preferential rates (Dunning 1993). In earlier chapters we noted that the Korean economy is strongly dominated by conglomerates ('chaebol'). Although the chaebol, as opposed to Japanese conglomerates (especially the so-called horizontal '*keiretsu*'; see Belderbos 1994), did until recently not include financial activities, it is known that members of the chaebol typically have had favourable access to additional capital. For a long time, preferred access to (often relatively cheap) credits provided by Korea's government-controlled banks was an important source for the chaebol (Sakong 1993). As discussed in Chapter 4, an additional financial advantage for chaebol *vis-à-vis* non-chaebol companies concerns the intragroup financial resources available through cross-lending and cross-guaranteeing of loans and the cross-holding of equity of member firms. Another, non-financial, advantage of belonging to a chaebol is that most of them also possess trading companies (see Chapter 4). As will be illustrated in Chapter 7, these trading companies often act as overseas advance guards that do not only sell the chaebol's products overseas, but also collect important information about overseas markets. *We therefore argue that a positive relationship exists between being a member of a chaebol and outward investment.* Since such conglomerates are not present in the Taiwanese electronics industry, this hypothesis is only tested for the Korean sample. If a sample firm is member of one of the conglomerates (CONG), a dummy variable will take the value 1 and 0 otherwise.

To sum up, the following equations are estimated in this chapter:

1 *All countries* Initially, the equation for investments in all countries (namely, industrialised as well as developing economies) will be estimated to investigate the common significance of part of the aforementioned determinants of FDI. For Korea, the size of the company (SIZE), human capital (HUMCAP), export orientation (EXP), foreign ownership (FOR), labour productivity (LABPROD1) and company structure (CONG) are assumed to be significant both in industrialised and in developing economies. In the case of Taiwan, the significance of company size (SIZE), human capital (HUMCAP), export orientation (EXP), foreign ownership (FOR) and labour productivity (LABPROD2) will be tested. In other words:

$$FDI_{KORi} = F(SIZE, HUMCAP, EXP, FOR, LABPROD1, CONG);$$

and

$$FDI_{TWNi} = F(SIZE, HUMCAP, EXP, FOR, LABPROD2)$$

2 *Industrialised countries* With regard to FDI in industrialised economies we hypothesise that, next to the variables listed under (1), for Korea marketing capabilities (MARKINT1) and protectionist measures (DUMP) have to be added to the equation. In the case of Taiwan, marketing capabilities (MARKTINT2) and technological capabilities (RDINT) are added. As a result, the equations look as follows:

$$FDI_{KORi} = F(SIZE, HUMCAP, EXP, FOR, LABPROD1, CONG, MARKINT1, DUMP);$$

and

$$FDI_{TWNi} = F(SIZE, HUMCAP, EXP, FOR, LABPROD2, MARKINT2, RDINT)$$

3 *Developing countries* Regarding FDI in developing nations, we hypothesise that for both economies, next to the explanatory variables included in equation (1), labour intensity (LABINT) is a significant determinant. As a result, the equations to be estimated look as follows:

$$FDI_{KORi} = F(SIZE, HUMCAP, EXP, FOR, LABPROD1, CONG, LABINT);$$

$$FDI_{TWNi} = F(SIZE, HUMCAP, EXP, FOR, LABPROD2, LABINT)$$

In Table 6.1, the independent variables included in the respective models and the expected signs of the relationships are listed.

As compared to the earlier empirical studies on Korean and Taiwanese FDI as discussed in the introduction, the present study shows a number of important improvements. As a matter of fact, by including only companies operating in the electronics industry, we control for the dominance of industry characteristics over firm determinants. Moreover, by including investment projects that were approved as late as 1993, the investigation is not limited to pioneering firms which may show deviant patterns from 'average' investing firms. In addition, new hypotheses are tested. Contrary to the previous studies on Taiwanese investments, the use of a tobit model also takes into account investment size.

Table 6.1 Independent variables and expected signs

	(1) All countries		*(2) Industrialised countries*		*(3) Developing countries*	
	Korea	*Taiwan*	*Korea*	*Taiwan*	*Korea*	*Taiwan*
SIZE	+	+	+	+	+	+
RDINT[a]	ni	ni	ni	+	ni	ni
MARKINT1	ni	ni	+	ni	ni	ni
MARKINT2	ni	ni	ni	+	ni	ni
HUMCAP	+	+	+	+	+	+
LABPROD1	±	ni	±	ni	±	ni
LABPROD2	ni	±	ni	±	ni	±
EXP	+	+	+	+	+	+
LABINT	ni	ni	ni	ni	+	+
FOR	–	–	–	–	–	–
DUMP	ni	ni	+	ni	ni	ni
CONG	+	ni	+	ni	+	ni

Notes:
ni = not included
a For Korea, Pearson's correlation coefficient between RDINT and FDI for a limited number of observations will be calculated

Empirical results

In this section, the results of the estimates will be discussed for the Korean and Taiwanese samples.

Korea

In Appendix 6.2, the correlation matrix shows that no high degrees of correlation can be found between the explanatory variables. The results of the empirical analysis are reported in Table 6.2.

Table 6.2 Results: Tobit estimates – Korea

Explanatory variables[a]	*(1) All countries*	*(2) Industrialised countries*	*(3) Developing countries*
SIZE[b, c]	27.981***	23.851***	9.438***
	(6.914)	(4.918)	(5.550)
MARKINT1	ni	−1,631.370	ni
		(−0.527)	
HUMCAP	11.804*	18.234**	1.520
	(1.476)	(1.845)	(0.423)
LABPROD1	−24.525**	−23.823*	−15.106***
	(−2.275)	(−1.898)	(−2.653)
EXP	17.710***	19.383***	6.023***
	(3.289)	(2.850)	(2.636)
LABINT	ni	ni	−6.708**
			(−1.672)
FOR	−35.082**	−25.349*	−18.281**
	(−2.188)	(−1.409)	(−2.182)
DUMP	ni	26.435**	ni
		(1.919)	
CONG	−14.015	−18.567*	−6.626
	(−1.080)	(−1.304)	(−1.165)
Observations	141	135	141
Positive observations	37.6%	23.7%	29.8%
L.L.[d]	−658.277	−402.547	−495.459
Goodness-of-fit[e]	$\chi^2(6)=100.990$***	$\chi^2(7)=88.232$***	$\chi^2(8)=82.178$***

Notes:

a The explanatory variables in the table correspond to the log-linear specification of the model

b *: significant at 10% level; **: significant at 5% level; ***: significant at 1% level

c Figures in brackets are asymptomatic *t*-values. A one-tailed test is used in all except the case of the coefficient for which no a priori prediction could be offered (*viz.* labour productivity)

d L.L. refers to log likelihood of the estimated model

e $\chi^2(n)$ indicates the likelihood ratio statistic on the null hypothesis that the coefficients of all explanatory variables except the constant term are zero, where *n* reflects the degrees of freedom (cf. Greene 1993)

(1) All countries

First, the variables were tested which we assumed to be important in explaining all investment projects, irrespective of their destination. Table 6.2 shows that, with the exception of the membership of a conglomerate (CONG), all explanatory variables have the expected sign. The size of the firm (SIZE) turns out to be a highly significant determinant in explaining the likelihood of and subsequent extension of overseas investment projects. Moreover, the hypothesis that the overall quality of the human capital embodied in a firm (HUMCAP) influences FDI in a positive manner is also supported – albeit at the 10 per cent level. In addition, a heavy reliance on markets outside the home country (measured by the export orientation of firms (EXP)) turns out to be positively related to outward investment. Likewise, the suggested negative effect of foreign ownership of companies (FOR) on overseas activities through FDI is confirmed here. Interestingly, the estimates indicate that indeed a significant relationship exists between the manufacturing capabilities of firms (LABPROD1) and outward investment. The sign of the relationship is negative rendering support to the proposition that a high level of productivity encourages a company to keep production activities in Korea and serve overseas markets through exports instead of FDI. No statistical evidence is found for the hypothesis that membership of a conglomerate positively influences FDI. The estimate not only has a low asymptotic t-value, but also a negative sign. A possible explanation for this poor result is that for the existing sample, the assumed (financial and non-financial) advantages of conglomerates in reality are only confined to those chaebol for which electronics is a core activity.[17] Since for these other conglomerates sales in the electronics sector are often primarily limited to the home market, overseas investment – as far as operations in electronics are concerned – is not a real alternative for them.

(2) Industrialised countries

Subsequently, we tested which variables determine FDI in industrialised economies. Again, most variables (with the exception of CONG and MARKINT1) showed the expected sign. As in the case of the first equation, SIZE, HUMCAP, LABPROD1 and EXP turned out to be significant. Whereas in the case of HUMCAP the statistical significance here is stronger than in the previous analysis (at 5 per cent instead of 10 per cent level), the opposite holds for LABPROD1 (at 10 per cent instead of 1 per cent level). The statistical relationship between foreign ownership and outward FDI is significant at the 10 per cent level. Confirming our expectations, the results show that protectionist measures (namely, antidumping cases; DUMP) taken by authorities in industrialised economies against Korean electronics products have indeed encouraged the companies involved to invest in these countries. As far as the importance of marketing capabilities (MARKINT1) is concerned, the results do not support the hypothesis of a positive relationship with FDI as the sign is negative. This outcome suggests that in the case of Korean electronics companies, marketing capabilities as such are not an O advantage that, *inter alia*, explains outward investment.[18] The negative (instead of the expected positive) relationship between being a member of a conglomerate and FDI even turns out to be statistically significant at the 10 per cent level.

We have already noted that, due to the limited number of observations, technological capabilities (RDINT) could not be included in the equation estimate. The Pearson's correlation coefficient between RDINT and outward FDI in industrialised economies amounted to −0.154. Although we have to be very cautious in drawing inferences, this result suggests that the empirical support for the hypothesised positive relationship between technological capabilities and FDI in these countries is not strong.

(3) Developing countries

As far as the determinants of FDI in developing economies are concerned, the estimates reveal that in all but two cases (CONG and LABINT) the variables show the expected signs. As in the foregoing estimates, SIZE, LABPROD1 and EXP register statistically significant relationships. Moreover, with regard to FDI in developing economies the results confirm the expected negative impact of foreign ownership of firms in the electronics industry on outward investment from Korea. As opposed to FDI in industrialised economies, the general quality of human capital for investments in developing economies – though of the expected sign – is much less important as is shown by a t-value substantially below the critical value. Surprisingly, labour intensity (LABINT) influences – at the 5 per cent level – FDI in a *negative* (instead of the expected positive) manner. In other words, the results suggest that the more labour-extensive (namely, capital intensive) firms in particular, shift production to developing economies. This outcome suggests that the assumed 'macro' relationship between labour intensity and FDI does not hold at the firm level for the sample. Moreover, this result suggests that the more advanced (namely, labour extensive) companies shift parts of their production activities to economies with relatively low industrial wage levels. A possible explanation for this is that FDI by these companies is not so much a result of the high wage levels in Korea (in their case, labour is a relatively small portion of total production costs), but rather should be seen as part of their strategies to pursue an international division of labour. Given the large distinction in Korea's industrial structure between the more sophisticated (namely, labour extensive) companies and the relatively unadvanced (namely, labour intrusive) small and medium sized firms, it would be no surprise if the former category of companies in particular, are able to pursue such a strategy.

Taiwan

Appendices 6.3 and 6.4 show the correlation matrices for the explanatory variables. In Appendix 6.3, the pair-wise correlation coefficients are depicted for all firms in the sample excluding the variable RDINT. In Appendix 6.4, the coefficients are listed for the explanatory variables of companies for which values for

RDINT are also available. In some cases, relatively high values of pair-wise correlation coefficients are shown which (in combination with high standard deviations) suggest the presence of multicollinearity – an inconvenience often observed in this kind of cross-section analyses (Pugel *et al.* 1996). In Table 6.3, the results of the equation estimates are presented.

(1) All countries

In column 1 of Table 6.3, the results are shown of the estimated equation including all the variables that are assumed to influence FDI both in industrialised and in developing economies. All variables have the signs that had been expected to apply in advance. The size of companies (SIZE) turns out to be a highly significant factor in determining whether a firm invests abroad or not and, if so, the magnitude of that investment. The hypothesis that the export orientation (EXP) of firms is positively linked to its outward investment behaviour is also supported here. The same holds for the nationality of ownership (FOR) of the companies in the sample: foreign ownership strongly influences the likelihood of investing abroad in a negative manner. Although labour productivity (LABPROD2), as in the Korean sample, shows a negative relationship with FDI, the asymptotic t-

Table 6.3 Results: Tobit estimates – Taiwan

Explanatory variables	*(1) All countries*	*(2a) Industrialised countries (excluding RDINT)*	*(2b) Industrialised countries (including RDINT)*	*(3) Developing countries*
SIZE	7.933***	4.944***	6.189***	6.505***
	(5.890)	(4.254)	(4.479)	(4.660)
MARKINT2	ni	7.972***	7.215***	ni
		(3.211)	(2.603)	
HUMCAP	4.975	2.809	−0.146	4.791
	(1.060)	(0.718)	(−0.033)	(0.889)
LABPROD2	−1.342	−2.362	−2.549	3.359
	(−0.601)	(−1.207)	(−1.035)	(1.273)
EXP	3.338**	3.312**	3.581**	0.890
	(2.089)	(1.959)	(1.810)	(0.654)
LABINT	ni	ni	ni	2.973
				(1.213)
FOR	−23.824***	−15.293***	−14.667***	−18.556***
	(−4.753)	(−3.804)	(−3.461)	(−3.241)
RDINT	ni	ni	1.823*/**	ni
			(1.620)	
Observations	160	160	111	160
Positive observations	33.1%	28.8%	30.6%	16.3%
L.L.	−596.449	−510.876	−373.694	−294.673
Goodness-of-fit	$\chi^2(6)=71.836$***	$\chi^2(6)=68.190$***	$\chi^2(7)=58.160$***	$\chi^2(6)=48.158$***

Notes: See Table 6.2

value is much too low to suggest statistical significance. The same holds for the overall quality of human capital (HUMCAP).[19]

Since, as we noted before, the information on RDINT is only available for a portion of the sample, two equations are estimated regarding FDI in industrialised economies: one using the whole sample excluding RDINT as an explanatory variable (2a) and a second one on the basis of the partial sample including RDINT in the equation (2b).

(2a) Industrialised countries (excluding RDINT)

Again, all explanatory variables showed the expected sign for the hypothesised relationships with FDI. Also in the case of FDI in industrialised economies the assumed positive impact of SIZE on FDI is confirmed. The same holds for EXP and FOR which both show a positive and negative statistically significant relationship with FDI. To explain FDI in advanced economies, we also included a proxy capturing the marketing capabilities (MARKINT2) of firms. Although the dummy variable used (namely, membership of the Brand international promotion association) merely indicates the *willingness* to invest in commercial capabilities rather than the actual expenditures made, the results in the table suggest that those companies that are serious about increasing their commercial capabilities are also more eager to invest abroad. Although HUMCAP is again of the expected sign, the assumed relationship statistically is not significant. Also LABPROD2 has the same sign as in the Korean case but again shows a low t-statistic.

(2b) Industrialised countries (including RDINT)

For the smaller sample including only those firms with observations regarding technological capabilities, the results are for most explanatory variables similar to those in the previous equation. Also for this sample, SIZE, MARKINT2, and EXP are statistically significant variables in explaining outward investment by Taiwanese companies in the electronics industry. In addition, the hypothesis that the technological capabilities of individual firms (proxied by R&D intensity, RDINT) influence FDI in a positive manner is confirmed at the 10 per cent level (and almost at the 5 per cent level). As in the previous equations for Taiwan, the coefficient estimated for LABPROD2 shows a negative sign but has a statistically insignificant asymptotic t-value. HUMCAP in this estimate has become insignificant, though still of the expected sign.[20]

(3) Developing countries

Finally, we estimated equation 3 to test the hypotheses related to investments in developing economies. As in the foregoing cases, all variables have the a priori expected signs. For our sample, SIZE is also a very significant variable in explaining FDI in this category of countries. Moreover, the results confirm the assumed negative relationship between FOR and outward investment. Although there seems to be some evidence for the hypothesis that a positive relationship exists between the labour intensity of individual Taiwanese electronics companies and the likelihood of investing in developing countries (to make use of prevailing lower wages), the statistical support is insufficient. Also in the case of FDI in developing countries, HUMCAP shows low statistical significance. Furthermore, the estimated coefficient of EXP, though of the correct sign, has a low asymptotic t-value. Contrary to the Korean sample and previous estimates for Taiwan, the estimate for LABPROD2 here has a positive sign (though a low asymptotic t-value) suggesting that the higher the productivity, the more likely a company will invest in developing countries. Given the relatively high pair-wise coefficient for LABPROD2 and LABINT and LABPROD2 and HUMCAP, as well as relatively high standard errors for these variables, the latter outcome could be caused by some degree of multicollinearity.[21] Multicollinearity is known to be a typical sample problem (Stewart 1991). As noted earlier, some bias in the sample indeed may be expected since small investment projects are often not registered and FDI in the PRC is also not included. Given the fact that multicollinearity probably exists, it is dangerous to delete these variables on the basis of these estimates.

Summary and conclusions

In this chapter, we investigated the importance of firm-level characteristics in explaining outward FDI from Korea and Taiwan in the electronics industry. In Tables 6.4 and 6.5 the results of the present study are summarised and compared with those of previous studies discussed in the Introduction.[22] What inferences can be drawn from this exercise?

In the case of Korea, the size of companies in all cases turned out to be a significant explanatory variable.[23] The same holds for their reliance on foreign markets (i.e. export orientation). The hypothesis that foreign ownership is negatively related to FDI was also confirmed in the present study. Moreover, although the quality of

Table 6.4 Summary of results on Korean FDI

	Present study			*Jeon (1992)*	
	All	*DCs*	*LDCs*	*DCs*	*LDCs*
Size	+	+	+	+	ni
Marketing capabilities	ni	I	ni	I	ni
Human capital	+	+	I	ni	ni
Productivity	–	–	–	ni	ni
Export orientation	+	+	+	ni	ni
Labour intensity	ni	ni	–	ni	+
Foreign ownership	–	–	–	ni	ni
Protectionism	ni	+	ni	+	ni
Company structure	I	–	I	ni	ni

Notes: ni = not included; I = insignificant

Table 6.5 Summary of results on Taiwanese FDI

	Present study				*Chen (1992)*	*Chen* et al. *(1995)*		
		*DCs**						
	All	*−RDINT*	*+RDINT*	*LDCs*	*All*	*All*	*DCs*	*LDCs*
Size	+	+	+	+	+	+	+	+
Technological capabilities	ni	ni	+	ni	ni	I	+	I
Marketing capabilities	ni	+	+	ni	ni	ni	ni	ni
Human capital	I	I	I	I	ni	I	+	I
Productivity	I	I	I	I	ni	ni	ni	ni
Export orientation	+	+	+	I	+	+	I	+
Labour intensity	ni	ni	ni	I	I	I	I	I
Foreign ownership	–	–	–	–	ni	–	–	–

Notes: ni = not included; I = insignificant

* −RDINT excludes technological capabilities; +RDINT includes them (see text)

human capital also registered as important for 'all economies', its significance especially held for FDI in advanced economies. The investigation further revealed that protectionist measures against Korean products make a difference in the decision to invest abroad or export. Being hit by antidumping measures taken by industrialised economies turned out to encourage Korean companies to invest there. Although Jeon (1992) looked at non-tariff barriers, his study found a similar impact of protectionist measures. Also interesting is the negative significant relationship found in the present study between labour productivity and FDI.

Not all hypotheses were confirmed in the present investigation. Marketing capabilities of individual firms turned out to be of no influence on the decision to invest overseas. This outcome suggests that (at least until recently) marketing has played a minor role for Korean electronics companies. This impression is confirmed by Jeon (1992). Moreover, no evidence was found that belonging to one of the conglomerates made a (positive) difference in investing abroad or not. Surprising, and contradictory to Jeon's study, was the negative relationship found between labour intensity and FDI in developing countries.

In the Taiwanese electronics industry, the size of companies also turned out to be a significant determinant for investments abroad – a result which had also been found in the previous studies by Chen (1992) and Chen *et al.* (1995). The same held for the hypothesised negative relationship between foreign ownership of companies and outward investment. This outcome parallels the findings by Chen *et al.* (1995). In addition, again confirming the study by Chen *et al.* (1995), it was found that technological capabilities positively influence FDI in industrialised economies. The assumed importance of sales outside Taiwan was confirmed when all investments were taken into account, but after having separated FDI in developed and developing countries, statistical significance only could be determined in the former case.[24] The present investigation further showed that for

Taiwanese electronics companies, the likelihood of investing in developed countries (and the subsequent magnitude of those investments) is also positively influenced by the *prospective* marketing capabilities of individual firms. Since the proxy variable in the Taiwanese case does not measure actual marketing efforts to date but instead the intention to put more efforts into marketing activities in the future, this outcome does not contradict the results of the Korean sample. For in the Korean case the significance of the actual marketing efforts were measured.

Although Chen *et al.* (1995) found a statistically significant positive relationship between human capital and FDI in DCs, this result could not be confirmed here – probably due to multicollinearity. Moreover, no evidence was found for the hypothesised influence of productivity on the likelihood of investing abroad. In addition, the impact of labour intensity on FDI in LDCs could not be proved convincingly.

Although the tobit analysis described in this chapter is a powerful research instrument, it also has its limitations. The (cross-section) analysis, for instance, was based on characteristics observed at one point in time. This implies that rapid changes in explanatory values (and therewith their influence on FDI) are not taken into account. As was shown in earlier chapters, the industrial transformation of both Korea and Taiwan takes place at a very rapid pace. Therefore a deeper insight into the dynamic aspects of the internationalisation of the companies involved is still needed. Moreover, the investigation for a major part is based on hypotheses derived from 'conventional' insights. The results show that at least part of these 'conventional' firm characteristics leading to outward FDI are also relevant in the case of Korean and Taiwanese electronics firms. In addition, however, it was proved that in the case of Korean firms more idiosyncratic features such as labour productivity (underlining the typical strength in manufacturing) and protectionist measures influence FDI behaviour. To investigate whether other determinants that are not described in the mainstream MNE literature play an important role in the outward investment behaviour of Korean and Taiwanese companies, in Chapter 7 the case study method will be used to investigate a limited number of Korean and Taiwanese firms in more detail.

Appendices

Appendix 6.1 Variable definitions and data sources

Symbol	*Variable definition*
SIZE	Total sales
MARKINT1	Direct and indirect selling expenditures/total sales
MARKINT2	Dummy: 1 if Bipa-member; 0 otherwise
HUMCAP	Average annual earnings/number of employees
LABPROD1	Value added/employee
LABPROD2	Total output/employee
EXP	Exports/total sales
LABINT	Number of employees/gross assets
FOR	Dummy: 1 if foreign ownership >50%; 0 otherwise
DUMP	Dummy: 1 if hit by antidumping measure; 0 otherwise
CONG	Dummy: 1 if member top-20 conglomerate; 0 otherwise
RDINT	R&D expenditures/total sales

Appendix 6.2 Correlation coefficients for independent variables – Korea

	SIZE	HUMCAP	EXP	FOR	MARKINT1	LABINT	LABPROD1	DUMP	CONG
SIZE	1.00000								
HUMCAP	−0.03385	1.00000							
EXP	0.04855	−0.25046	1.00000						
FOR	−0.06329	0.27050	−0.02271	1.00000					
MARKINT1	0.15076	0.23711	−0.20721	−0.00618	1.00000				
LABINT	−0.13723	−0.26124	0.05374	0.15368	−0.10898	1.00000			
LABPROD1	0.17521	0.21618	0.0026181	0.084051	−0.06784	−0.33689	1.00000		
DUMP	0.46091	−0.092481	−0.003410	−0.11437	0.15544	−0.13981	0.17173	1.00000	
CONG	0.41247	−0.029261	−0.021159	−0.052841	0.20782	−0.13749	0.25525	0.36802	1.00000

Appendix 6.3 Correlation coefficients for independent variables – Taiwan I

	SIZE	LABPROD2	RDINT	MARKINT2	FOR	LABINT	HUMCAP	EXP
SIZE	1.00000							
LABPROD2	0.13700	1.00000						
RDINT	0.0093	0.11444	1.00000					
MARKINT2	0.0920	0.085903	0.26196	1.00000				
FOR	0.16637	−0.00085388	−0.15891	−0.20221	1.00000			
LABINT	−0.15216	−0.44903	−0.34250	−0.13599	0.36267	1.00000		
HUMCAP	0.20867	0.50160	0.42501	0.082011	0.050746	−0.51225	1.00000	
EXP	−0.00495	0.093585	−0.25848	0.041529	0.14387	0.29832	−0.37428	1.00000

Appendix 6.4 Correlation coefficients for independent variables – Taiwan II

	SIZE	LABPROD2	MARKINT2	FOR	LABINT	HUMCAP	EXP
SIZE	1.00000						
LABPROD2	0.076930	1.00000					
MARKINT2	0.14414	0.057380	1.00000				
FOR	0.089928	−0.029995	−0.19809	1.00000			
LABINT	−0.14594	−0.44131	−0.13972	0.38261	1.00000		
HUMCAP	0.22919	0.46198	0.099628	0.031938	−0.49883	1.00000	
EXP	−0.059102	0.054732	−0.045409	0.13604	0.33355	−0.38623	1.00000

7 Four case studies of Korean and Taiwanese multinationals

Introduction

Although the econometric analysis discussed in the previous chapter revealed interesting insights about the firm-level determinants of Korean and Taiwanese outward FDI, not all questions can be analysed by means of such techniques. Since the tobit models that were used are based on cross-sectional data, dynamic aspects of the internationalisation of the companies were not taken into account. Moreover, new elements that are not brought forward by received MNE literature but might be relevant in explaining the internationalisation of Korean and Taiwanese companies could not be tested. In addition, as is also voiced by the Investment Development Path (Dunning and Narula 1996), outward investment patterns to a large extent are influenced by contextual conditions that could not be included in the econometric analysis. In this chapter, we therefore use the case study method to investigate the internationalisation behaviour of Korean and Taiwanese companies more in detail.

The following questions are addressed in this chapter:

- How can the internationalisation patterns of Korean and Taiwanese MNEs be characterised? Or, more specifically:
  - What are the (potential) O advantages on which their investments abroad are based and how are these accumulated?
  - What entry modes and expansion strategies can be determined?
  - What are their motives to invest abroad?
- What problems do these companies face in operating in industrialised economies and to what extent are these hurdles related to their late industrialisation background?
- What differences in the outward investment pattern can be discerned between and among the Korean and Taiwanese firms and how can they be explained?

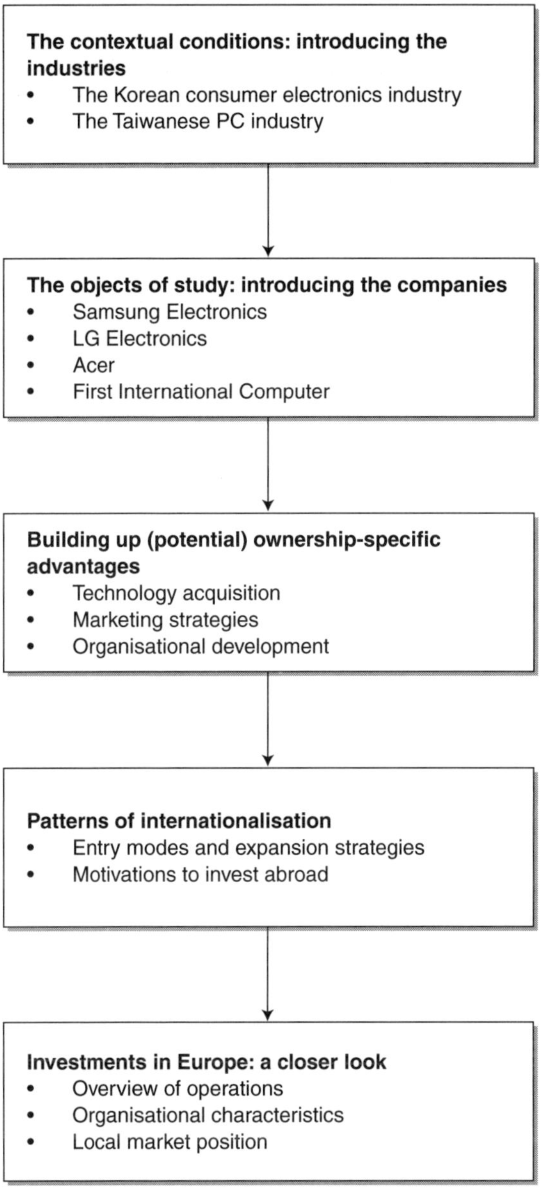

Figure 7.1 Set-up of case studies

In this chapter, so-called 'embedded multiple case studies' (Yin 1994) are constructed, which bear the following characteristics:

- *More levels of analysis are distinguished.* Although the emphasis is on the internationalisation of individual firms, an adequate analysis also requires examining the environment in which these companies operate. Since both in Korea and Taiwan, electronics companies are the most important source of outward FDI (representing no less than 54 and 16 per cent, respectively, of total outstanding FDI by the end of 1995),[1] subsectors from this industry were selected. In the case of Korea, firms operating in consumer electronics are studied whereas in the case of Taiwan, companies active in the PC industry are the central focus of attention. Both lines of business show a high degree of internationalisation in terms of outward investment.

- *More than one case is studied.* To analyse differences and similarities between the companies, (sub)sectors as well as between the home economies, two companies from each country that operate in the same sector are investigated. For Korea, Samsung Electronics and LG Electronics are studied in more detail, whereas in the case of Taiwan, the internationalisation process of Acer and First International Computer are analysed in-depth.

There were various reasons for selecting these companies for the case studies. First, they are among the leading companies in their national industries and all are engaged in overseas investment. Second, since also their operations in industrialised economies (namely Europe) will be investigated, investment projects should not be limited to the Asian region. Third, the companies had to be willing to participate in the investigation.

The set-up of this chapter is as follows (see also Figure 7.1). After having elucidated the data collection, the industries in which the companies operate are discussed. Subsequently, the companies are introduced, followed by an examination of their (potential) O advantages. Next, their internationalisation patterns are compared, after which the companies' operations in Europe are investigated in more detail. Conclusions will be drawn in the final section.

The data collection

Multiple sources were used to construct these cases. For the industry section, besides existing literature and statistical publications of various government bodies, especially the (published and unpublished) material provided by the Electronics Industry Association of Korea (EIAK) and the Market Intelligence Centre (MIC) of the Institute for Information Industry (III) in Taiwan, was relied on.

The frame of reference for the company studies were questionnaires filled in by the global headquarters in Korea and Taiwan as well as the regional headquarters in Europe. In Table 7.1, an overview is given of the subjects about which information was collected through the questionnaires.[2] Besides the questionnaires, multiple open interviews were held with senior managers to gain a deeper understanding of the most important issues. These interviews were held in Korea and Taiwan in 1993 and 1995 and in Europe during the period 1994–96. In addition, throughout the research period additional specific information – if needed – was gathered via contact personnel in the companies. The information compiled from the questionnaires and interviews was supplemented with other

Table 7.1 Subjects of questionnaires

Company as a whole (parent company)	
Subject	*Items*
General company information	Year of establishment, main activities, employees and turnover per region
Product groups & competition	Main products, competitors, strengths and weaknesses in Korea *vis-à-vis* competitors
International co-operation	Importance licensing agreements, management contracts, turnkey contracts, international subcontracting, joint ventures, OEM, M&A
Research & development	R&D budgets, nature R&D, results R&D, problems encountered
Marketing strategy	Marketing budgets, brand name policy
Internationalisation patterns	Information on investment projects abroad, entry and expansion strategies, investment motives, location choices
European operations (regional headquarters)	
General information	Per subsidiary: year of establishment, greenfield/M&A, main activities, employees, nationality management, ownership shares, production capacity
Product groups & competition	Main products, competitors, strengths and weaknesses in Europe *vis-à-vis* competitors
Internationalisation patterns	Breakdown of stages of internationalisation
Research & development	Nature of R&D in European units
Difficulties	Problems faced in Europe

material (e.g. company reports, press releases) provided by the firms. Moreover, external sources (e.g. business magazines, specialised newspapers) were studied to acquire a more complete and balanced picture of the companies. To avoid the risk of misinterpretation, draft versions of the case studies were subsequently sent to the four companies allowing them to render feedback.

Before analysing the case studies, it is important to draw attention to the restrictions of this research strategy. First of all, the table shows that the information requested from the companies covers a wide range of subjects. As a result, data had to be gathered from different departments within the firms. Given the limited time available within the companies for the research project, it was impossible to get accurate answers on every question. In particular, information on situations in the past could not always be retrieved completely. Furthermore, it should not be forgotten that the companies in their home economies are highly regarded and often referred to as examples to follow. As a (non-native) researcher, basing his analysis for an important part on information provided by these companies, it was not always easy to distinguish between truly 'objective' facts rendered and statements that rather express pride about what has been achieved. Cultural differences also – to some extent –

determined the way in which data were collected. For example, the answers to the questions referring to 'opinions' rather than facts (e.g. on investment motives and perceived strengths and weaknesses *vis-à-vis* competitors) clearly had to match the 'official' (top management) notion on these matters making a 'survey'-type of approach posing the same question to a large number of persons within the same firm useless.[3]

Another limitation relates to the implications of using case studies as a research strategy. By choosing an in-depth analysis of a limited number of companies, it is clear that we do not intend to enumerate *statistically sound generalisations*. Instead, the aim in this respect is to refine the existing theories brought forward in Chapter 2 ('*analytic generalisation*') (Yin 1994).

The contextual conditions: introducing the industries

The Korean consumer electronics industry

Although the consumer electronics (CE) industry in Korea was initiated only three-and-a-half decades ago, it has already gone through various stages of development. In the 1960s, the CE industry was almost completely domestic as imported parts and components were assembled into finished products such as radios and black and white television sets by purely indigenous firms (Michell 1988). At that time, the local firms had very few ownership-specific (O) advantages and products were primarily sold on the local market.

Since those early days, the performance of Korean CE industry by all standards has been impressive. Total output went up from US$ 679 million in 1977 to US$ 13.9 billion in 1995. In the same period, the export value rose from US$ 436 million to US$ 7.9 billion. South Korea is now the second largest producer of CE products in the world behind Japan, realising about 9.8 per cent of global production (EIAK 1996).

Who are the major players in present-day Korea's electronics industry? From Table 7.2, which lists the top 20 companies in terms of total sales, some interesting observations can be made.[4] The industry is dominated by a limited number of companies. Total sales of the biggest individual enterprise, Samsung Electronics, for instance, is almost twelve times higher than that of number 5, LG Cable and Machinery. Moreover, many companies listed in the table are members of one of the conglomerates (denoted by an asterisk). Of the top 10, only three companies do not belong to a conglomerate. Samsung, LG and Daewoo all have three companies in the top 20. These three companies no doubt dominate the Korean CE industry to a large extent and are also the most active overseas investors in this sector. In addition, only four companies are partly-owned by foreign MNEs. Three of these foreign firms are in the top 15. As compared with the situation in 1972, this is a remarkable achievement that illustrates the rapid indigenisation of the CE industry. In that year, Motorola Korea realised the largest sales volume in the electronics industry; in addition, seven other foreign companies had joint ventures that were in the top 15 (Bloom 1992).

Table 7.2 The top 20 electronics companies in Korea (1995 figures)[a]

	Company name	*Sales (billion won)*	*Employees*	*Main products*[b]
1	Samsung Electronics*	8,714	47,600	DRAMs, CTVs, VCRs, MWOs, fax machines, PCs, phones, ICs, washing machines
2	LG Electronics*	3,677	31,000	CTVs, VCRs, fax machines, A/Cs, car radios, copy machines
3	Daewoo Electronics*	1,640	12,348	Radios, phones, fax machines, washing machines, computers
4	Samsung Display Device*[c]	1,181	11,000	Monitors
5	LG Cable and Machinery*	741	5,600	Cables
6	Samsung Electro-Mecanics*	730	7,050	Storage units, convertors, transformers
7	Anam Industrial[c]	545	6,225	Memory ICs
8	Trigem Computer	388	1,500	PCs, workstations, monitors, terminals, printers
9	Orion Electric	384	4,927	Monitors, LCDs
10	Daewoo Telecom*	361	2,467	Telephones, LAN systems, computers, printers
11	LG Information & Communication*	273	2,700	Modems, bar code readers, LAN systems, keyphones
12	Tae Il Media[c]	147	700	Modems, PCs, monitors, printers, motherboards
13	Maxon Electronics	109	2,000	Phones, answering machines
14	Inkel	104	2,000	Monitors, keyboards, car radios, speaker systems
15	Daewoo Electronics Components*	101	1,200	ICs, capacitors
16	Haitai Electronics	100	1,360	Vending machines, PCs, workstations, car stereos
17	Korea Data System	100	620	Fax machines, monitors, terminals, power supply equipment
18	Anam Electronics	90	2,100	Word processors, compact disk players, speakers, CTVs
19	Samyoung Electronics[c]	74	1,850	Sensors, electronic capacitors
20	Korea Computer	72	900	Automatic cash dispensers, computers, monitors

Sources: Business Korea (October 1996), EIAK (1996)

Notes:
* Member of top 20 conglomerates
a Fiscal year June 1995–June 1996
b In most cases, other goods are produced as well
c Joint ventures with foreign companies

The simultaneous government policy of 'infant industry' protection and export promotion has undoubtedly contributed to the oligopolistic nature of the CE industry. As was noted in Chapter 4, a number of selected companies have for a long period been protected from foreign competition. Imports of major electronic goods were prohibited from entering the Korean market up to the early 1980s. In addition, successful exporters received preferential access to loans at subsidised interest rates for their export activities. Only in the mid-1980s did the government gradually start to liberalise imports of electronics products (Bark 1991).[5] Membership of OECD and especially the pressure of the IMF to liberalise the Korean economy (which was one of the key conditions for receiving financial support during the recent financial crisis) are expected to result in a further increase of the presence of foreign electronics companies in Korea.

From the 1960s onwards, the Korean government has also played a crucial role in creating the L advantages required to upgrade the CE industry. Important was the establishment of institutes aimed at building an adequate technological infrastructure (see Table 7.3). The Korea Institute for Science and Technology (founded in 1966) was a multidisciplinary industrial research institute trying to bring more foreign technology to Korea and support companies that commercialised this knowledge. It also played an important role in bringing back foreign-trained scientists to Korea. Responding to the need for more specialised institutes in key industrial sectors, among others, the Korea Institute of Electronics Technology was established in 1976. Instead of ('pure') fundamental research there is a strong emphasis on research that yields knowledge that can be commercialised relatively easily – among others in IC design and wafer production.

Table 7.3 Institutes supportive to the technological capabilities of the electronics industry in Korea

Institute	*Established*	*Tasks*
Electronic Industries Association of Korea (EIAK)[a]	1966	Overseas promotion of Korean electronics products, provide training courses in electronics, enhance product quality
Korea Institute for Science and Technology (KIST)/Korea Advanced Institute of Science and Technology (KAIST)[b]	1966	Bringing foreign technology to Korea and assisting in commercialisation of these technologies
Korea Institute of Electronics Technology (KIET)/Electronics and Telecommunication Research Institute (ETRI)[c]	1976	Collaborative research between state and private sector aimed at yielding technologies for private firms in such fields as semiconductors and computers

Sources: Bloom (1992); Hong (1993); Michell (1988)

Notes:
a Merged with Electronics Industries Department of the Korea Fine Instruments Centre in 1979
b Merged with KAIS (postgraduate training institute) in 1980
c Merged with Korea Electrotechnology and Telecommunication Research Institute in 1985

Although its contribution was not technological in a strict sense, the Electronic Industries Association of Korea (founded in 1966) has also played a pivotal role in the development of an indigenous electronics industry. The Association has various functions, such as the promotion of Korean electronics products in overseas markets and the provision of various training courses in the field.

As opposed to most other industries in Korea, foreign MNEs played an important role in the development of the electronics industry (Das 1992). To attract foreign companies, the Foreign Capital Inducement Act was passed in 1960. In addition, Export Processing Zones (EPZs) were established in Masan in 1971 and in Iri in 1974.[6] Foreign investors were offered various incentives such as the free remittance of profits, tax holidays for five years and a halving of corporate tax for a further three years (Oh 1993). The first foreign investments consisted of the establishment of a few assembly lines by US companies. The early operations were mainly of an enclave nature with minimal local production linkages.[7] Only very limited technical know-how was diffused to Korean companies, although some spin-off effects took place as a result of the presence of foreign engineers and production managers who 'exposed' their technical expertise (Hong 1993). In 1968 MNEs were still responsible for 71 per cent of Korean electronics exports (Ernst 1994a). After 1970 when the government lifted the ban on Japanese companies setting up wholly-owned subsidiaries, Japan emerged as a major investor (Bloom 1991). As will be illustrated in the company cases, many joint ventures were established with local companies which also resulted in substantial technology transfers. With the building up of an indigenous electronics industry and deteriorating traditional L advantages (e.g. rising wages, and the expiration of special tax benefits for MNEs), the relative importance of MNEs declined in the 1980s. By the end of 1995, in total US$ 1,731 million had been invested in Korea's electronics industry (on approval basis). Of this cumulative amount, 41.5 per cent was invested by US companies, whereas Japanese firms accounted for 39.7 per cent of total FDI.[8]

Unfortunately, annual outward FDI figures from the electronics industry are only available for the years since 1989. We do know, however, that Korean electronics companies only started to invest abroad in the 1980s. As is shown in Figure 7.2, especially in the 1990s, outward FDI increased rapidly. By the end of 1995, no less than US$ 4,882 million had been invested abroad (approval basis). The most important geographical targets have been North America (41.3 per cent), Asia (32.5 per cent) and Europe (19.8 per cent). The figure suggests that a substantial positive net outward investment position (depicted by the line in the figure) exists in the Korean electronics industry.

Next to 'conventional' FDI, Ozawa (1996) and others (cf. Oman 1984) have emphasised the importance of non-equity investments such as licensing agreements. As Sakong (1993) points out, Korea has generally preferred such arrangements to FDI for foreign technology transfer. In Figure 7.3, therefore the 'balance of technology trade' (namely import and export of technology) of the electronics industry is shown. Whereas in the beginning, technology imports primarily reflected knowledge purchased by joint ventures from their foreign parent

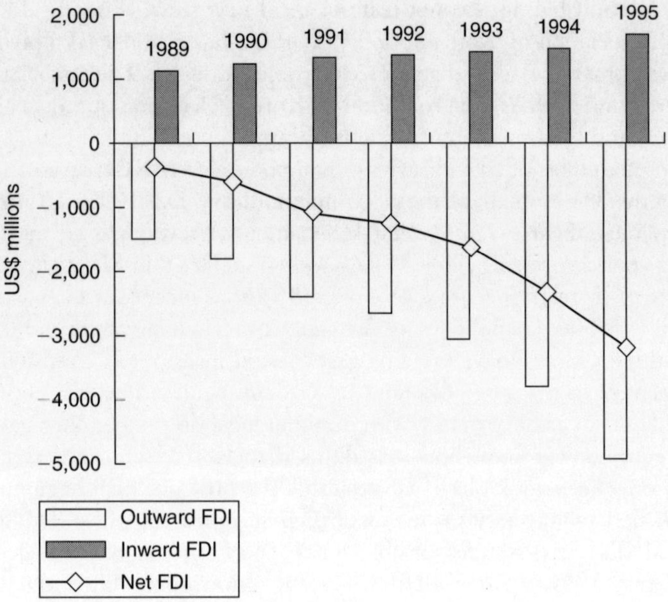

Figure 7.2 Cumulative inward and outward FDI – Korean electronics industry
Source: EIAK (1996), MFE (1996)

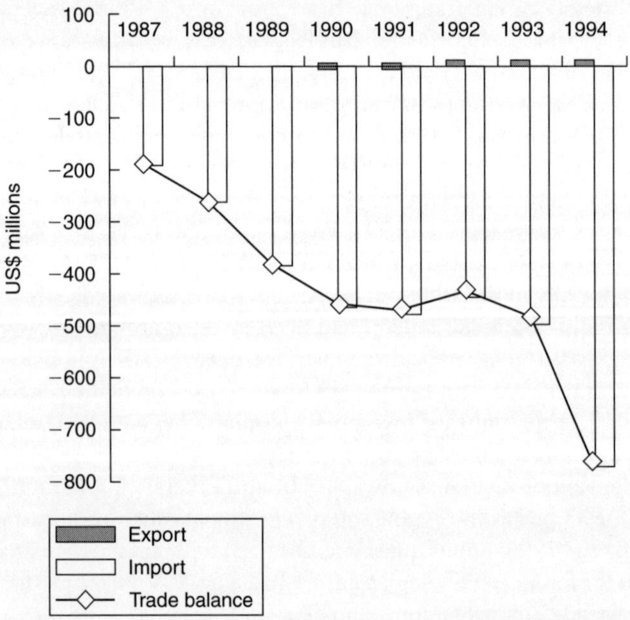

Figure 7.3 Balance of technology trade – Korean electronics industry
Source: KITA (1995)

company, in time technology imported by indigenous firms became more important. Especially since the 1980s, the imports of technology increased rapidly. Figure 7.3 clearly illustrates that the import of technology by the Korean electronics industry far exceeds the export of knowledge. In fact, in recent years, the gap has become even wider. By the end of 1994, total technology imports had gone up to US$ 3,836 million based on 2,385 projects.[9] The US and Japan were by far the most important sources contributing 64.6 per cent and 27.0 per cent, respectively, of all Korean technology imports in electronics. Total Korean electronics technology sold abroad amounted to a very modest US$ 48 million (or 107 projects).

Comparing Figures 7.2 and 7.3 reveals a considerable discrepancy between the positive net outward investment position and the negative technology trade balance of the Korean electronics industry. As noted earlier, the traditional L advantages in Korea (such as the industrial wage levels) have deteriorated making the country a less favourable location for labour-intensive production. This development – at least partly – explains the shift of such production activities to developing countries. As will be illustrated in the company case studies, large-scale investments have also been made in industrialised economies. At the same time, the balance of technology trade suggests that the O advantages (at least, those related to technology) have not improved substantially – an observation among others shared by Jun and Yoon (1995). To gain a better understanding of this paradox, the nature of the internationalisation behaviour of Korean electronics companies needs to be investigated more closely.

The Taiwanese PC industry

The history of Taiwan's PC industry[10] is even more recent than that of the consumer electronics industry in Korea. Although American National Cash Register (NCR) established an office in Taiwan as early as 1954, there were hardly any local computer producers before the end of the 1970s when a domestic PC assembly industry emerged (Chang 1992). An important boost for the growth of the Taiwanese PC industry was the outsourcing of (parts of) their production process to local companies by 'established' foreign companies.[11] The pressure to lower production costs in the PC industry became increasingly urgent when, in the second half of the 1980s, leading companies such as IBM and Compaq switched to price competition as an important sales strategy. Because the production techniques had become easily accessible and only a relatively small amount of capital was required many local entrepreneurs were attracted.

In less than a decade-and-a-half the PC industry has developed into one of the most successful business sectors in Taiwan. The total output of PC products increased between 1985 and 1995 from US$ 1,260 million to US$ 14,156 million. Yet, Taiwan has not achieved an *overall* leading position in the global computer industry, but is rather focused on a limited range of products in which it has gained a very strong position. In 1995, 29 per cent of all computer monitors, 27 per cent of portable PCs, and 9 per cent of all PC systems built in the world were produced in Taiwan (III/MIC, various issues). Moreover, Taiwan's shares in global production of motherboards amounted to 41 per cent, whereas 64 per cent of all scanners and 55 per cent of all computer mice were produced on the island of Taiwan.[12]

In Table 7.4 the top 20 of Taiwan based PC producers (based on total sales) are listed. From the table, a number of interesting observations can be made. Compared to the Korean electronics companies, the industry is much less dominated by a limited number of companies. Whereas in the Korean electronics industry total sales of the largest company was almost 12 times higher than that of the fifth largest company, in Taiwan, total sales of the largest PC company (namely, Acer) in 1995 was only about eight times higher than that of the twentieth largest firm (namely, Clevo).[13] Moreover, with only one exception (namely, Digital Equipment), all companies are locally owned, which illustrates the rapid transformation of the PC industry into a truly indigenous activity. Even in 1989, only four companies in the top 10 were still foreign owned. This is in contrast with other parts of the electronics industry in Taiwan where no less than nine companies in the top 20 were foreign owned (CCIS 1996).

It was only in 1982, that the information industry was designated as a 'strategic industry' making it eligible for government subsidies and other assistance.[14] The decision by the government to develop a computer industry was partly based on the successful experience Taiwan had in building up a consumer electronics industry. More importantly, however, Taiwan's L advantages and future policy orientation matched very well with the typical characteristics of the computer industry (van Hoesel 1996). First, Taiwan's shortage of low-cost labour and its abundance of well-trained manpower could serve the knowledge (and capital-) intensive computer industry very well. Second, the information industry requires only little energy, which fitted Taiwan's scarcity of energy resources. Third, by the beginning of the 1980s, Taiwan had strong intentions to shift from low-value-added to higher-value-added production operations and the computer industry, especially if vertical integration took place, would answer this objective. Fourth, the computer industry would not further deteriorate the environment, which was already seriously polluted.

The accumulation of O advantages by indigenous PC companies would not have been possible without continuous substantial investments in the improvement of Taiwan's knowledge infrastructure (see Table 7.5). The Electronics Research and Service Organisation (founded in 1974) and linked with the Industrial Technology Research Institute (ITRI), is the island's premier research centre in terms of Taiwan's current and future advancements in microelectronics. The Organisation initiated numerous extensive R&D programmes in collaboration with universities and the private sector. A well-known example is the establishment of a VLSI (very large-scale integration) company (namely, Taiwan Semiconductor Manufacturing Corporation) in which also Philips and the National Chiaotung University are involved (Simon 1992). The Computer and Communication Laboratories, another leg of ITRI, for instance, launched a consortium with 47 firms to develop a notebook PC (Wang 1994). Knowledge dis-

Table 7.4 The top 20 computer companies in Taiwan (1995 figures)

Company name	*Sales (NT$ m)*	*Employees*	*Main products*[b]
1 Acer	62,517	3,788	PCs, electronic notebooks, monitors, motherboards
2 First International Computer	26,399	2,581	PCs, electronic notebooks, motherboards
3 Acer Peripherals	26,311	1,478	Monitors, keyboards, laser printers, fax machines
4 GVC	20,811	1,870	Disk drives, printers, motherboards, modems, LAN products, mice
5 Winbond Electronics	18,220	2,352	ICs, semiconductor devices
6 Lite-On Technology	17,499	1,011	Monitors, palmtop computers
7 Digital Equipment International[a]	16,217	1,636	Printers, terminals, business computers, PCs
8 MAG Technology	16,139	1,357	Monitors
9 Chuntex Electronic	15,547	1,408	Monitors
10 ADI	14,676	1,419	Monitors, terminals, transceivers, keyboards, PCs, laser printers
11 Texas Instruments – Acer	14,613	1,419	ICs
12 Inventec	12,759	1,945	Electronic diaries, electronic notebooks, electronic translators
13 Mitac International	12,361	1,150	Monitors, electronic notebooks, motherboards, PCs
14 Compal Electronics	10,941	1,462	Electronic notebooks
15 Hon Hai Precision Ind.	10,805	1,025	Plugs and sockets, power strip sockets, IC memory card connectors
16 Twinhead International	10,023	1,023	Electronic notebooks, work stations
17 Elitegroup Computer Systems	9,587	1,400	Motherboards, LAN stations
18 Quanta Computer	8,763	904	Electronic notebooks
19 Asustek Computer	7,872	530	Motherboards
20 Clevo	7,684	826	Keyboards, electronic notebooks

Sources: CCIS (1996) and China Productivity Centre (1996)

Notes:
a Foreign owned company
b In most cases, other goods are produced as well

Table 7.5 Institutes supportive to the technological capabilities of the computer industry in Taiwan

Institute	*Established*	*Tasks*
Industrial Technology Research Institute (ITRI):	1973	Develop technologies and new products related to industrial development and transfer these research results to local firms
• Electronic Research and Service Organization (ERSO)	1974	Develop technologies for micro-electronics and diffuse technology through licensing agreements and establishment of new spin-off companies
• Computer & Communication Laboratories (CCL)	1990	Research in computer hardware and communications
Institute for Information Industry (III):	1979	Introduce and develop software technology, assist government agencies and public enterprises in their computerisation projects, train and educate professionals, form alliances with MNEs to promote application of computer technologies
• Market Intelligence Centre (MIC)	1979	Gather and distribute market information to computer industry

Source: Simon (1992)

semination from the Institute for Information Industry (III, founded in 1979) to local firms takes place on a royalty basis or by some sort of licensing scheme. The Market Intelligence Centre of III collects commercial information about the industry. The economies of scale realised by these institutes, to some extent at least, compensate for the small size of most Taiwanese PC companies.

MNEs played an important role in initiating the PC industry in Taiwan (van Hoesel 1996). In the first half of the 1980s, foreign companies such as Atari, Digital, Philips and Wang were attracted to Taiwan. Most of these firms initially invested in offshore production facilities in the Export Processing Zones (EPZs) of Kaohsiung and Nantze that were established in 1966 and 1969, respectively. The EPZs offered a well-developed physical infrastructure, tax incentives and no import duties on equipment and parts used to produce exports. In the beginning, the interaction between these MNEs and the local Taiwanese companies was limited. Although difficult to measure, it appears that in the longer run the presence of foreign MNEs did bring spillover effects. Advanced managerial and production technology knowledge, for instance, were diffused to local enterprises mainly through labour mobility. The maximisation of potential spillover effects of MNEs was further enhanced by the establishment of Hsinchu Science Based Industrial Park (see also Chapter 3), where most computer companies and research institutes are nowadays located. On the one hand, companies are offered an excellent physical infrastructure, vast resources of human capital, as well as a five-year tax holiday, import duty exemptions and certain financial and foreign exchange freedoms (Lim and Pang 1991). On the other hand, companies that operate in the industrial park also have to meet certain obligations, such as the establishment of a sizeable research department and training of local personnel in advanced technologies. These requirements have not deterred MNEs from investing in Taiwan. On the contrary, as can be derived from Figure 7.4, inward FDI in recent years has grown rapidly and by the end of 1995, more than US$ 5.8 billion of approved cumulative FDI in the electronic and electric appliances industry had been registered – a much higher figure than, for instance, in Korea.[15] More important, companies like IBM, Matsushita, Hewlett Packard and NCR all have established R&D centres in Taiwan. By the end of 1995, US firms were the largest investors in the sector (50.0 per cent), followed by Japan (26.8 per cent) and Europe (9.7 per cent).

Figure 7.4 also shows that in the 1990s the size of outward FDI from Taiwan's electronics industry has increased substantially. By the end of 1995, in total no less than US$ 1.6 billion had been invested abroad. The most important regions in which Taiwanese electronics companies invested were Asia (51.4 per cent) and the USA (32.5 per cent). Because of the continued growth of inward FDI, however, Taiwan's electronics industry is still a net recipient of FDI.

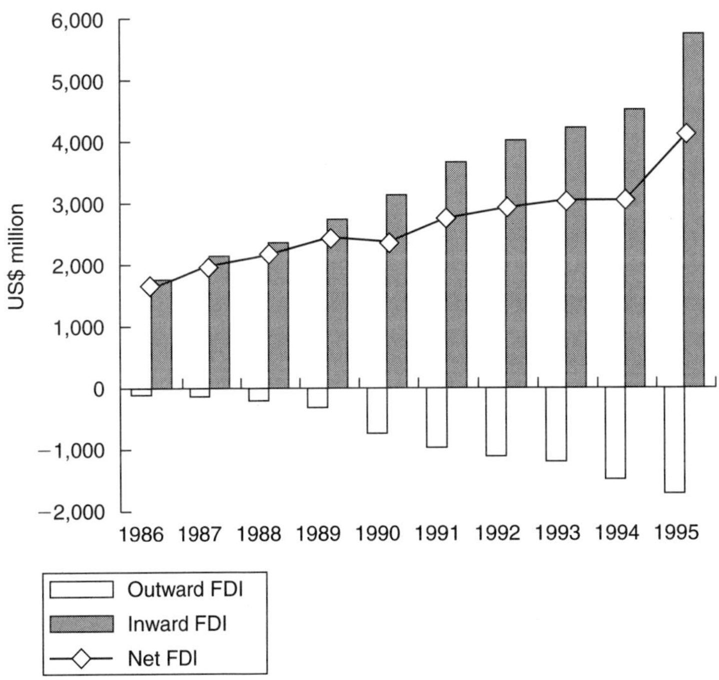

Figure 7.4 Cumulative inward and outward FDI – Taiwanese electronics industry
Source: MOEA/IC (various issues)

In Figure 7.5, the technology imports and exports (unfortunately only available in units) in the electronics industry are depicted. As inward FDI started to increase, so did the number of approved technical co-operation agreements that were often a requirement in the approval procedure for foreign investments. Whereas in the period 1972–81 the total number of approved technology imports in the electronics sector amounted to 240, by the end of 1995 this number had gone up to 1,246. With a growing local capacity to absorb technology, the number of agreements with indigenous firms has grown in recent years at the expense of intra-firm transfers between parent and subsidiary. In terms of the number of agreements, Japan (56.8 per cent) is the biggest foreign contributor to Taiwan's technological capabilities in this field, followed by the USA (33.6 per cent).

The Taiwanese authorities have been less successful in stimulating indigenous R&D efforts than the authorities in Korea.[16] By the end of 1995 a scant total of 12 cases of technology exports from the electronics sector were registered. As a result, a large gap still exists between the number of technology imports and exports (see also Figure 7.5).

Although not as extreme as in the case of the Korean electronics industry, a divergence also appears to exist here between the 'meso IDP' and the balance of technology trade. We observed that deteriorating L advantages have made labour-intensive production in Taiwan less attractive, which stimulated large-scale FDI in developing economies. However, as in the case of Korea, Taiwanese

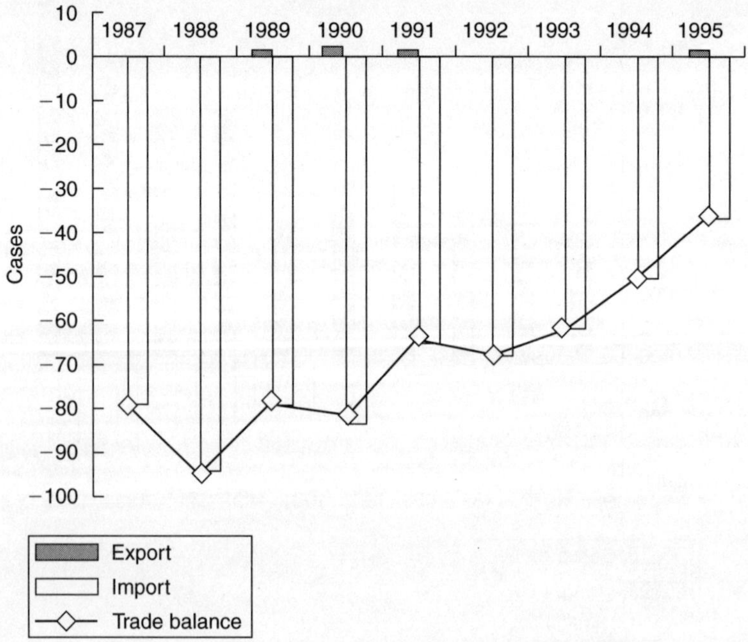

Figure 7.5 Balance of technology trade – Taiwanese electronics industry
Source: MOEA/IC (various issues)

firms also invest in industrialised economies. The limited number of technology export arrangements suggest that the (technological) O advantages in the industry are still very limited. These observations justify a more detailed study of the internationalisation of Taiwanese PC companies.

The objects of study: introducing the companies

Samsung Electronics

Samsung Electronics is one of the most prominent members of the Samsung Group, one of Korea's oldest conglomerates. The Samsung Group was founded in 1938 by the late chairman Lee Byung-Chull to trade in agricultural commodities with neighbouring countries. Following the Korean War, in the 1950s, the company provided basic necessities for the war-devastated country (Steers *et al.* 1989). Cheil ('number one') Sugar & Co., Ltd. was founded in 1953, followed one year later by Cheil Wool Textile Co., Ltd. As Korea's welfare grew, Samsung entered the service sector with businesses in insurance, broadcasting, securities and a department store. In 1975, Samsung Co. was designated as one of the first Korean General Trading Companies. By the end of the 1960s, the Samsung Group had an annual turnover of US$ 100 million; in 1994 total sales amounted to no less than US$ 63.9 billion. In 1994, approximately 206,000 persons were employed by the Group. In Appendix 7.1, an overview is provided of their wide range of business activities.

Samsung Electronics Co., Ltd. (SEC) was founded in 1969. In 1972, the company started to produce black and white TVs. Since then, the product line has been extended substantially. The company went public in 1975. In the early 1980s, the company overtook Goldstar, the pioneer of consumer electronics in Korea, as the biggest CE enterprise in the country. In 1980, Korea Telecommunication Co. was acquired by Samsung (renamed 'Samsung Semiconductor & Telecommunication Co., Ltd.') and in 1988 merged with SEC.

As is illustrated in Table 7.6, Samsung Electronics has expanded substantially. From 1985 to 1994, total sales grew from US$ 1,491 million to US$ 14,617 million. In the same period, the total number of employees increased from 15,700 to 63,500. At present, Samsung Electronics is involved in a wide range of activities covering semiconductors and computers (46.5 per cent of total sales in 1994), information systems (15.9 per cent) and consumer electronics (37.6 per cent).

Table 7.6 Selected core data, Samsung Electronics

	Sales (US$m)	*Net income (US$m)*	*Employees*	*Sales/employee (US$000)*
1985	1,491	7	15,700	108
1990	6,291	284	43,500	145
1994	14,617	1,199	63,500	231

Source: Samsung Electronics

SEC's success in the semiconductor business has resulted in a sharp increase in the relative importance of the semiconductor and computer segment; its share went up from 37.9 per cent in 1993 to 46.5 per cent in 1994.[17]

LG Electronics

LG Electronics, formerly known as Goldstar Co., is the most important electric and electronics company of the Lucky Goldstar (LG) Group. Lucky Goldstar was launched in 1947 when Mr. Koo In-Whoi founded the Lucky Chemical Company. Initially, the company specialised in cosmetics for women. Although by 1958 Lucky dominated the domestic market for various personal care products, its growth opportunities were rather limited (McDermott and Young 1989). To avoid stagnation, in that year Goldstar Co. was founded, Korea's first electronics manufacturer. As opposed to many other chaebol (including Samsung), Lucky Goldstar started in manufacturing and only later on diversified into trading activities instead of vice versa. Like Samsung, the LG Group diversified into many lines of business, which are listed in Appendix 7.2. Basically, the company has followed a dual-track growth path primarily stressing the development of chemicals (through Lucky Ltd.) and electronics (primarily through LG Electronics), although later service and public welfare activities were added. In 1976, Lucky Goldstar International was designated a General Trading Company – one year after the first batch of companies. In 1994, the LG Group realised total sales of US$ 47.5 billion, employing approximately 103,500 persons world-wide.

LG Electronics (LGE) often acted as a pioneer in Korean electronics business. The company was the first in Korea to produce transistor radios (1959), refrigerators (1965), and television sets (1966), followed by washing machines, air conditioners and elevators in 1969 (LG Electronics 1994). In 1970, Goldstar became a publicly owned company. Goldstar established the first private R&D facility in Korea in 1975, the Goldstar Central Research Laboratory. Also in terms of outward direct investment, Goldstar was a pioneer, establishing Korea's first overseas TV plant in the United States in 1982. As was indicated earlier, Samsung Electronics overtook Goldstar's leading position in the Korean electronics industry in the 1980s.[18] The two companies are still by far the two most important electronics companies in Korea.

Table 7.7 illustrates that LG Electronics has grown considerably with total sales going up almost fivefold between 1985 and 1994 from US$ 1,407 million to

Table 7.7 Selected core data, LG Electronics

	Sales (US$m)	*Net income (US$m)*	*Employees*	*Sales/employee (US$000)*
1985	1,407	13	28,300	51
1990	4,168	47	33,500	124
1994	6,526	133	33,200	197

Source: LG Electronics

US$ 6,526 million. The growth in the number of employees was more modest during the same period and the total number even decreased somewhat between 1990 and 1994. In 1994, the most important product category of LG Electronics was home appliances (contributing 31.9 per cent of total sales), followed by audio and video equipment (20.6 per cent), colour TVs (17.4 per cent), computer and office equipment (14.0 per cent) and other products such as video displays, telecommunication parts, and components (16.1 per cent). As compared to Samsung Electronics, consumer electronics represents a much more sizeable part of total sales. The configuration of LGE's product line is, however, expected to change substantially in the near future. By the end of 1994, the reunification of LG Electronics with Goldstar Telecommunication (meanwhile renamed 'LG Information and Communications') had been announced.[19] A major difference with Samsung Electronics' product line remains the semiconductor business, which, as we saw earlier, has become the most important product group of Samsung Electronics. In the case of the LG Group, semiconductor business is handled by LG Semiconductor of which 59.7 per cent is owned by LGE.[20] Recently, plans were also announced to merge LG Electronics and the semiconductor division into one electronics company.

Acer

In 1976, Mr. Stan Shih and five other shareholders in Taiwan founded the company – at that time it was called Multitech – with a capital of only US$ 25,000 and 11 employees. Between 1976 and 1981 the company's core activity was to render design services in Taiwan to manufacturers of (among others) electronics games. In addition, it represented overseas companies like Texas Instruments and Rank Xerox in the Taiwanese market. Its first PCs were produced in 1981.

The growth of the company since then has been dazzling. By 1987, the company already ranked fourth among the exporters of own brand name products in Taiwan's computer industry, behind three foreign manufacturers. In 1988, when the company name was changed to Acer, its stock was traded for the first time on the Taiwan Stock Exchange. Table 7.8 shows that total sales went up from US$ 28 million in 1983 to US$ 531 million in 1988 and US$ 3,220 million in 1994. During the same period, the total number of employees increased from 592 to 9,700.[21] By 1994, the company had become the tenth largest PC vendor in the world (with a global market share of 2.5 per cent) and had also reached the tenth

Table 7.8 Selected core data, Acer Group

	Sales (US$m)	*Net income (US$m)*	*Employees*	*Sales/employee (US$000)*
1983	28	–	592	48
1988	531	27	1,263	126
1994	3,220	119	9,700	332

Source: Acer

position in the global electronic notebook market with a share of 1.6 per cent (FT 05.04.95 and 03.05.95).[22] In Taiwan, Acer is not only the biggest computer company, but is also often regarded as an example that other companies should follow. Or, as one senior government official put it, the international marketing strategy of Taiwan is to 'have 100 "Acers" by the year 2000' (McDermott and Young 1989: 126).

During the 1990s, the Acer group evolved into a network of more or less autonomous business units, subsidiaries and affiliated companies all operating in subsectors of the PC industry (see Appendix 7.3). In 1994, PC and notebook systems represented 60 per cent of the total sales of the Acer Group. The second most important product category was peripherals (24 per cent), followed by semiconductors (14 per cent). Other products realised 2 per cent of total sales.

First International Computer

In 1980, First International Computer (FIC) started as an automation department within the Formosa Plastics Group (FPG), one of Taiwan's leading industrial conglomerates. The company was founded by one of FPG chairman Y.C. Wang's daughters, Charlene Wang. She and her husband, Dr Ming Chien, now run the company as president and chairman, respectively. In 1983, FIC for the first time assembled PC systems – initially only for use within FPG. From 1986 onwards, FIC commenced to sell PCs outside the company. In 1987, FIC took the strategic decision to move into the mass production of motherboards and add-on cards. In the same year, the company also entered the international market. First International Computer went public in 1992.

Although the company is internationally less well-known than Acer, FIC has also experienced an extraordinary fast expansion: its total sales went from a modest US$ 90 million in 1990 to US$ 600 million in 1994 (see Table 7.9). According to *Datamation Magazine* (01.12.92), FIC ranked number 4 among the world's fastest growing PC manufacturers. Another illustration of its fast rise is its position in the Taiwanese computers and peripherals industry. From an eighteenth position in 1988 it moved into the second position, i.e. immediately behind Acer, in 1993 (CCIS 1994).

Although FIC is not a formal subsidiary of the Formosa Plastics Group it no doubt belongs to what sometimes is called the 'Wang Web' (see Appendix 7.4).

Table 7.9 Selected core data, FIC

	Sales (US$m)	*Net income (US$m)*	*Employees*	*Sales/employee (US$000)*
1987	10	1	10	1,000
1990	90	3	200	450
1994	600	10	2,000	300

Source: FIC

The strong ties with other member companies in related industries has led to a, for Taiwanese standards, relatively well-developed stage of vertical integration.

FIC itself consists of a number of departments, of which the personal computer division is by far the most important one, realising 90 per cent of total sales.[23] This division is headed by Cher Wang, the sister of president Charlene Wang. Next to the production and marketing of PCs and computer components, FIC also acts as the distributor and takes care of the servicing of a number of international hardware and software companies in Taiwan, such as Computervision, Stratos, Sun Microsystems and Fujitsu. For an overview of the various divisions as well as their activities see Appendix 7.4.[24]

Because of the continuous drop in motherboard prices, FIC has begun to shift its attention to new growth products, such as electronic notebooks. Eight months after production commenced, the company had already become the fourth producer of them in Taiwan. In 1994, 40 per cent of total sales was realised by motherboards – of which FIC is one of the biggest producers in the world. Notebook systems took about 30 per cent of total sales, followed by PC systems which realised 15 per cent of total revenues. Other products (such as add-on cards) in that year represented about 15 per cent of total turnover.

Building up (potential) ownership-specific advantages

Before analysing the actual internationalisation of the companies, we first attempt to gain a better understanding of (potential) ownership-specific advantages on which their operations abroad are based.[25] To this end, we shall examine their technology acquisition, marketing strategies and organisational development. This section primarily intends to get an impression of the companies' strengths in the various fields, and does not claim to render an all encompassing objective evaluation.

Technology acquisition

First, the various sources of technology acquisition that were important for the four firms are discussed. Subsequently, an assessment is made of the technological level that the companies have reached.

Samsung Electronics

For a long time, SEC's own technological base was very weak. An important source that enhanced its technological capabilities have been the joint ventures established with technologically advanced foreign companies. Only a year after the establishment of Samsung Electronics in 1969, for instance, a joint venture had already been formed with Japanese NEC,[26] followed in 1973 by a joint venture with Japanese Sanyo Electric.[27] In more recent years, however, SEC again established a number of joint ventures in Korea with foreign companies, that primarily are related to its semiconductor business. Together with DNS

(Japan), SEC (with a 29 per cent stake) produces semiconductor manufacturing equipment. A silicon wafer plant was established with Korean POSCO and American MEMC; SEC's stake is 20 per cent. In addition, SEC owns 40 per cent of a joint venture established together with TOWA (Japan) and Hanyang Precision (Korea) to produce semiconductor fabrication facilities. In the beginning of 1995, SEC and NEC agreed to co-produce semiconductors for the EU market.

Another important technology input for SEC have been licensing agreements bought from leading companies in various lines of business. In 1992, for instance, the company had about 40 licensing agreements that among others provided knowledge to produce digital compact cassettes, TV tuners, and car phones.

Especially in the beginning, Original Equipment Manufacturing (OEM) contracts have been another crucial technology source. These OEM contracts do *not* imply the joint development of new (or improved) knowledge, but instead reflect a clear division of tasks whereby the OEM buyer markets the product produced by SEC. (The product can be developed by the OEM buyer in an earlier stage, but that is not necessarily the case.)[28] In 1993, SEC still had about 160 OEM contracts with buyers from a wide variety of countries, such as the USA, Japan, Germany, France, and Argentina. The OEM clientele consists of both leading global companies – which often require narrow product specifications – as well as much less well known companies that only want their own brand name on an existing product. Learning effects among others arose in terms of product quality, design, and managerial processes. At present, the importance of OEM as a technology source for its CE business has become limited.

To build up technological capabilities, SEC also acquired companies abroad. When US Amperex, one of the three manufacturers of microwave ovens (MWOs) in the world was closed, for instance, Samsung bought its equipment and transferred it to Korea (Bloom 1991). This, in combination with technical assistance from Toshiba, one of the two remaining MWO manufacturers, provided Samsung with the technology needed to produce MWOs successfully. In recent years, SEC has become increasingly active in mergers and acquisitions – although only partly in consumer electronics. In 1993, for instance, SEC acquired US Harris Microwave Semiconductor, the world's leading maker of optical semiconductors. In 1994, 51 per cent of Lux, the Japanese hi-fi audio producer, was taken over. In the same year, a similar stake was acquired in a US-based CAD/CAM software technology company, named Control Automation. In addition, Integrated Telecom Technology, a US company specialising in ATM technology was bought. In February 1995, SEC acquired a 40.25 per cent stake in US computer producer AST Research at US$ 378 million.[29]

In the 1990s, the in-house accumulation of technological capabilities became a major objective of the company. In 1994, approximately 9,000 out of a total of more than 63,000 workers were engaged in R&D activities. About US$ 750 million or 5.1 per cent of total sales was spent on R&D as compared with only about 2 per cent (or US$ 30 million) in 1985. At present, the company employs ten R&D centres of which five are located abroad (See Table 7.10). Most of the

Table 7.10 Overseas R&D centres of SEC as of Spring 1995

Location	*Activity*	*Established*
San Jose (USA)	Telecommunication, semiconductors	1983
Brussels (Belgium)	Telecommunication	1991
Frankfurt (Germany)	A/V products & home appliances	1992
Tokyo (Japan)	Product design	1987
Osaka (Japan)	A/V products	1991

Source: Samsung Electronics

R&D activities, however, are centred around the Samsung Advanced Institute of Technology (SAIT), Korea's largest private research institute.[30]

Next to joint ventures with (mostly foreign) partners, SEC also increasingly seeks non-equity strategic alliances for the joint development of new products.[31] In recent years, SEC, for instance, concluded such agreements with Qualcom (digital cellular systems), Micron Technology (memory devices) and ISD (multi-level storage sound processing ICs), all from the USA. In addition, agreements with Toshiba (ICs in CE products) and Fujta (TFT-LCDs) from Japan and UK-based ARM (3–bit risk processors) were concluded.

In technological terms, the semiconductor business is SEC's strongest draw-card. One of the most impressive results of the company's technology accumulation was accomplished in September 1992 when it developed – as the first company in the world – a 64 Mbyte DRAM memory chip.[32] In 1994, Samsung Electronics was the world's seventh largest manufacturer of semiconductors and the largest manufacturer of memory products. In the same year, Samsung's OEM notebook produced for AT&T (marketed under the name 'Globalyst2000') was selected as the best notebook computer by the US-based computer magazine, *PC World*. An example of the *overall* quality of the company's products is that all major products attained ISO 9000 certification, including microwave ovens, colour TVs, refrigerators, VCRs, camcorders, audio equipment, PCs and monitors. In the field of consumer electronics, however, Samsung Electronics cannot yet be labelled as an innovator – the company usually follows top companies in developing and producing new product items.

LG Electronics

As in the case of Samsung Electronics, LGE's initial technology base was extremely weak and therefore strongly dependent on external sources. Two transistor plants, for instance, were set up in 1961–62 under the technical guidance of Japanese companies (namely, Matsushita and Sanyo) with whom joint ventures by that time were not yet legally permissible (Michell 1988). In 1970, Goldstar formed a joint venture with NEC.[33] Also in more recent years, LGE invested considerably in affiliated companies located in Korea. These joint ventures no doubt are an important input in enhancing the company's technological capabilities – albeit primarily in other areas than consumer electronics. Together with Japanese

Alps Electronics (the company in which LGE has a 50 per cent share) produces and sells electronics parts. Software is developed in a joint venture (70 per cent share) with Olivetti from Italy. LG-Honeywell, in which LG Electronics has a 25 per cent stake, produces automatic control systems; the company of which 40 per cent is owned by American Honeywell. Goldstar Micronix produces picture tubes and is owned 50 per cent by LG Electronics, 40 per cent by Dai Nippon Screen and 10 per cent by Kanematsu Kosho, both from Japan. In addition, LG Electronics holds shares in companies such as LG Information & Communications (25 per cent) and LG Cable & Machinery (9.5 per cent) which both are partly owned by foreign companies (Siemens: 3 per cent in Goldstar Telecom and Hitachi Cable: 33.5 per cent in LG Cable & Machinery).

OEM contracts, especially in the beginning, served as an important source of technology. Although the importance in terms of learning effects over time has decreased considerably, in 1993 LG Electronics still had about 120 OEM contracts with a wide variety of companies originating in countries such as the USA, Canada, Japan, China, Germany and Argentina. As in the case of SEC, for LGE the requirements of the OEM clients vary considerably. In the future, the company intends to focus its OEM business increasingly on higher value-added products and decrease the share of low quality, low value-added products.

With the accumulation of technological know-how within the company and an increased capacity to absorb new knowledge, 'independent' licensing agreements (i.e. not included in a joint venture with a foreign company) have become another important source of technology acquisition. From 1988–92, for instance, LGE had about 35 licensing agreements to acquire knowledge in the fields of, among others, DCC, CD-I, and video compressors.

In the past, acquiring companies abroad has not been an important source of technology for LGE. In the second half of 1995, however, LG Electronics announced the acquisition of a 57.7 per cent stake, for about US$ 350 million, in Zenith the last US-owned manufacturer of TV sets (FT 04.08.95, 18.09.95, 19.09.95).[34] Zenith, which reported losses in nine of the previous ten years, is expected to provide LG Electronics with important complementary technology in the field of TV manufacturing.

Nowadays, LG Electronics engages in substantial R&D efforts itself, spending US$ 381 million (about 6 per cent of total sales) in 1994, and employing some 4,500 persons out of a total of more than 33,000. In 1985, for instance, R&D represented about 4.5 per cent of total sales, which by that time amounted to approximately US$ 65 million. Most R&D efforts are carried out in the four central and seven regional laboratories in Korea. LG Electronics established three research centres abroad, namely in Japan, USA and Ireland (see Table 7.11). As is also the case for SEC, these overseas R&D centres are mainly complementary to the core R&D activities conducted in Korea and aim at absorbing locally available knowledge in the various fields.

Non-equity strategic alliances with leading companies meant to jointly generate new technology are expected to become increasingly important inputs for LGE's technological enhancement. In 1994, a 50/50 joint research co-operation

Table 7.11 Overseas R&D centres of LGE as of Spring 1995

Location	*Activity*	*Established*
Tokyo (Japan)	Consumer electronics, office automation	1981
San Jose (USA)	Computers, office automation	1984
Dublin (Ireland)	Product design	1990

Source: LG Electronics

(called 'Frontech') was established with Japan's Alps Co. in the latter's home country. The set-up is rather unusual as neither a transfer of knowledge nor a joint production takes place; instead the intention is to collaborate on next generation high technology in the field of TFT–LCDs (thin film transistor–liquid crystal displays), which are expected to be a key component of next-generation image displays.

Although we noted that in the 1980s SEC overtook LGE as a domestic leader in electronics technology,[35] in more recent years LG Electronics has again shown itself able to be 'the first in Korea' at a number of important occasions in consumer electronics, such as the introduction of the compact disk player, camcorder, digital compact cassette and interactive compact disk.[36] All LGE factories received ISO 9000 certification. Nevertheless, as remarked in an interview at the corporate headquarters, the technological gap with leading global companies in some areas is still very substantial and in some cases even amounts to 5–10 years.[37] Therefore, it can be concluded that, as in the case of Samsung Electronics, LG Electronics is not yet an innovator in the global consumer electronics industry.

Acer

In the beginning of the 1980s, when Acer built its first PCs, production activities were confined to the assembly of components produced by others. The technological capabilities that were required at that time were minimal. As pointed out earlier, in more recent years the pressure in the Taiwanese PC industry to produce components locally has increased considerably. Acer has responded to this challenge by expanding its business to these technologically more complicated activities. Various sources were tapped to build up the necessary technological capabilities.

Forming joint ventures with leading companies in specialised fields no doubt have enhanced Acer's technological capabilities. In 1989, Acer entered into a joint venture (58 per cent stake) with Texas Instruments in Taiwan (named 'Texas Instruments-Acer Inc.') to produce DRAMs and ASICs (application-specific integrated circuits).[38] The aim is that the factory, provided with TI technology, will secure Acer's supply of these vital PC components.[39] In 1990, Acer also formed a joint venture with the German aerospace company MBB, named 'Ambit Microsystems Co.'. This joint venture, in which Acer has a 50 per cent stake, produces hybrid microelectronics for packing and assembly processes.

Acer has also acquired existing companies to speed up the accumulation of technological capabilities. Already in 1987, the company bought US-based Counterpoint Computers (for US$ 17 million) to gain entry into the minicomputer business. This venture turned out to be a failure, however, resulting in the closing down of Counterpoint Computers. In the same year, Acer purchased Taiwan's largest independent industrial design firm, Union Design and Development. In 1989, Princeton Publishing Labs (PPL) was bought to enhance its position in high-end sophisticated peripheral and monitor markets.

The first OEM contracts were granted in 1984 from ADDS (a subsidiary of NCR) and ITT. Also in the case of Acer, an important spin-off of OEM contracts has been knowledge gained in such fields as quality control and design. At present, joint developments with OEM buyers still enhance the company's technological capabilities. In addition, an important advantage of OEM business for Acer is that it expands the size of the company's production activities thus realising economies of scale. The recent shift in importance from OEM to ODM ('Original Design Manufacturing'), however, illustrates a more advanced level of technological capabilities. In the case of ODM, the design of the product is also carried out by the producer.[40]

To improve its product design capabilities outside sources also are used. The recently introduced 'Aspire' computer, for instance, was created in close co-operation with a Californian design firm.

The company established a R&D facility in Silicon Valley in the USA where the latest technological developments are scanned ('environmental scanning'). This unit is considered an important input of technology acquisition.

In recent years the company has clearly stepped up its own R&D efforts. Spending US$ 58.7 million on R&D (or 1.82 per cent of total sales)[41] in 1994 and employing in total about 800 persons in R&D activities, it outdoes probably all Taiwanese competitors. However, these figures are still well below that of most of its global competitors.[42] In 1990, a non-equity strategic alliance was formed with National Semiconductor to jointly develop new ASIC technology.[43]

In Taiwan, Acer is at present considered the technological leader in the PC industry. The company is also frequently praised in international computer magazines and won prizes at computer shows for the technological performance of its products. In 1992, manufacturing facilities in Taiwan, Malaysia, and the USA were granted ISO 9002 quality certificates. In 1993, Acer's Hsinchu and Taoyuan manufacturing plants earned compliance with ISO 9001. A landmark regarding the recognition of its technological capabilities was the decision of Intel in 1993 to licence Acer's chip-up technology. This was the world's first single-chip CPU upgrade to be sold on the retail market, allowing users to improve the performance of their system by inserting a CPU chip directly onto the motherboard.

FIC

Like Acer, FIC started out by assembling PCs which required only limited technological know-how. The technology embedded in the components that the company began to produce later on (e.g. motherboards, add-on cards) was also rather mature. Gradually, according to the company, the level of technological sophistication of these components increased. What sources were tapped to strengthen the company's technological capabilities?

Although FIC and the Wang family acquired a firm in the US (named Everex) which was considered an innovator in high-end multi-user systems, accumulating technological know-how was not the prime motivation for this investment. Other mergers or acquisitions have not yet taken place.

An important source of technological knowledge for FIC have been their OEM contracts. About 30 per cent of the global top 20 PC producers are OEM customers of FIC. The contracts encompass a broad range of production activities carried out by FIC. OEM clients sometimes only subcontract the production of motherboards to FIC whereas at other occasions complete PC or notebooks systems are produced for them.[44] In the case of motherboards, in particular, which are often designed in close co-operation with its customers, important knowledge transfer has taken place. According to FIC, OEM implies an exchange of information with the buyer which always results – to a greater or lesser extent – in a learning effect.

In recent years, FIC has substantially increased the in-house efforts to enhance its technological capabilities. At present, the company spends about 5 per cent of total sales on R&D with some 180 persons engaged in this activity. While 100 persons work on the development of motherboards, 80 are engaged in the development of notebook computers. The focus of R&D is not so much on developing new technologies but more on mastering and improving existing knowledge in these fields. Internal R&D is not only carried out in Taiwan. The overseas R&D centre located in Silicon Valley is also considered an important source of technology.

Although FIC possesses no patents, it has frequently been the first company in Taiwan to introduce new types of motherboards. FIC apparently is particularly strong in the design of motherboards and in high-volume manufacturing. Its plants all have ISO 9001 certification. Although in 1994 FIC invested almost US$ 100 million in an IC testing and IC packaging facility in Hsinchu Science and Industrial Park, the company does not yet consider itself to be an innovator.

Summary

Samsung Electronics and LG Electronics have only in recent years started to invest on a substantial scale in building up new technological O advantages – beyond their shopfloor operations. New developments (such as the emergence of new OEM producers from cheaper countries) as well as increased government pressure certainly stimulated this development (Ernst 1994a). In the past, because of a heavy reliance on OEM sales among other factors, these firms did not feel the urge to extend their technological capabilities beyond production capabilities. In an attempt to catch up with leading electronics companies in the world, both SEC and LGE recently not only increased their budgets allocated to research and development but also tapped complementary sources such as the acquisition of knowledge intensive companies abroad (namely, strategic asset-seeking FDI) and various kinds of equity and non-equity collaborative agreements with leading firms in particular fields. Although Samsung Electronics in particular, realised some important successes in related fields (especially memory chips), in consumer electronics both firms can still be considered technological followers.[45] In Figure 7.6, the various sources of technology acquisition are depicted.

Although the two Taiwanese companies have also clearly increased their efforts to improve their technological capabilities,[46] in the case of FIC the efforts are still relatively modest and do not pretend to result in innovations of any kind. This can be partly explained by the smaller size of the company which, for instance, makes it a less attractive partner in strategic alliances for leading foreign companies. In addition, the assembly nature of the PC industry makes it possible (if other requirements are fulfilled, such as an efficient logistical network) to operate in the world market without accumulating crucial firm-specific knowledge. In view of the fact that the company was established only a decade-and-a-half ago, combined with its relatively small size, this strategy appears quite rational in the short term. Acer, however, did attempt to break new technological grounds and indeed achieved a technological edge – albeit in a specific field – even over conventional leading PC companies. Its technological efforts as well as overseas commercial successes made Acer an increasingly interesting partner for joint technological developments with leading PC companies.

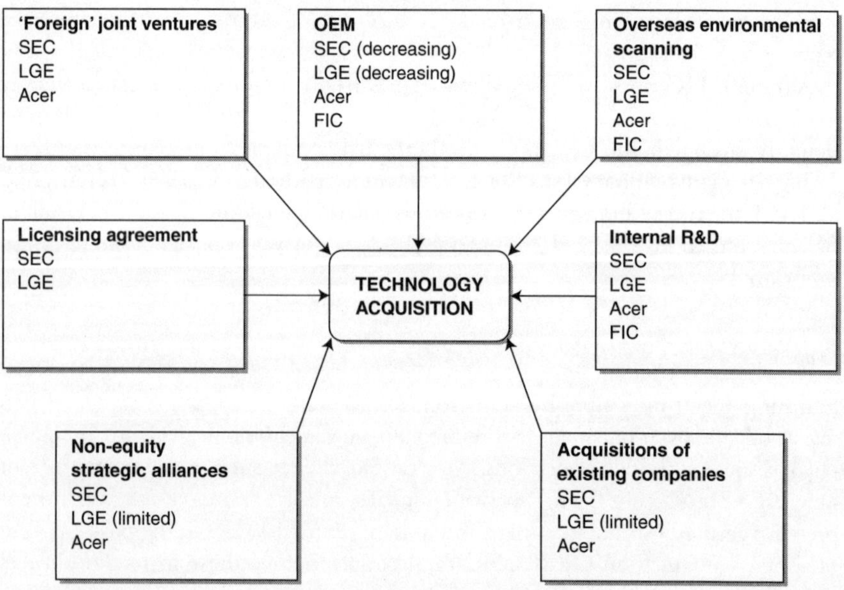

Figure 7.6 Sources of technology acquisition

Marketing strategies

In this section, we focus primarily on the building up of own brand names and company images. Other aspects of marketing strategies, such as the development of distribution channels, are addressed later on in this chapter.

Samsung Electronics

In Korea, most of Samsung's CE products – sold under the company's brand name – have a market share of 40 per cent or more. Although the company has sold products under the Samsung name for a long time (for an extensive analysis of SEC's advertising policy, see Cho *et al.* 1994), the company's sales abroad until the beginning of the 1990s were mainly based on OEM contracts. With the strategic focus being on production operations, the importance attached to marketing – and the budget allotted to this activity – was accordingly limited. Especially in SEC's major markets in the industrialised world, the awareness of the Samsung brand name was until recently very low and the corporate image weak. According to an image survey conducted by London-based Research International, the perceived image of the company's products ranked far below those of Sony, Matsushita and Sharp (FT 20.10.95).

In 1992, Samsung took the strategic decision to increase its efforts to market its own products. According to the company, it is impossible to become a global top electronics company and, at the same time, remain dependent upon OEM contracts.[47] In 1994, 4.8 per cent of SEC's brand sales was spent on marketing activities. The introduction of the new Samsung logo one year earlier illustrates Samsung's determination to strengthen its global marketing position.[48] By 1994, 65.7 per cent of total sales were already realised under the Samsung brand name. Although the company is determined to increase this share to 100 per cent, it acknowledges that at this stage, its marketing position in the global market is still not strong enough.

To minimise the dependence on OEM activities SEC also acquired a major part of the American-based firm AST Research. This move can be seen as an attempt to strengthen SEC's weak position in the information systems market which is a result of a poor international brand name recognition and inadequate sales channels abroad (*Business Korea*, April 1995).

LG Electronics

In the domestic market, LG Electronics has, in almost every line of business, a very strong position with market shares for the Goldstar brand averaging 40 per cent. The position of its own brand name in developing countries such as Peru, Egypt and Indonesia is also very strong, as opposed to its share in many industrialised countries. In the latter countries, consumers used to purchase LGE produced goods primarily under other brand names. In recent years, LGE stepped up its efforts to improve its global marketing position. In 1994, marketing expenditures represented between 1.5 and 2 per cent of total sales (or roughly 4 per cent of own brand name sales). In imitation of Samsung, on 1 January 1995, the Lucky Goldstar Group introduced a new logo to improve the company image which was considered too conservative. In addition, as there was confusion among consumers about the fact that Lucky and Goldstar are two legs of the same company, the names had to be combined. Since the full name was considered too long, the company chose to use the initials, LG. As a result, the company name 'Goldstar' was converted into LG Electronics – a change more far-reaching than Samsung's image makeover where the original name remained intact. In 1994, the share of sales under the own brand name was about 45 per cent, a percentage that has increased only a little during the last few years. In the coming years, LGE hopes to increase this proportion to 75 per cent.

Acer

As early as in 1984 the company started to sell under its own brand name – by that time under the name 'Multitech'. By mid-1986, it was realised that 'Multitech' did not possess the 'distinctiveness . . . necessary . . . to make a real impact on the world market' (McDermott 1991: 131). This led to a change of name in 1988 into 'Acer', the Latin root for acute and sharp.

Acer decided to step up its marketing efforts in the late 1980s and increase the importance of sales under the Acer name on the international market. In 1994, 'selling expenses' amounted to no less than 6.6 per cent of total sales (or US$ 212 million). By emphasising Acer's image as a 'global brand name with a local touch' Acer had become the number one PC brand name in eight countries (namely, Mexico, Chile, Uruguay, Panama, Thailand, Malaysia, Philippines and Indonesia), with a top three position in 15 other developing countries in 1994. Own brand sales in industrialised economies were until recently, however, not so successful. In the beginning of 1994, even less than 5 per cent of American PC consumers had heard of the Acer brand name (*Business Week* 04.07.94).

The attempts that were made at the end of the 1980s to sell exclusively under the company's own brand name failed. As a result, this goal was subsequently abandoned. To generate sufficient funds to finance its marketing expenses, OEM (as well as ODM) orders turned out to be of crucial importance. Next to the income aspect, OEM orders from well known companies like Apple Computers, Compaq, Fujitsu and NEC, enabled the company to realise economies of scale and lower production costs. Acer's commitment to building a brand name among others means a fight against prejudices left over from the days when Taiwan was merely a high-volume, low-cost producer. Or, as Shih puts it, 'The made-in-Taiwan image is still a burden. People still have the idea that a Taiwan company must be worse than a US company' (FT 24.12.94).[49] Whereas in 1994 OEM orders represented about 30 per cent of total sales, the aim is to increase this share to 50 per cent in order to realise a certain balance between OEM and own brand sales.

Notwithstanding these major drawbacks, the company's image has certainly improved in recent years. The manifold interviews with Acer's chairman in international magazines and newspapers and the many product awards no doubt also contributed to the general positive reputation of the company.

Another method used to improve its commercial position overseas was the acquisition of companies with strong distribution channels and/or brand names. In 1989, Acer decided to buy Kangaroo Computer in the Netherlands, followed in 1990 by the acquisition of US Altos computers for US$ 94 million. The existing elaborate distribution channels were a prime motivation in the purchase of both these companies. The same motivation holds for the 19 per cent stake Acer in 1992 took in Computec, its distributor in Mexico.

FIC

Eventually, FIC wants to become a major global computer company with a strong own world-wide brand name. Yet FIC took the strategic decision to focus primarily on the industrial market. In the present situation, the company considers the vast commercial and financial efforts needed to build up a strong own brand name and extensive distribution channels in advanced economies not worth the risk. An additional reason mentioned by FIC for focusing on the production and sales of intermediate products instead of on final products is that too much emphasis on the marketing of own brand name PCs and notebook systems may deter buyers of motherboards who are not willing to, indirectly, support one of their competitors.[50] A consequence of this focus on industrial consumers, is that FIC spends a modest 1 per cent of total sales on marketing activities. FIC also receives much less attention in the media than Acer.[51]

A different situation exists in the Asian region, where the company is very successful in selling PC systems and notebooks under its own Leo brand name. Leo is among the top selling brands in Taiwan, Thailand, Malaysia, Singapore and Indonesia.[52] Recently, its motherboards were also given a brand name. Overall, the company ranks eleventh in Asia – a position that is still improving.

For FIC, OEM sales are of crucial importance. Examples of FIC's OEM clients are Escom and Peacock (both located in Germany), ICL (UK), AT&T and Texas Instruments (both in the USA). In 1994, 50 per cent of PC systems were sold under the Leo brand name whereas 50 per cent was sold to OEM buyers. In the case of electronic notebooks, no less than 90 per cent was sold to OEM customers and only in 10 per cent of the cases was the Leo name used. The disadvantage of slim profit margins, characteristic for OEM business, is compensated for by the large volumes produced by FIC.[53] An additional advantage of OEM contracts is that this mode of business means long-term contracts which are considered highly desirable in a dynamic market like the PC industry.

Although we noted that building up its own brand name in industrialised economies does not have the highest priority, attempts are increasingly made to strengthen the company's commercial capabilities. Next to 'internal' efforts, by the end of 1993, FIC and the Wang family had acquired American-based Everex Systems for US$ 2.3 million.[54] The well-established brand name, the government contracts,[55] and the Everex factory in Fremont were the main motivations for FIC taking over the company. The company is now run by Cher Wang.

Summary

The econometric analysis in Chapter 6 revealed that, at the time, marketing capabilities were not yet a significant determinant of outward investment by Korean electronics companies. Here we have shown that building up internationally accepted brand names is not an easy task. As Wee (1994) showed, it took Japanese companies about 40 to 50 years to achieve their present standing. Because of their heavy reliance on OEM arrangements in the past, the building up of global brand names was ignored by SEC and LGE for a long time. In recent years, however, the companies have clearly stepped up their endeavours to improve their marketing capabilities and become less dependent on OEM sales. These efforts were not confined to increased advertising budgets, but also included far-reaching measures such as changes of logos and – in the case of LGE – even its corporate name. Samsung Electronics, in particular, which seems to be somewhat more ambitious in minimising its OEM business than LG Electronics, has been increasingly successful in this respect.

The goal of the two Korean companies to minimise their OEM dependence forms an interesting contrast with the policies of Acer and FIC. Although especially Acer in recent years has made substantial progress in terms of building up its own brand name (also in advanced economies),[56] the rapid increase of operations overseas does not mean that these companies intend to minimise OEM business. On the contrary, for both companies OEM contracts will remain important in the future. This cannot merely be ascribed to the learning effects involved in OEM contracts but primarily because these contracts enable the companies to realise the necessary scale advantages which both companies, as opposed to the chaebol, lack.

Organisational development

In this section, an attempt is made to depict the 'organisation culture' in the companies. Also discussed are the organisational changes that have been introduced to prepare for their increased presence overseas.

Samsung Electronics

Various types of links with Japan have been in existence for a long time. Founder Lee was briefly educated at Japan's Waseda University, as was his third son who took over on his death in 1987. One of Samsung's first major capital projects, the building of a fertiliser plant in 1964, was financed by the Japanese government and designed and built by Japanese companies. Many key organisational and strategic features of Samsung are modelled after successful Japanese electronics companies (Bloom 1992).

In 1993, at the fifty-fifth anniversary of the company, the Samsung Group launched an ambitious reform programme in which 'all employees are being urged to stop dwelling on Samsung's being the top in Korea and to focus on making the group a major player in the world' (*Samsung Newsletter* 1993). Recognising the need for more emphasis on product quality, chairman Lee noted that the making of defective products is 'cancerous, and a criminal act on the part of management'. As a result, a new defect control system was introduced implying that as soon as a defect is observed, the assembly line is stopped until the problem is solved. Another major change was the adoption of working hours from 7 a.m. to 4 p.m. In Korean society, where staying in the office till (very) late in the evening is common, this was considered a radical change. According to Samsung, the new office hours allow employees to use their free time for 'self development through physical exercise, language studies, and participating in company sponsored clubs' (*Samsung Newsletter* 1993). Furthermore, all Samsung executives are required to spend more time away from the office visiting the production plant or construction site (*Business Korea* August 1993).

Of particular relevance here are the changes introduced in the light of the company's globalisation policy. Every year, several hundred employees are selected and sent abroad for a year of study in a foreign country of their choice. These 'regional specialists' are encouraged to study that country's language, customs, and culture and report on the accumulated knowledge at the corporate headquarters. In addition, the number of local managers will be increased and overseas employees are brought to Korea for training and to make them more familiar with Samsung's corporate culture.

LG Electronics

LG is often described as a conservative, stable company that avoids risky ventures and stresses group harmony in all of its endeavours (cf. Steers *et al.* 1989). The present chairman of the Group, Mr Koo Cha-Kyung is the oldest son of founder Koo In-Whoi. According to Steers *et al.* (1989), the organisation of LG differs in important respects from most other Korean enterprises. There is much less of a personality cult than in most other conglomerates. Top management is supposed to limit itself to corporate strategy and leave considerable autonomy to the lower levels of management. Whereas in the case of Samsung Japanese management practices have acted as a role model, at LG American management models have been more dominant. In 1992, with the help of the US-based consulting firm McKinsey, LG started a large-scale reorganisation. As a result, the company was subdivided into various strategic business units (namely, audio, video, information systems, magnetech, TVs, video-display, kitchenware, livingware, appliance components) and operational business units based on the unique demands of each industry. Each of these operational units have managerial and profit responsibility in their respective fields. The employees were also given greater responsibility in managing the assembly lines. More attention is also paid to the professionalisation of management. Recently, Lee Hun-Jo, who is not a family member, was named LG Electronics' chairman.[57]

Although organisational changes have attracted less (media) attention than those launched at Samsung, similar changes also took place at the LG Group. Since 1994, employees are able to choose their time schedule during which they want to come to work. Furthermore, every other Saturday has become a day off, while Korean employees in most other companies usually work on Saturdays. As a result, according to LG, 'the company's work environment has become more relaxed and creative' (LG Electronics 1994).

LG has also initiated programmes specifically related to its goal to become a global company. The status of locally-hired personnel and expatriates dispatched from the home office, for instance, has become equal. In the coming years, the number of local presidents is expected to grow. Moreover, personnel at the home office are trained in international business practices and individual overseas markets. To improve the relations between overseas personnel and the staff working at the parent company, a 'work exchange programme' was introduced. One element of the programme consists of an invitation for foreign employees working abroad to visit the corporate headquarters and factories in Korea to get better acquainted with the company and to boost corporate culture globally. To prepare foreign employees for international managerial positions, in 1994 the first group of employees who were hired in their home country (Europe, USA, or Asia) were invited to the Korean headquarters for a longer period (6 months–1 year).

Acer

Acer undoubtedly benefits from the quality and influential position of its leadership. Chairman and CEO Shih is a very prominent person in Taiwan and increasingly also abroad. *Business Week* (08.01.96) even selected Shih as one of the 'Top 25' managers of the year, naming him 'Asia's paramount entrepreneur'. Unlike many other leading figures in Taiwan's PC industry, he was not only born in Taiwan, but was also educated there, which makes him even more of a role model.[58] Notwithstanding Shih's charismatic leadership style, the strategic influence by other managers is considerable. Illustrative in this respect is that Acer's employees remain the biggest shareholders of the company.[59]

The beginning of the 1990s witnessed major organisational changes in the company. Convinced that vertical integration was no longer a profitable strategy, Acer was transformed into a 'client–server' structure in which each business unit acts as a server for its sister clients. Business units are not required to source materials from other units of the Group if components of the same quality can be bought elsewhere more cheaply. Tight controls are prescribed to ensure that workers follow the same testing procedures wherever they are located.

Of special relevance for the internationalisation of Acer was the transformation of the company into a 'federation of stand-alone units' (*Business Week* 04.07.94). Whereas Acer in Taiwan produces key components and provides design services to Acer plants as well as to other (OEM) buyers, the main goal of the overseas subsidiaries (which are co-ordinated by the regional business units) is to push sales of the Acer brand. To combine both goals, the Group decided to spin off no less than 21 business units that will be listed on local Stock Exchanges by the end of the decade. The parent company intends to keep shares ranging from 19–40 per cent, so that ultimate control will be retained. These overseas business units will be run by local managers. At present, the Acer Group consists of six strategic business units that are responsible for R & D, manufacturing and OEM business. In addition, five regional business units are responsible for assembly, distribution, service and marketing of the Acer-brand products (see Appendix 7.3). This organisational set-up has to ensure that conflicts of interest do not take place.

To speed up delivery and minimise inventory costs, a so-called 'fast food franchise' strategy was introduced. Unfinished PCs are shipped to overseas assembly sites located in the markets of the buyers. Key components that quickly lose value (so-called 'perishable' components such as microprocessors and hard disk drives) are sent by air transport to ensure fast delivery. 'Non-perishable' components (such as computer cases and floppy disk drives) are shipped via slower sea transport.[60]

FIC

To be a member of the 'Wang Web' is considered one of FIC's most important strengths. Although there is no formal vertical backward integration, FIC's ties with other members of the Web assure a solid supply of a number of crucial components – especially for the production of motherboards. Petrochemicals for printed circuit boards (PCB) production, for instance, originate from the Formosa Plastics Group. Sister company Nan Ya Plastics produces the copper-clad laminate base as well as PCBs for FIC.

Responding to the limits of petrochemical expansion,[61] the Wang group intends, in the long run to convert itself into a computer industry conglomerate, with all individual companies performing their specific task.[62] In view of the dynamic nature of the PC market, it is impossible to foresee the division of tasks among the various members of the Wang Web in the more distant future. In the coming years, however, FIC will retain its focus on more downstream activities such as the production of motherboards and assembly of final systems. FPG and other members, on the other hand, are expected to expand their activities more on the raw materials end where scale advantages are crucial. Although the corporate family ties do not allow complete autonomy, competition between suppliers is considered important. This implies that outsiders who sell components of the same quality level against lower prices will in principle be preferred to more expensive Web members.

Next to these 'technical links', access to capital is another major asset to FIC of being a member of the Wang Web which no doubt has helped in pursuing domestic and overseas ventures. Although Charlene Wang claims she has never accepted investment money from her father, his guarantees were critical in raising loans from Taiwan's risk-adverse banks (*Business Week* 15.08.94). Moreover, the ability to continuously cut costs, a necessary condition for survival in this highly price-sensitive line of business, is a strongly entrenched tradition in the family.[63]

Although both Charlene Wang and Dr Ming Chien were trained at American universities, FIC considers itself to be a Chinese family company (cf. Chen 1995). Top positions in the firm, as was mentioned before, are indeed occupied by family members and FIC allows little autonomy to the lower levels of the organisation. It is not expected that the Group will follow the trend of some Korean conglomerates in decentralising and promoting non-family as their founders retire (FT 24.07.95). Although in the longer run this 'informal' organisational set-up may become less appropriate, the existing absence of a heavy bureaucracy is valued positively by the employees, mainly because the communication lines between employees and top management are relatively short. To strengthen the commitment to the company, about 10 per cent of the company shares are owned by the employees.

An important recent organisational change that is also relevant for the company's activities overseas was the introduction of the so-called 'barebone system'. Acknowledging that end users have different preferences regarding various features of systems (especially with regard to the CPU, internal memory and hard disk), PCs are initially assembled without these components.[64] An additional advantage of this set-up is that components such as CPUs, DRAMs and hard disks can be purchased in local markets (if available), thus speeding up delivery and avoiding import duties.

Summary

As Dunning (1993: 216) rightfully remarked 'once a firm engages in foreign transactions it is faced with new organisational challenges'. In fact, this holds for every step a company takes on its way to becoming a full-fledged multinational enterprise. Various examples were given of major organisational restructurings as well as the adaptation of 'new' human resource policies that took place at Samsung Electronics and LG Electronics in the 1990s. These observations confirm the results from the econometric analysis described in Chapter 6 that showed a positive significant relationship between human capital embedded in Korean electronics companies and their propensity to invest abroad. Given the typical characteristics of the Korean paternalistic management style (cf. Chen 1995), many of the changes concerning human resource policy carried out by the companies studied here, are no doubt even quite unique in the local context. Considerable emphasis is given to simultaneously 'teach' Korean managers about the rest of the world and teach overseas employees about Korea.

In Taiwan, Acer is known for its 'progressive' (namely, 'non-Chinese') human resource policy allowing for open debates among personnel and the willingness to introduce changes. The fact that Acer is often seen as an exception to the rule in Taiwan might explain that no significant relationship was found between human capital and FDI in the econometric analysis in Chapter 6. To optimise their international logistical networks and become or remain competitive in the global market place, both companies have introduced ways (the franchise and barebone systems, respectively) that safeguard flexibility and speed of delivery. For Acer this was one of the main considerations in breaking up its formerly integrated structure in the 1990s. In the case of FIC, the (relatively) loose organisational ties within the Wang Web served this purpose well. Interestingly, these network structures, which now are an important base for their international success, show a strong resemblance to the strengths typically ascribed to Chinese family firms, as were discussed in Chapter 4.

Patterns of internationalisation

In this section, first the entry modes in overseas markets of the four companies and their subsequent expansion strategies are examined. Subsequently, the motivations that have led to these investments are investigated.

Entry modes and expansion strategies

To describe the broad path companies follow in becoming a MNE, commonly a certain sequence is identified (see Chapter 2). It begins with the assumption that, initially, the firm is purely a domestic firm, both in terms of production and markets. When the limits of the domestic market are reached, overseas markets are penetrated through independent overseas sales agents. To exert closer control in a next stage, sales subsidiaries are set up. Eventually, the company may become engaged in production facilities overseas.[65] As Dicken (1992) emphasises, however, a firm may bypass intermediate stages and, for instance, set up an overseas production facility as a first step. Other variations are also possible. Firms, for example, may get involved in overseas markets through licensing agreements with local firms that are permitted to use a particular technology or manufacture a specific product for a defined market on payment of fees and royalties. Which steps were taken by the four companies in entering and expanding overseas markets?

Samsung Electronics

The first export activities by Samsung Electronics resulted from OEM sales and were not initiated by the Group's General Trading Company (GTC) (namely, Samsung Co.) which was only established in 1975 (Kang 1993). After Samsung Co. was set up, its commercial network was utilised to export Samsung branded CE products. Also after the establishment of the GTC, OEM remained important in entering overseas markets. In 1980, for instance, microwave ovens were sold for the first time in the USA through an OEM contract with storehouse J.C. Penny. When in the 1980s, sales volumes became sufficiently large, the company began to set up its own sales offices abroad. Although the first office abroad was established in the USA in 1978, it was especially in developing areas (e.g. Central and South America, Southeast Asia and the Middle East) that SEC was successful in increasing the proportion of its own brand sales (Cho *et al.* 1994). In more advanced economies SEC primarily sold under the brand names of its OEM buyers.

In the course of the 1980s, SEC also started to establish manufacturing operations abroad. The first overseas (CTV) factory was established in Estoril in Portugal (this, however, is no longer in operation). As can be derived from Table 7.12 the number of overseas production facilities has grown rapidly, especially in the 1990s. In these years, the network of overseas sales subsidiaries and branches (that are co-ordinated by regional headquarters), also expanded rapidly. In addition, as was noted earlier, the company established a number of R&D centres overseas.

In 1994, the North American continent was the most important region outside Korea realising 37 per cent of total sales abroad (OEM and own brand sales/OBS), followed by Asia and Oceania (34 per cent), Europe and the CIS (21 per cent), Central and Latin America (5 per cent) and the Middle East and Africa (3 per cent).

Table 7.12 Overseas production units of SEC as of Spring 1995[a]

Region	*Country*	*Year established*	*Products*	*Ownership (%)*	*Employees*[b]
America	USA	1984	MWO	100	1,700
	USA	1993	Semiconductors	100	
	Mexico	1988	CTV	100	
Asia	Thailand	1988	CTV, REF, W/M, VCP	51	7,500
	Indonesia	1989	REF	50	
	Indonesia	1991	AUDIO, VCR	80	
	Malaysia	1989	MWO	100	
	PRC	1992	VCR	50	
	PRC	1993	AUDIO	90	
	PRC	1993	Telecommunication switching systems	46	
	PRC	1994	CTV	50	
Middle East & Africa	Turkey	1989	CTV	20	370
Europe	Portugal	1994	Semiconductors	38	2,500
	UK	1987	CTV	100	
	Spain	1989	VCR	100	
	Slovakia	1991	REF	44.8	
	Hungary	1989	CTV	95	

Source: Samsung Electronics

Notes:

a It should be stressed that the table only shows existing units. By Spring 1995 new investment plans were already announced for China, India, Mexico, Malaysia, Vietnam, Brazil and the UK.

b Total overseas employment in the region (including other subsidiaries).

LG Electronics

For a long time, LGE exported primarily through its OEM buyers. Its first export sales, for instance, were to a New York mail order firm in 1962 (Michell 1988). By the end of the 1970s and the beginning of the 1980s, LGE began to establish sales offices in major markets to promote its own brand name. LG International, the General Trading Company of the Group that was established in 1976, sells LGE products in countries where LGE has no sales subsidiaries. Although the Goldstar brand name has been particularly successful in developing countries, its first overseas office was launched in 1978 in the USA.

In the 1980s, LGE expanded its operations abroad by setting up factories in various regions of the world (see Table 7.13). The first factory abroad was established in 1982 in the USA. In the 1990s, the number of overseas factories

Table 7.13 Overseas production units of LGE as of Spring 1995[a]

Region	*Country*	*Year established*	*Products*	*Ownership (%)*	*Employees*[b]
America	USA	1982	Video tapes	100	1,600
	Mexico	1988	CTV	100	
Asia[c]	Thailand	1988	CTV, AUDIO, W/M	49	2,500
	Thailand	1988	Telephones, keyphones	60	
	Philippines	1988	AUDIO, CTV, W/M	40	
	Indonesia	1990	CTV, AUDIO, REF	49	
	Indonesia	1994	Pumps	49	
	PRC	1993	Keyphones	35	
	PRC	1993	AUDIO	65	
	PRC	1994	Colour picture tubes	51	
	PRC	1994	CTV	70	
	PRC	1995	Dolby systems	60	
	PRC	1995	VCR	70	
	PRC	1995	A/C, MWO	80	
	PRC	1995	Pumps	100	
	PRC	1995	W/M	60	
	PRC	1995	REF	50	
Middle East & Africa	Egypt	1990	CTV components	51	130
Europe	Germany	1986	CTV, VCR	100	980
	UK	1988	MWO, CTV	100	
	Italy	1990	REF	100	

Source: LG Electronics

Notes:

a As in the case of Samsung Electronics, this only represents a snapshot of LG Electronics operations abroad. By 1996 ambitious plans existed for additional capacity in Indonesia, Malaysia, and even North Korea (FT, 14.04.96).

b Total overseas employment in the region (including other subsidiaries).

c Since most factories in the PRC only commenced in 1995, the number of employees in Asia is still relatively modest.

expanded considerably. Table 7.13 shows that in recent years LGE has invested on a large scale in Asia in particular – especially in China. Like SEC, also in the 1990s under the umbrella of regional headquarters, LGE rapidly expanded the geographical coverage of its sales network by opening a large number of sales subsidiaries and branches. Next to the production and sales facilities, we have already noted that LG Electronics established three research centres abroad.

In 1994, the American region was by far the most important market outside Korea realising no less than 42 per cent of total LGE sales abroad (OEM and OBS), followed by Asia (29 per cent), Europe (19 per cent), and the Middle East and Africa (10 per cent).

Acer

Like most Taiwanese PC companies, from the start Acer sought its major customers abroad; however, this occurred not only in the form of OEM contracts. From the beginning it also promoted its own brand name – albeit on a modest scale. Although a small office was opened as early as 1977 in North America, 1985 saw Acer setting up a fully-fledged sales subsidiary for the first time, in Germany. In the second half of the 1980s and the 1990s, the company's network of sales offices was further extended in North America and Europe as well as in Japan and Australia. In markets in which Acer has no sales subsidiaries, local distributors sell its products.

In the 1990s, Acer entered the next stage in its internationalisation. In 1992 a large factory was established in Malaysia to manufacture peripherals primarily for the export market.[66] In the PRC, Acer acquired a company which in 1996 started to produce keyboards.[67] In addition, as a result of the 'fast food' business model described earlier, a substantial and rapidly growing number of sites have been set up in target markets where PCs and notebooks are assembled (see Table 7.14).[68] The assembly site in the USA is located in Silicon Valley where Acer has also established a R&D facility – which was opened even before production started.

Although most sales subsidiaries are located in Triad markets (namely, Western Europe, North America and Japan), in the developing world in particular, Acer has built up top positions using its own brand name. In North America and Europe, a substantial part is still sold through OEM contracts. In terms of own brand sales (OBS), North America nevertheless is by far the most important market realising 50.6 per cent of total OBS turnover in 1994. Acer's sales in Europe represented 17.1 per cent of total OBS turnover and Latin America 12.6 per cent. Other parts of the world (namely, Asia, Pacific, Africa, Russia and the Middle East) contributed 19.7 per cent to Acer's total OBS sales.

FIC

Like Acer, FIC from the beginning, primarily targeted the international market to sell its products. The year in which FIC started to sell PCs outside the Wang

Table 7.14 Overseas production units of Acer as of Spring 1995

Region	*Country*	*Year established*	*Products*	*Ownership*[a] (%)	*Employees*[b]
N. America	USA	1992	PCs, notebooks	na	1,098
Asia	Malaysia	1992	Peripherals	100	2,679
	PRC	1995	Keyboards	100	
	PRC	1995	PCs, notebooks	JV	
	PRC	1995	PCs, notebooks	JV	
	Japan	1993	PCs, notebooks	na	
	Singapore	1995	PCs, notebooks	na	
	Indonesia	1992	PCs, notebooks	na	
L. America	Mexico	1993	PCs, notebooks	JV	402
	Chile	1994	PCs, notebooks	JV	
	Miami[c]	1993	PCs, notebooks	na	
	Brazil	1992	PCs, notebooks	na	
Other regions	UAE	1994	PCs, notebooks	na	1,259
	Australia	1993	PCs, notebooks	na	
Europe	Netherlands	1992	PCs, notebooks	na	320
	Germany	1993	PCs, notebooks	na	
	UK	1994	PCs, notebooks	na	

Source: Acer

Notes:
a JV: joint venture;
na: not available (confidential)
b Total overseas employment in the region (including other subsidiaries)
c Production intended for the Latin American market only

Group, 1986, was also the first time the company presented its products outside of Taiwan, at Cebit in Germany, the world's largest computer exhibition. Initially, besides direct contacts with industrial customers, FIC sold its final products (namely, PC systems) through local agents. In 1990, the company internationalised one stage further by setting up sales offices in the USA, Europe and Hong Kong – following on with offices in Australia and Japan in more recent years.

At present, the company has a distribution network in more than seventy countries. In most Asian countries, the distribution is not managed by FIC itself, but instead is taken care of by local (usually ethnic Chinese) agents. According to FIC, the similarities of these Asian markets to the home market and the cultural background shared with the local agents make the establishment of own subsidiaries unnecessary.[69]

In the 1990s, FIC's international presence was further consolidated by the setting up of production lines overseas (see Table 7.15).

In the PRC, FIC has established a factory where less-advanced motherboards are produced and final systems are assembled. A second factory is planned near

Table 7.15 Overseas production units of FIC as of Spring 1995

Region	*Country*	*Year established*	*Products*	*Ownership (%)*	*Employees*[a]
N. America	USA	1993	PCs, notebooks	100	100
Asia	PRC	1994	PCs, motherboards	100	500
Europe	Netherlands	1993	PCs, notebooks	100	75

Source: FIC

Notes:
a Total overseas employment in the region (including other subsidiaries)

Shanghai, where monitors will, for the first time, be manufactured.[70] In addition, assembly lines have been established in the USA and the Netherlands. The site in the USA is located in Silicon Valley where a R&D facility has also been opened.

In 1994, no less than 40 per cent of total sales were realised in Europe. North America was the second largest market representing 30 per cent of total sales, followed by Asia (20 per cent) and other countries (10 per cent).[71] It should be underlined, however, that in Europe and North America most sales were realised through OEM contracts whereas in Asia own brand sales were far more important.

Summary

In Figure 7.7 the internationalisation patterns of the four companies are schematised. It is important to keep in mind that this scheme merely represents a simplification of reality. On some occasions, companies from LDCs may also have concluded OEM contracts with one of the four companies studied here, but this has not been a major mode of doing business between these countries and Samsung Electronics, LG Electronics, Acer or FIC. Moreover, although exceptions prove the rule, the use of local importers ('AGENTS') has not been a major entry or expansion strategy in the case of the Korean companies. In addition, the figure does not show the relative importance (or any changes in it) of the various modes of international business. Furthermore, overseas R&D or design units are not included in the scheme.

Nevertheless, some interesting observations can be made on the basis of the cases studied. As has already been shown in Chapter 6, Korean and Taiwanese outward investors tend to realise an important share of their turnover abroad. The case studies, however, yielded important additional insights in this respect. First of all, the figure shows how rapidly the companies have gone through various stages of internationalisation, i.e. from export to overseas production. Increasing international competition requiring faster establishment in leading markets and the growing importance of M&As as a means of entering foreign

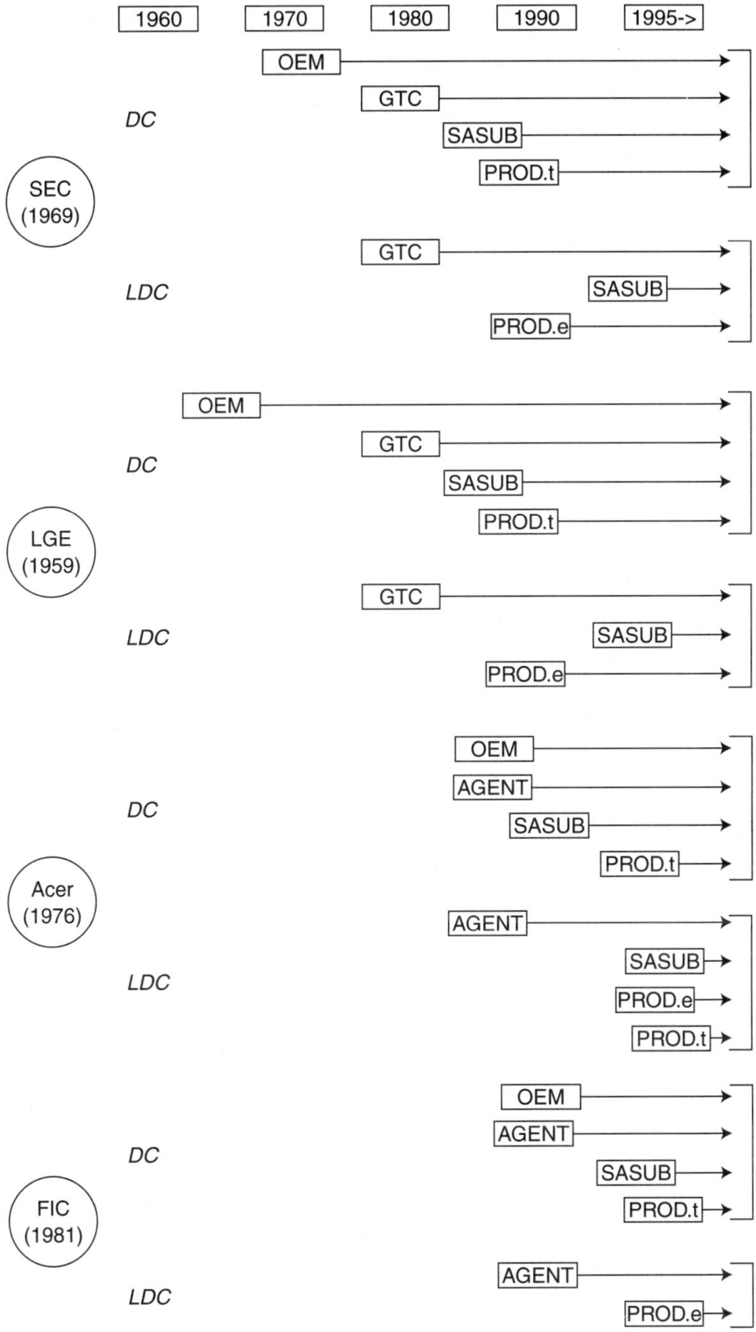

Figure 7.7 Internationalisation patterns

markets and accessing foreign technology appear to have speeded up this development (cf. Cantwell and Tolentino 1990, Tolentino 1993, Zander 1994). In fact, we noted that, unlike 'conventional' images of internationalisation (such as put forward by the Scandinavian School – see Chapter 2), from the beginning the most important markets for the four companies were located outside their home economies; only in later stages of company development were important market shares gained at home. This can be explained by the relatively backward state of the home economies by the time these companies started to expand. We saw that from the initial stages, OEM contracts have been a major mode of cross-border business for these companies. Since most of these OEM contracts were concluded with major (manufacturing and other) companies originating in industrialised economies, again contrary to the internationalisation patterns described by the Scandinavian School, from the start, the latter were the most important markets.

For the two Korean companies, especially in the second half of the 1970s and the first half of the 1980s, the international networks that their affiliated General Trading Companies (namely, Samsung Co. and LG International) rapidly built up throughout the world have certainly contributed to their successful entry into foreign markets. At present, their role has become very limited in promoting the goods of SEC and LGE. Both companies have established a network of own sales subsidiaries or branches in all regions of the world. The figure also nicely illustrates that the 'follower strategies' for which Korean chaebol are known (cf. Ghoshal 1988, Kang 1993) and which, according to the late industrialisation paradigm, are typical for the oligopolistic markets found in many late industrialised economies also extend to their internationalisation behaviour. SEC and LGE almost simultaneously advanced from OEM to the establishment of sales offices and, subsequently, to production facilities abroad. This practice also confirms the 'follow-the-leader' hypothesis brought forward by Knickerbocker (1973).

The short period in which Acer and FIC have advanced through various stages of internationalisation is truly striking. In fact, the actual time span is even shorter than may be derived from the figure. As pointed out earlier, in the first years after they were established Acer and FIC were not engaged in selling PCs they produced themselves. The highest pace of internationalisation was pursued by Acer so that in a few years time it had built up a global presence. The progress of FIC, the 'youngest' of the four companies, has been somewhat more modest in this respect and no FIC sales subsidiaries have yet been established in developing countries.

In Figure 7.7, a distinction is made between overseas production targeted at the local market or region ('PROD.t') and overseas production meant for export to other countries or regions ('PROD.e'). As far as FDI in advanced economies is concerned, the production operations of all four companies are meant for local sales. In the case of FDI in developing countries, we observed that in the cases of SEC, LGE and FIC, the goods produced are meant eventually for other markets. This clearly reflects the distinct motivations to invest in industrialised and in developing countries which are discussed in the next section. Interestingly, in the

1990s, Acer started to open assembly units in developing countries whose output is targeted at the local market. This development illustrates the growing importance of these new target markets.

Motivations to invest

We asked the four companies to indicate how important the various motives were in deciding to invest in factories in the EU, USA and (developing) Asia. In Table 7.16, the importance of these investment motivations are listed.

For SEC and LGE, avoiding trade barriers was a highly important motive for investing in the EU and USA. This observation confirms our findings in Chapter 6 on the significance of protectionist measures against Korean electronics companies. Both companies, for instance, have been frequently hit by antidumping measures in Europe (see Appendix 7.5). In the case of LGE, avoiding trade barriers was also a highly important motivation for investing in the Asian region. This holds not only for the production countries but also for exports from these production locations to other industrialised nations – especially the USA. In 1989, the USA discontinued the Generalised System of Preferences (GSP) for Korean imports.[72] By exporting from developing economies that do possess GSP status, Korean firms can still benefit from the reduction of normal customs duties. For Acer and FIC, protectionism as such, e.g. antidumping measures, has not played a role in their decision to produce in the EU and USA. The importance attached to circumventing trade barriers can be explained by the fact that the local assembly of PCs or electronic notebooks reduces import duties.

The possibility of collecting commercial information in the EU and the USA played a role in the decision of SEC (important), LGE (highly important) and Acer (important) to invest in these important target markets. An example in the case of SEC is the 'Euro-lab' in France where among other things, European food habits are investigated – information that is used to adapt existing products or develop new ones. LGE established a design centre in Ireland to tailor products to specific local needs. Acer is motivated by their desire to be 'close to the market'. The PC business is a highly dynamic market in which the required computer configurations change continually; this characteristic requires close monitoring of local demand. The fact that FIC, on the other hand, did not indicate that this motive was considered important reflects their focus on industrial clients. Recently however, a shift towards selling under the own brand name has been observed at FIC so that we expect that this motivation will become increasingly important in the future. Only LGE indicated that gathering commercial information was of some importance in Asia. The relative insignificance can be explained by the fact that – certainly until recently – the production facilities established in Asian developing countries did not sell their output in these countries but it was exported to advanced economies instead. In view of the continuous growth of Asian economies, their importance as markets increases rapidly – a development also illustrated by the establishment of sales subsidiaries (SEC, LGE and Acer) and assembly lines (Acer) in these countries in the last few years.

Table 7.16 Motivations for investing

	SEC			*LGE*			*ACER*			*FIC*		
	EU	*USA*	*Asia*	*EU*	*USA*	*Asia*	*EU*	*USA*	*Asia*	*EU*	*USA*	*Asia*
Circumvent trade barriers	3	3	1	3	3	3	0	0	0	2	2	1
Gather commercial information	2	2	0	3	3	1	2	2	0	0	0	0
Technology scanning	1	1	1	2	3	1	0	2	0	2	2	0
Quality labour	1	1	1	2	2	1	2	2	2	2	2	0
Host country incentives	1	1	1	3	0	1	2	2	2	2	0	2
Access raw materials	0	0	0	3	3	1	0	0	0	0	0	0
Low labour costs	0	0	2	0	0	3	0	0	3	0	0	2
Low capital costs	0	0	0	2	2	3	0	0	3	0	0	3
Limited size of the home market	3	3	3	1	1	1	0	0	0	2	2	2

Sources: SEC, LGE, Acer and FIC

Note:
0 = no importance; 1 = of limited importance; 2 = important; 3 = highly important

For SEC, the possibility of scanning the local environment for new technological developments in all three regions was of limited importance. In the case of LGE, this consideration was more relevant in the EU (important) and especially the USA (highly important). For Acer, making use of locally available 'national technological capabilities' (see Chapter 2) was an important reason to invest in the USA. The company selected San Jose in the heart of Silicon Valley where it also established a R&D facility. FIC, as noted already, often develops the motherboards in close co-operation with its OEM clients. Since many clients are located in Europe, scanning the latest technological developments was important for the company both in the USA and in the EU.

For the companies studied, the quality of labour also played a role in their decision to invest in most regions. For SEC, it was only of limited importance, whereas LGE indicated that it played an important role in the EU and the USA but was of limited importance for their investments in Asia. Acer indicated that it was an important consideration for all investments. We like to stress, however, that whereas in the USA the high level of technological expertise of labour was an important attraction for the company, in Malaysia the quality of blue collar labour was meant. For FIC, the quality of labour was important in deciding to invest in the EU and the USA.

According to SEC, the provision of incentives (such as tax exemptions, free office space, and cheap credit) played only a limited role in their decision to invest in the three regions. Interestingly, both in the case of LGE and FIC such incentives were not relevant as far as the investment in the USA is considered whereas in the EU they were considered highly important and important, respectively. This probably reflects the competition between host countries and even regions in the EU which, in their attempts to attract foreign investors, offer a wide range of incentives. FIC indicated that also the investment in Asia was influenced by incentives. For Acer, host country incentives were important for investment in all three regions.

Table 7.16 shows that only in the case of LGE did access to raw materials play a role in its decision to invest abroad. With regard to their investments in the EU and the USA this was even a highly important motive – this in view of the company's strategy to localise its operations.

For all four companies, the low cost of labour in the Asian region was an important (SEC and FIC) or highly important (LGE and Acer) motivation to set up a factory there. The shift of production operations to these countries has been a response to worsened factor conditions at home and are partially intended to maintain export competitiveness. In the case of Acer, the low labour cost argument holds for the investment in the peripherals factory in Malaysia and the keyboards factory in the PRC. As was noted earlier, for the assembly sites in Asia targeting at the local markets, this was not an important consideration.

As was discussed in Chapter 4, regular interest rates both in Korea and Taiwan are relatively high. It is therefore no surprise that for LGE, Acer and FIC, the relatively low cost of capital was considered highly important in deciding to set up factories in the Asian region. For LGE, the costs of capital also played a role – though of less importance – in deciding to invest in the EU and the USA.

And finally, according to SEC (highly important), LGE (limited importance) and FIC (important) the limited size of the home market necessitated their expansion abroad.[73]

To sum up, we conclude that for all four companies, the motivations to invest in the EU and the USA are largely the same.[74] Moreover, we observed that the Korean investments in advanced economies were primarily defensive in nature. Although at first instance the investments by Acer and FIC may appear much more offensive, we also noted that the nature of their lines of business more or less forces them to invest in their target markets. The processes investigated are highly dynamic, however, which is also reflected in the motivations to invest abroad. As a result of the NAFTA agreement, SEC and LGE shifted CTV production to Mexico in the second half of the 1990s. Similar examples of such efficiency-seeking FDI are provided in the next section on their operations in Europe. In addition, as a result of SEC's and LGE's efforts to further globalise their business operations, since the 1990s we observe a shift towards a more offensive attitude to expand production operations in advanced economies. This, *inter alia*, implies that access to local raw materials will also become more important for SEC and LGE. The latter also holds for Acer and FIC.

Investments in Europe: a closer look

In this section, the investments by the four companies in Europe are discussed in more detail. First, an overview will be provided of their operations in the region. Subsequently, some organisational characteristics are addressed, followed by a discussion of their market position in Europe.

Overview of operations

Samsung Electronics

The CTV factory in Portugal, established in 1982 was the first production facility established by SEC outside Korea and the very first Korean investment in Europe. (As noted earlier, this factory is no longer in operation.) By 1995, SEC had built up an impressive physical presence in the region with investments in five production facilities (see Table 7.17). In the UK factory in Billingham (established in 1987) as well as at the Hungary site (1989), CTVs are produced. In Spain (1989), the company manufactures VCRs. In Slovakia, in 1991 SEC formed a joint venture with Caltex to produce refrigerators. In 1994, Samsung Electronics, together with American Texas Instruments (TI), invested in a factory in Portugal to produce memory chips for the European market.

Next to these factories, SEC established sales offices in Germany, UK, France, Spain, Italy, Sweden and Portugal. As we saw earlier, small R&D centres are located in Belgium and Germany doing research on telecommunication systems

Table 7.17 SEC's major operations in Europe – some factual data as of Spring 1995

	Year of establishment	*Greenfield or M&A*	*Ownership (%)*	*Activities*[a]	*Employees (fte)*	*Korean employees*	*Nationality of GM*	*Annual production*	*Capacity (units)*
Portugal	1994	M&A	38	P, SE	165	8	Korean	4M DRAM	1M 16M DRAM 0.3M
UK	1987	GF	100	P, SE	425	7	Korean	CTV	700K
Spain	1989	GF	100	P, SE	210	5	Korean	VCR	700K
Slovakia	1991	GF	44.8	P, SE	942	5	Korean	REF	360K
Hungary	1989	GF	95	P, SE	201	6	Korean	CTV	120K
Germany	1982	GF	100	W, SA	112	12	Korean		
UK	1984	GF	100	W, SA	124	9	Korean		
France	1988	M&A	100	W, SA	81	6	Korean		
Spain	1990	GF	100	W, SA	67	6	Korean		
Italy	1991	GF	100	W, SA	36	5	Korean		
Sweden	1992	GF	100	W, SA	21	4	Korean		
Portugal	1993	GF	100	W, SA	30	2	Korean		

Source: Samsung Electronics

Notes: The Wynyard facility is not yet included

a Production (P), warehousing (W), sales (SA), service (SE), research & development (R&D), regional headquarters (RHQ), mergers & acquisitions (M&As)

and A/V systems and home appliances, respectively. These centres primarily focus on scanning local market developments in these fields.

In the autumn of 1994 the company announced its plans to develop a fully integrated electronics and home appliance production complex in Wynyard (Stockton-on-Tees), England.[75] In the first of the three stages of construction Samsung intends to invest US$ 700 million. By its completion in 1999, the complex is expected to have an annual local production volume of approximately US$ 2 billion employing 3,000 local workers. Samsung received about US$ 90 million of regional selective assistance for this project. In addition, infrastructural spending by the government-sponsored agency English Partnership amounted to more than US$ 18 million. Support from Cleveland local authorities, Teesside Training and Enterprise Council and Teesside Development Corporation is estimated to be worth US$ 26 million over six years (FT 18.10.94, 10.12,94). In 1995, the production of MWOs and monitors commenced, followed in 1996 by CTVs.

LG Electronics

As is illustrated in Table 7.18, LGE's operations in Europe are also quite substantial. In 1986, LGE invested in its first production facility in Europe, a CTV factory in Germany. At the beginning of 1995, this factory, which then produced colour TVs and VCRs, was on the verge of losing its colour TV production to the UK, where it was to be integrated within existing operations. The plant in the UK, established in 1988, produced microwave ovens and colour TVs. In the Italian factory, set up in 1990, refrigerators are produced.[76]

LGE also operates sales subsidiaries in the UK, France, Germany and Hungary. At LG Design Technology in Dublin (Ireland), products are designed which are tailored to European lifestyles and local consumers. The design centre is supported by two design companies, one British and one Swiss. LGE's commitment to gaining a strong position in the Eastern European market is underlined by a permanent exhibition hall of its home appliances opened in Prague at the end of 1992. (The first one established by a Korean appliance maker in Europe.)

In 1995, LGE commenced the construction of an integrated complex in the North of England (the Newcastle area) to produce colour TV sets and microwave ovens; total investment is put at about US$ 40 million (*Business Times* 01.03.95). Approximately 15 per cent of this amount was granted by the local authorities as regional selective assistance (FT 08.07.94). Products will be exported from Newcastle to the whole of Europe; with up to 500 people employed. However, LG has more ambitious plans for Europe. In 1996 it announced a US$ 2.65 billion investment in Wales (the biggest Korean investment in Europe to date) where the company will establish two factories producing semiconductors and colour TV components. The enormous support package (including a 250–acre site, infrastructural support and cash grants) amounting to some US$ 45,000 a job, resulted in an investigation by the comptroller and auditor general. Not only was

Table 7.18 LGE's major operations in Europe – some factual data as of Spring 1995

	Year of establishment	*Greenfield or M&A*	*Ownership (%)*	*Activities*	*Employees (fte)*	*Korean employees*	*Nationality of GM*	*Annual production*	*Capacity (units)*
Germany	1986	GF	100	P	282	8	Korean	CTV	300K
								VCR	700K
UK	1988	GF	100	P	334	9	Korean	MWO	700K
								CTV	200K
Italy	1990	GF	100	P	180	–	Korean	REF	200K
Ireland	1992	GF	100	R&D (Design)	7	8	Korean		
Germany	1980	GF	100	RHQ, W, SA, SE	70	5	Korean		
France	1990	GF	100	RHQ, W, SA, SE	40	4	Korean		
UK	1987	GF	100	W, SA, SE	52	3	Korean		
Hungary	1992	GF	100	W, SA, SE	10	2	Korean		

Source: LG Electronics

Notes:

GF Greenfield
M&A Mergers & acquisitions
RHQ Regional headquarters
P Production
W Warehousing
SA Sales
SE Service
R&D Research & development

this amount substantially higher than most other investors had received until then, it was also considered by some to be a waste of money which created unhealthy economic distortions. It also underlined the substantial differences that exist between the budgets available in Wales in comparison with the English regions. The Northern Development Company (the North-east English agency), for instance, could 'only' offer some US$ 125 million to attract LG (FT 11.12.97, 08.01.98). Notwithstanding the generous support, the economic problems in Korea, however, have slowed down the preparations for the actual production activities (FT 24.03.98).

As compared to Samsung Electronics, which has two production facilities in Eastern Europe (in Hungary and Slovakia), LG Electronics' presence in the former Eastern bloc was, by the beginning of 1995, still rather modest and confined to the sales subsidiary in Hungary.

Acer

During the last few years, Acer has considerably expanded its physical presence in Europe. In Table 7.19, an overview of the most important European operations is given. At the beginning of 1995, no less than three assembly lines and eight sales offices had been established. Although Acer established its first office in Europe as early as 1985, it was only in 1992, 1993 and 1994 that assembly operations began in the Netherlands, Germany and the UK, respectively. Next to PC systems, Acer in 1994 started to produce notebook computers in the Netherlands. In 1996 notebook production was launched in Germany. By the end of 1997, Acer Peripherals announced the establishment of a factory in United Kingdom (Wales) to manufacture televisions and computer monitors. As in the case of LG's investment in the region, the Northern Development Company claimed that Acer has been lured away by a higher support offer from Wales (FT 22.01.97). The US$ 40 million investment is expected to generate 1,000 jobs.

Acer's sales offices are located in the Netherlands, Germany, the UK, France, Italy, Austria, Belgium, Denmark, Norway, Spain and Hungary. Since Europe is not considered a technological centre of computer know-how, Acer did not establish a R&D facility here. This, however, does not mean that no knowledge transfer (besides market knowledge) takes place as a result of the company's local presence. As a response to the strict environmental EU regulations, for instance, Acer has developed new packaging material which was subsequently introduced world-wide.

FIC

Table 7.20, which lists the operations of FIC in Europe, shows that an assembly line was established in 1993 in the Netherlands for PC systems as well as notebooks. In 1995, FIC moved to a much larger plot of land to allow for a future expansion of production. In 1990, it had already opened a sales office in that country after having sold PC systems and motherboards through local agents

Table 7.19 Acer's major operations in Europe – some factual data as of Spring 1995

	Year of establishment	*Greenfield or M&A*	*Ownership (%)*	*Activities*	*Employees (fte)*	*Taiwanese employees*	*Nationality of GM*	*Annual production*	*Capacity (units)*
Netherlands	1990	M&A	100	RHQ, P, W, SA, SE	110	12	Danish	PCs & notebooks	180K
Germany	1985	GF	100	P, W, SA, SE	120	2	German	PCs	120K
France	1988	GF	100	W, SA, SE	26	0	French		
UK	1988	GF	100	P, W, SA, SE	43	3	Taiwanese	PCs	60K
Italy	1990	M&A	100	W, SA, SE	15	0	Singaporean		
Austria	1992	GF	100	W, SA, SE	16	2	Taiwanese		
Belgium	1993	GF	100	SA, SE	6	0	Danish		
Denmark	1990	M&A*	100	SA, SE	15	1	Taiwanese		
Norway	1994	GF	100	SA, SE	3–4	0	Norwegian		
Spain	1995	GF	100	SA, SE	1–2	0	Spanish		
Hungary	1994	GF	100	SA, SE	3	0	Hungarian		

Source:: Acer

Notes:

GF Greenfield
M&A Mergers & acquisitions
RHQ Regional headquarters
P Production
W Warehousing
SA Sales
SE Service
* Former subsidiaries of acquired company 'Altos'

Table 7.20 FIC's major operations in Europe – some factual data as of Spring 1995

	Year of establishment	*Greenfield or M&A*	*Ownership (%)*	*Activities*	*Employees (fte)*	*Taiwanese employees*	*Nationality of GM*	*Annual production*	*Capacity (units)*
Netherlands	1990	GF	100	RHQ, P, W, SA, SE	55	5	Taiwanese	notebooks	40K
								PCs	20K
Spain	1990	GF	100	SA	6	1	Taiwanese		
Czech Republic	1992	GF	100	W, SA, SE	9	1	Taiwanese		
France	1994	GF	100	W, SA, SE	3	1	Taiwanese		

Source: FIC

Notes

GF	Greenfield
P	Production
W	Warehousing
SA	Sales
SE	Services
RHQ	Regional headquarters
PC	Personal computers

since 1986. Moreover, FIC established a European service centre in the Netherlands. In addition, the company set up sales offices in Spain, France and the Czech Republic – the latter targeting at the Eastern European market. At the end of 1997, FIC announced that it also intends to set up a computer factory near Prague. The US$ 100 million plant is scheduled to employ about 15 people and produce 50,000 computers a month, 95 per cent of which will be exported. FIC has not opened a R&D centre in Europe which is not considered a centre of technology in the computer field.

Next to the formal subsidiaries listed in the table, FIC Europe has also invested in two joint ventures that do not bear the FIC name. One of these firms is located in Western Europe, realising a total turnover of about US$ 180 million and employing some 240 persons. The other company is located in Eastern Europe and has about 200 employees who realise a turnover of US$ 50 million. (By special request of the company, the host countries are not specified.)

Organisational characteristics

To get an impression of the way in which the European operations are organised, the location choice of the regional headquarters, the nationality of management and the global logistical networks are addressed here.

Regional headquarters

The main task of the regional headquarters is to co-ordinate the sales activities of own-brand products in the various countries of Europe. Moreover, smaller investments are usually made from the regional headquarters. Large OEM clients are mostly handled by the corporate headquarters in Seoul and Taipei, respectively.[77]

Where did the four companies base their regional headquarters? SEC originally selected Germany for the establishment of its European regional headquarters. (As a result of the large-scale investments in the UK, it was decided in 1995 to move the European headquarters to London.) LGE, interestingly, has two regional headquarters in Europe. Initially, the company had only a single regional headquarters (namely, in Germany) for the whole of Europe. Partly in view of the different nature of the advanced markets in Western Europe and the Eastern European transformation economies, it was decided to create two separate regional offices, one focusing on the German and Eastern European markets and the other one (located in Paris) dealing with consumers in Western Europe. Both Acer and FIC selected the Netherlands as the location for their European headquarters.

What criteria played a role in selecting these countries? In Table 7.21, the importance attached by the companies to the various location factors listed are given. The table shows that incentives granted by the local governments (such as tax concessions) played an important (SEC, Acer and FIC) or very important (LGE) role for all four companies in deciding where to establish their regional headquarters. The same holds for the quality of the infrastructure in the host

Table 7.21 Criteria for selection of regional headquarters

	SEC (Germany)	*LGE (Germany)*	*Acer (Netherlands)*	*FIC (Netherlands)*
Political stability	2	2	0	1
Host government incentives	2	3	2	2
Good infrastructure	3	3	2	2
Relatively cheap land	1	2	0	0
Relatively cheap labour	1	1	0	0
Loan availability	1	3	2	0
Availability of advanced technologies	1	3	0	0
Availability of technical and skilled manpower	1	3	0	2
Availability of raw materials	1	3	0	0
Size and potential of host country market	3	3	0	0
Geographical location	2	3	0	3
Business & family connections	1	1	0	0
Absence of FOREX control & possibility of repatriating profits	1	2	0	1
Language & cultural affinity	2	3	1	2

Sources: SEC, LGE, Acer and FIC

Note:
0 = no importance; 1 = of limited importance; 2 = important; 3 = highly important

country (highly important for SEC and LGE, important for Acer and FIC). The geographical location within Europe was rated as important by SEC and highly important by LGE and FIC. Language and cultural affinity played an important role in selecting the host country in the case of SEC and FIC and was even a highly important motivation in the case of LGE. For SEC and LGE, the motive is limited to 'cultural affinity' whereas in the case of FIC it also refers to the widespread use of English in the Netherlands. Although the national cultures (in a narrow sense) of these home and host countries formally have little in common, it was noted by the companies that, for instance in terms of the 'way of doing business', Korea and Germany on the one hand and Taiwan and the Netherlands on the other are relatively similar. For Acer, this argument was only of limited importance. In addition, for SEC and LGE the (potential) size of the host market was a highly important motive for selecting Germany[78] It is not surprising that in the case of Acer and FIC, which are based in the Netherlands, this was of no relevance.

Other motives that for one or two companies were considered 'highly' relevant in selecting the host country, were: loan availability (LGE); availability of advanced technologies (LGE); availability of technical and skilled manpower (LGE) and availability of raw materials (LGE). 'Important' selection criteria mentioned, were: political stability (SEC, LGE); relatively cheap land (LGE); loan availability (Acer); availability of technical and skilled manpower (FIC); and absence of foreign exchange ('FOREX') control and possibility to repatriate profits (LGE). And finally, of limited importance were: political stability (FIC); relatively cheap land (SEC); relatively cheap labour (SEC, LGE), loan availability (SEC); availability of technical and skilled manpower (SEC), availability of raw materials (SEC); business and family connections (SEC, LGE); and absence of foreign exchange control and possibility to repatriate profits (SEC, FIC).

Some of the selection criteria mentioned by the companies as being important in selecting a host country in Europe appear to apply more to their manufacturing operations rather than the activities of the regional headquarters *per se*. Given the fact that the head offices are all indeed located in countries where the companies also operate major production activities, this is no surprise. Although it need not necessarily be the case, this does suggest, however, that the importance of the regional headquarters is still rather limited, implying that more important decisions are still made by the global headquarters in the home economies.

Nationality of the management

During the interviews with the companies, all four underlined the importance of the participation of local management in the European activities. Tables 7.17 and 7.18 showed, however, that all the subsidiaries of SEC and LGE in Europe were still run by Koreans. The same held for the nationality of the European 'president'. Given the training programmes that were discussed earlier, it is expected that in the near future a growing number of non-Korean managers will be appointed.[79] The 'official' European ventures of FIC (Table 7.20) as well as the overall European operations are also run by Taiwanese managers. The two other ventures in Europe owned by FIC (that are not specified in the table) are, however, completely managed by locals. Acer is by far the most 'localised' company in terms of its management (Table 7.19); although the European president is Taiwanese, most ventures in Europe are managed by local and other non-Taiwanese persons.

Global logistics

We noted that in the PC industry, an efficient global logistics network is considered to be of crucial importance. For the consumer electronic products produced and sold by SEC and LGE in Europe this is less relevant since the embedded technology has reached maturity and the rate of technological change in the industry is fairly low. Given the importance attached by Acer and FIC to the global logistics system, a further elucidation is given here.

We have seen already that, to speed up delivery and minimise inventory costs, Acer introduced a 'fast food franchise' strategy. How is this strategy applied for its operations in Europe? To maximise flexibility, Acer Europe has in principle, been granted full autonomy by the parent company regarding the purchase of components, i.e. as long as they comply with the quality guidelines as issued by the headquarters in Taiwan. For the European operations, CPUs, for instance, are bought in the USA (from Intel), while most hard disks come from Singapore. Memory chips are bought from the Intel plant in Scotland. The motherboards are flown in directly from the Acer plant in Taiwan. Other computer parts, such as computer cases, for which speed of delivery is less crucial, are shipped to Europe.

FIC's production operations in Europe are also limited to the final assembly of PC and notebook systems. In their case, non-crucial parts (such as keyboards and computer cases) are centrally purchased by FIC Taiwan and shipped to Europe. Components that rapidly lose value (such as CPUs and memory chips) are bought on the European continent. With the shipment of barebone systems, production operations in Europe in the future may be expanded. If, for instance, the headquarters of Texas Instruments in the USA concludes a deal with FIC in Taipei to produce PC systems for the European market, barebone systems will be manufactured in Taiwan (and the PRC in the near future). These are subsequently shipped to Den Bosch in the Netherlands where final assembly and testing will take place, after which the final PCs are sent to Texas Instruments in Utrecht (also in the Netherlands). From there, the latter company will take care of the marketing and distribution.

Local market position

In this section, the position of the four companies in the European market and the problems they face in improving their role in the local CE and PC market are discussed.

Samsung Electronics

By the beginning of 1995, SEC had gained a strong position in Europe for some of its products. This, however, holds only for some markets and for some products. Especially in less competitive markets such as the Czech and Slovakian Republics, Hungary and Romania, Samsung Electronics performed very well. Examples of market shares built up (by 1995), are: 31 per cent for refrigerators and 21 per cent for CTVs in the Czech and Slovakian Republics; 21 per cent for microwave ovens and 25 per cent for CTVs in Hungary; 29 per cent for CTVs in Romania. In large markets such as Germany and France, SEC had not yet been able to capture substantial market shares for its Samsung brand products.

We asked senior executives to indicate the *perceived* strengths and weaknesses of Samsung Electronics *vis-à-vis* its competitors in the Korean and in the EU market with regard to consumer electronics products. It is important to underline here that, since SEC in most European markets does not yet directly compete with top brands (such as Sony and Philips), the perceived position in Europe indicated in Table 7.22 is compared with more mid-range brand names.

Although the figures in Table 7.22 only represent an overall personal judgement of the interviewees, it nevertheless revealed a number of interesting insights. With regard to technological capabilities, patents, cost structure and product quality no or only a small deviation exists between the perceived positions in Korea and Europe. This appears to confirm the conventional strength of LIE companies in shopfloor operations. Also with regard to the quality and attitude of its work force, access to capital and managerial capabilities, no major differences appear to exist between the home market and Europe. This, however, is where the similarities end.

The production capacity of SEC in Europe – as compared to the capacity available in Korea – was relatively weak. As noticed before, one of the prime reasons for commencing production operations in Europe was to overcome problems of market access. Especially at the start of the production operations in the region, the company imported as many parts as possible from Korea and its traditional suppliers and these inputs were subsequently assembled in the European Union. When in 1989 local content requirements in Europe were raised from 20 to 40 per cent, the company had no other choice but to localise production operations to a larger extent. To realise this goal, in 1994, FGT, a company producing TV glass bulbs, was acquired in the former East Germany by SEC's sister company Samsung Corning. In 1990, Samsung Electro-Mechanics, another company closely affiliated with Samsung Electronics, had already established a (wholly-owned) CTV and VCR parts plant in Portugal. The investments in the Wynyard facility also illustrate SEC's determination to increase its production capacity and localise production in Europe.

What is striking is the perceived relatively weak position in Europe in terms of the company's commercial capabilities. With regard to all the items listed related to marketing operations (namely, product differentiation, brand names, distribution networks, advertising/promotion skills, sales force, service, knowledge of customers' needs and loyalty of customers) the company perceived its position as (considerably) weaker in Western Europe than in Korea. The biggest difference exists with regard to the brand name which is considered to be relatively weak in Europe whereas in Korea the Samsung brand name is very well established. In 1992, for instance, the average brand awareness of the Samsung brand name in Europe was still only 3 per cent. At the time, most sales in Europe were on an OEM basis. As part of SEC's policy aim to become a global leader, a three-year programme was introduced in 1993 to convert the company in Europe from a production into a more marketing-oriented enterprise. A direct result of all this was the decision to substantially expand the distribution network in Europe. In Italy, for instance, a new sales organisation was established while in France, a well-performing distributor was acquired.

Two years after the introduction of the new strategic orientation, the first results became visible. The importance of OEM business has decreased sharply and the position of own brand sales had gone up considerably. In 1994, the OEM share (excluding semiconductors) amounted to only 17.5 per cent of total European sales.

Table 7.22 Strengths and weaknesses of SEC and LGE in Korea and Europe with regard to CE products

	Samsung Electronics		*LG Electronics*	
	Korea	*Europe*	*Korea*	*Europe*
Technological capability	+3	+2	+3	0
Patents	+3	+2	+3	0
Cost structure	+2	+3	+2	+2
Work force and attitude	+3	+2–3	+3	+1
Production capacity	+3	+1	+3	+3
Access to capital	+2	+3	+3	+2
Managerial capability	+2	+2	+3	0
Product quality	+2	+2	+3	+1
Product differentiation	+2	+1	+3	0
Brand names	+2	−1	+3	−1
Distribution channels	+2	0	+2	+1
Advertising/promotion skills	+2	0	+2	0
Sales force	+2	+1	+2	0
Service	+3	+1–2	+2	−1
Knowledge of customers' needs	+2	0	+3	−1
Loyalty of customers	+2	0	+3	−2

Source: Interviews held at parent companies and European headquarters
(−3 = very weak position; −2 = weak position; −1 = relatively weak position; 0 = neutral; +1 = relatively strong position; +2 = strong position; +3 = very strong position)

LG Electronics

Unfortunately, no detailed information is available on LGE's market shares in individual European countries. According to company sources, in the beginning of 1995, the Goldstar brand name had an overall market share of about 2 per cent for colour TVs, 3 per cent for video recorders and 6 per cent for microwave ovens in the European region covered by the German regional office. Unfortunately, these market shares do not distinguish between the company's position in Eastern Europe (where it had acquired a relatively strong position with own brand sales) and, for instance, in Germany (where the majority of LGE products is still sold through OEM contracts).

It should be realised that LGE perceives companies like Samsung and less-high-profile Japanese electronics companies like Panasonic as its major competitors. According to LGE, the company does not directly compete with the top brands in Europe. How is the *relative* position of LGE in Western Europe (*vis-à-vis* these competitors) perceived in comparison with the situation in its home country? With the exception of two items (namely, production capacity and cost structure), LGE perceives its position as weaker in Europe than in the home market (see Table 7.22). With regard to technological capabilities and patents LGE thinks the company takes a neutral position in Europe, whereas in Korea its position was labelled as 'very strong'. Also the scores with regard to 'work force and their attitude' and 'product quality' are considerably lower in Europe than in Korea. Interestingly and unlike Samsung Electronics, LGE's production capacity in Europe is not considered a problem; this notion is confirmed by the fact that the German factory, for instance, in 1995 had excess production capacity.[80] This does not mean that production activities in Europe are shielded from any hurdles. As a result of stricter local content requirements, the initial strategy of keeping costs low by assembling imported inputs locally has become less feasible in recent years. The company's intention of localising its overseas operations, as voiced in the 1990s, is therefore not surprising. Eventually, LGE intends to source almost 100 per cent of parts and materials from the region. Because of the perceived erection of high walls against non-European products by the European Union, the plant in the UK is meant to become a purely European venture.[81] In addition, as was noted earlier, one of the two plants that LG announced it plans to build in Wales will produce TV components which will further increase the local content of CTVs.

As in the case of Samsung Electronics, Table 7.22 clearly underlines the company's relative weakness in terms of commercial capabilities. With respect to all relevant items mentioned, LGE considers its position in Western Europe as weaker than in Korea. With regard to brand names, services rendered and knowledge of customers' needs, LGE perceives its position in Europe as 'relatively weak', whereas the loyalty of customers with regard to LGE's products was even labelled as 'weak'. According to an interviewee,[82] 'customers in Germany perceive our products as third rate (with local products and Japanese being first and second rate, respectively), being low priced but also of low quality'. At present, however, the company does not invest as extensively as Samsung Electronics on improving its marketing performance. A continued focus on the production side of operations and a relative disregard of the marketing function may force the company to maintain its low price (and therefore cost-cutting) strategy. Until now, LGE has not been able to break out of this deadlock.

Acer

The market shares of Acer products in the individual countries of Europe vary considerably. In 1994, for instance it had acquired 1.9 per cent of the German PC market (being the tenth biggest company), as compared with 3.3 per cent in France (ninth position) and 3.9 per cent in the Netherlands (seventh position). In view of Acer's aspirations to become a global leader in the computer business, the company is not yet satisfied with its present position in the region.

In view of this aim, the top global companies are considered as its major competitors, in Europe as elsewhere. How does Acer perceive its relative position in Taiwan and in Europe? From Table 7.23, it is obvious that Acer is quite confident about the quality and differentiation of its product line – both at home and abroad. Given its technological efforts and achievements, this is not surprising. A large gap, however, exists with regard to Acer's marketing position in Taiwan and in Europe. With regard to all the commercial capabilities listed (namely, brand name, distribution channels, advertising/promotion skills, sales force, service rendered, knowledge of customer's needs and the loyalty of customers) Acer perceives its position in Europe as substantially weaker than in Taiwan. In most countries, the brand awareness, for instance, is still too low. Since the costs involved in building up brand awareness and a positive image are very high, this is considered a long-term process in which efforts are made in individual countries to increase the awareness of the Acer brand. Or, as Shih puts it, 'it takes perhaps two years to see returns from manufacturing. Technology development may take three to four years and require 10 times as much effort. Image building takes one hundred times the effort. It takes a very long time' (FT 08.10.93). Not only do substantial resources have to be allocated to advertising, the company's distribution channels in Europe also have to be expanded. Acer therefore continually tries to increase the number of retailers selling their PCs – a costly and time-consuming process. In France, Acer agreed to sell its PC systems through Carrefour, the country's biggest wholesale distributor. Although a necessary condition, these efforts are not only a matter of financial resources. Equally crucial is the quality of local commercial and creative people. Since the company is still relatively unknown in Europe, it has less choice in recruiting new employees – the same goes for managerial positions. The lack of flexibility of the labour market in Western Europe (e.g. where it is much more difficult to fire employees than in Taiwan) is considered another serious handicap.

Table 7.23 Strengths and weaknesses of Acer and FIC in Taiwan and Europe with regard to PC systems

	Acer		*FIC*	
	Taiwan	*Europe*	*Taiwan*	*Europe*
Technological capability	+3	+3	+2	+1
Patents	+3	0	0	0
Cost structure	0	+1	+2	+2
Work force and attitude	+3	+1	+2	+2
Production capacity	+3	+2	+2	+1
Access to capital	+3	+2	+3	+1
Managerial capability	+3	+1	0	+2
Product quality	+3	+3	+1	+1
Product differentiation	+3	+3	0	+2
Brand names	+3	+1	+2	0
Distribution channels	+3	+1	+2	+1
Advertising/promotion skills	+3	+1	+1	+1
Sales force	+3	+1	+1	0
Service	+3	+1	+1	+2
Knowledge of customers' needs	+3	+1	0	+1
Loyalty of customers	+3	+1	0	+2

Source: Interviews held at parent companies and European headquarters
(−3 = very weak position; −2 = weak position; −1 = relatively weak position; 0 = neutral; +1 = relatively strong position; +2 = strong position; +3 = very strong position)

FIC

To assess the position of FIC on the European market, a distinction is made between the OEM business activities of the corporate headquarters with European PC companies and the operations of FIC Europe. Since FIC Europe assembles only PC and notebook systems, we focus on the latter's position. With whom does FIC compete in Europe? According to the company, the so-called top brand names in Europe (e.g. IBM, Compaq) are not direct competitors. The same holds for no-name PCs that are often assembled by very small local companies offering varying degrees of quality levels. The most direct competition comes from European companies (such as Vobis and Escom) which are pure assemblers but who maintain a constant level of quality. Although no concrete market shares are known, FIC indicated that in general – as far as PC systems and electronic notebooks are concerned – the company has only gained very modest market shares in Europe.

Table 7.23 indicates important differences between FIC's position in Taiwan and in Europe. The technological capability is rated stronger in Taiwan. The same holds for FIC's production capacity as well as for its access to capital. As in the case of Acer, the relative position of the Leo brand name and the distribution channels are considered stronger in Taiwan than in Europe. In view of the heavy (and risky) investments involved in positioning its own Leo brand, the company decided to emphasise primarily the expansion of OEM business. An additional complication is that FIC has until now avoided direct competition with the buyers of FIC motherboards and PCs by not actively promoting the Leo brand in Europe. As indicated by the company, changing this situation may imply the loss of clients. Similarly, the relative position of FIC's sales force in Taiwan is considered stronger than in Europe. Although it has been less 'aggressive' than Acer, the opening of sales subsidiaries in Europe, the assembly line in the Netherlands and the recent announcement of an assembly line in the Czech Republic, however, suggest that also FIC aims at improving its position as a producer and marketer of PC systems and notebooks. At present, in Europe 0.5 per cent of total sales is spent on marketing activities and this share is expected to grow rapidly in the coming years.

FIC perceives its after-sales service rendered in Europe to be stronger than in Taiwan. This can be explained by the relatively large service centre that the company operates in the Netherlands. Interesting to note and perhaps contrary to the general expectations is the perceived better knowledge of customer needs in Europe. This, as well as the relatively stronger customer loyalty in Europe, can be attributed in part to the narrowly defined target group of FIC customers who know exactly what technical requirements their PC has to comply with; the barebone set-up is able to fulfil this need. Dealer loyalty is also very high. The relative position in terms of product differentiation is also considered stronger in Europe than in Taiwan. The reason being that, as compared to its direct competitors in Europe (such as Vobis and Escom), the barebone system offers many options to satisfy the specific needs of customers. Finally, the managerial capabilities are considered to be relatively stronger in Europe than in Taiwan.

We conclude by stressing that these scores do not reflect the perceived position vis-à-vis the top brand names in Europe. In that case, the perceived position of FIC in Europe in many respects would be considerably weaker.

Summary and conclusions

In this chapter, we have looked more closely at the internationalisation pattern of two Korean consumer electronics companies (SEC and LGE) and two Taiwanese PC companies (Acer and FIC). First, the emergence and rapid development of both industries were described. In Korea, under strong guidance of the authorities, the L conditions were created under which the CE industry could develop at a very fast pace. To this end among others, research institutes were set up to build a technological base, an oligopolistic market structure was created to realise economies of scale and – at a later stage of development – incentives were provided to stimulate the building up of indigenous technological capabilities within individual firms. Moreover, inward FDI – especially joint ventures with leading indigenous companies – contributed considerably in this respect.

In Taiwan, the PC industry was carefully selected by the authorities as one of its key industrial sectors. This choice was based on Taiwan's L advantages at the time and the notions of the government about the future of Taiwan. A restructuring of the global PC industry (leading to *inter alia*, a price war), an active government policy aimed at maximising spillover effects resulting from the presence of MNEs, as well as considerable investments in an adequate research environment, led to the emergence of a successful indigenous PC industry. As opposed to the Korean CE industry, however, the industry was kept open to newcomers, initially resulting in hundreds of firms trying to gain a share in the home market.

In the company case studies the characteristics of the internationalisation patterns of Samsung Electronics, LG Electronics, Acer and FIC were discussed in detail. To draw conclusions based on these findings, we return to the questions formulated in the beginning of this chapter, namely:

- How can the internationalisation patterns of individual Korean and Taiwanese companies be characterised?
- What differences in the outward investment patterns can be discerned between and among the Korean and Taiwanese firms that were studied in this chapter and how can they be explained?
- What problems and challenges do companies face in operating in industrialised economies and to what extent are these hurdles related to their late industrialisation background?

Internationalisation patterns: the role of OEM

Both Korean companies invested in overseas production operations for the first time, at the beginning of the 1980s, and during the late 1990s both of them have stepped up their efforts to localise overseas operations in an attempt to become truly global companies. The latter development represents a new stage in the internationalisation strategies of these firms in which the motivations to invest have also partly changed. Whereas in the first phase of outward investment circumventing trade barriers in industrialised economies was the dominant reason for investing abroad (namely, market seeking/maintaining FDI), strategic asset seeking (aimed at tapping locally available technological or building up commercial O advantages) and rationalisation considerations (e.g. in the UK complexes) have become additional motives to invest in Triad economies. In developing nations, at present, the traditional natural resource seeking argument (namely, cheap labour and/or raw materials) is still primarily the most important reason to invest in factories there, although market seeking is expected to grow rapidly in importance.

The investments made by Acer and FIC in assembly units in major markets (primarily in industrialised economies) do not result from unfavourable domestic L conditions but rather follow from the nature of the PC industry in which rapidly changing demands require a highly efficient global logistics network and a close contact with clients. Therefore, these projects can also be labelled as market seeking/maintaining FDI. Similar considerations have led Acer to set up assembly lines outside the Triad countries in promising new markets. The shift of certain production activities to cheaper developing countries in both cases is a response to the deteriorated traditional L conditions in Taiwan that no longer allow labour-intensive production (namely, cheap natural labour resource seeking FDI).

The study of the internationalisation patterns of the four companies revealed some interesting characteristics. The (sequence of) the geographical direction of their overseas business activities differed from what might have been expected (namely, first advanced and subsequently developing economies). Perhaps most revealing, however, was the crucial role Original Equipment Manufacturing (OEM) has played in the development and internationalisation of all four companies.

Surprisingly little has been written about the (potential) role OEM can play in overcoming the technological and commercial disadvantages with which companies from LI economies are typically faced (Hobday 1995). The literature dealing with various forms of (non-equity based) international co-operation between corporations (cf. Hagedoorn 1993, Narula 1996b) does not address these questions. Moreover, this strand of research focuses primarily on agreements between firms within the Triad. In addition, *long-term* relationships ('strategic alliances') are usually investigated whereas OEM contracts are not necessarily long term in nature. As Wong (1992) notes, most literature on inter-organisational technology development also does not render the desired insights since the focus of attention is mostly either on commercial technology transfer or inter-firm co-operative R&D efforts where the technology flow is the primary (often contractually based) transaction activity. In the present study, however, the possible knowledge transfer and other effects of OEM contracts were not the primary transaction activity – at least from the buyer's perspective. Instead, the outsourcing of (part of) the production process to another (in our case: Korean or Taiwanese) company was the prime transaction.[83] The case studies nonetheless showed that for the local Korean or Taiwanese producer, various important if not crucial spin-offs have been realised through these contracts. Based on our findings in this chapter, an attempt will be made to further clarify the nature of these OEM relationships and the impact these agreements have had on the development and internationalisation of the four companies.

Before recapitulating the possible impact of OEM contracts on the development of the producing companies, it is important to underline the heterogeneous character of these contracts. In principle, OEM can be considered a form of subcontracting where a firm manufactures finished goods for a foreign buyer, often a leading company (Hobday 1995). The case studies illustrated, however, that 'OEM' has many dimensions:

- The buyers of OEM products may be either *producers or retailers/wholesaling firms*. Examples were given of American mail order and retail firms that ordered final goods from Korean consumer electronics companies. Also in the case of Taiwanese companies, similar orders were placed to buy PCs that are sold under the label of a retail company. Although no detailed data are available, it appears that at present, however, the vast majority of OEM contracts are concluded with industrial companies that outsource (part of) their production activities.
- From the interviews it became clear that the *frequency and intensity of the relationship* between the OEM buyer and the Korean and the Taiwanese producers differs widely. In some cases, incidental OEM orders are placed with local producers by buyers to offset temporary production capacity limitations in the buyer's factory(ies). In other cases, the OEM relationship between buyer and producer have evolved into a more lengthy affiliation. Although frequently the buyer may have relationships with various producers, 'monogamous' relationships for certain products are also no exception. In other words, the nature of OEM relationships varies widely ranging from something close to an arm's length transaction to a kind of strategic alliance.
- The actual activity carried out by the producer can either be confined to the production of a good in accordance with the specifications of the buyer (*'Original Equipment Manufacturing'*) or may also include the design of the product involved (*'Own Design Manufacturing'*). Especially in the case of the two PC companies, it appears that a shift in importance from OEM to ODM activities has taken place, one which illustrates their increased technological sophistication.[84]
- Although some authors limit 'Original Equipment Manufacturing' to the production of final goods (e.g. PC systems), in the Taiwanese PC industry OEM business is conceived somewhat more widely to include not only PCs, notebooks and peripheral products (such as monitors, scanners and mice), but also the production of such products as motherboards and add-on cards. On the one hand, to final consumers, the brand name of the motherboard

or add-on card is mostly less crucial (and is usually not purchased as a separate item) than that of the system bought. This suggests that labelling these products as OEM may seem inappropriate. On the other hand, most of the possible effects resulting from OEM contracts to produce final goods, also appear to hold for contracts confined to, for instance, the production of motherboards.[85]

- Finally, the exact nature of the relationship may further differ depending on whether the buyer: provides materials and/or components; finance; machinery and equipment; design specifications; and technical and/or general advice to the producer.

The company case studies clearly showed that OEM contracts have served as an important source to upgrade the technological capabilities of the Korean and Taiwanese firms. As Wong (1992) also found in his study on the technological development of indigenous firms through subcontracting linkages in Singapore,[86] the exact nature of the technology transferred varies widely. Among the most important spin-offs appear to be knowledge about quality control systems, product designs and more general managerial capabilities, such as MNE systems of managing and organising manufacturing activities. It should be noted, however, that these benefits are not all obtained through the same mechanism. An OEM buyer might provide a direct transfer of knowledge – for instance through the training of technical personnel of the producer. Instead, knowledge may also be gained through 'exposure' (e.g. the blueprints of product design) or by means of the feedback received from the buyer – for example on the quality of the goods.

The gains derived from OEM contracts are, however, not limited to technological spin-offs. Since the goods produced are marketed by the buyer, the OEM producer gains access to foreign markets that otherwise, certainly in the earlier stages of its development, would not have been possible. In fact, a (functional) division of tasks emerges in which, broadly speaking, the OEM buyer takes care of R&D and marketing whereas (part of) the production is carried out by the OEM producer.[87] As a result, R&D expenditures as well as the costs associated with building up international distribution networks and brand names are thus bypassed by the OEM producer enabling the latter to concentrate on the further efficiency improvement of its shop-floor operations and the quality of its output.

Experience has shown that, in the long run, a heavy dependence on OEM is not without costs, however. It should be stressed that the possible negative effects discussed hereafter do not necessarily occur in each case and are – at least partly – interrelated. Especially in the case of the PC industry, the heavy competition between the numerous Taiwanese firms to win OEM contracts has resulted in very narrow profits. Since building up own marketing capabilities is very costly, these thin profit margins have excluded many companies from reducing their dependence on OEM by increasing international sales under their own brand name.[88] As was shown in the case of FIC, the decision to promote an own brand name more aggressively in important markets of OEM buyers may also lead to the withdrawal of OEM buyers who feel threatened by the new competition. The case studies also illustrated that the heavy reliance on OEM does not stimulate companies to develop their own innovational capabilities. Partly because of the uncertainty about future OEM contracts, longer term financial budgeting is quite difficult. In the case of Samsung Electronics and LG Electronics, for instance, we saw that only during the 1990s – under strong pressure of the Korean authorities – have these firms stepped up their efforts to accumulate technological capabilities. In the case of the two Taiwanese companies, the latter process was further hampered by their relative small size (certainly in the beginning) combined with the thin profit margins and a more 'distant' government.

As noted earlier, not all benefits and disadvantages of OEM procurement will occur in every case or to the same extent. The actual effects in a particular case depend on a multitude of factors, that again are partly inter-dependent. In Figure 7.8 an attempt is made to depict some of the most important determining factors as derived from the case study findings – some of which may seem quite trivial. It should be stressed that the figure by no means intends to provide a complete picture and is meant for illustrative purposes only. The actual transfer of knowledge about product designs to the producer depends to an important degree on the capabilities possessed by the buyer: a retail store or a technologically

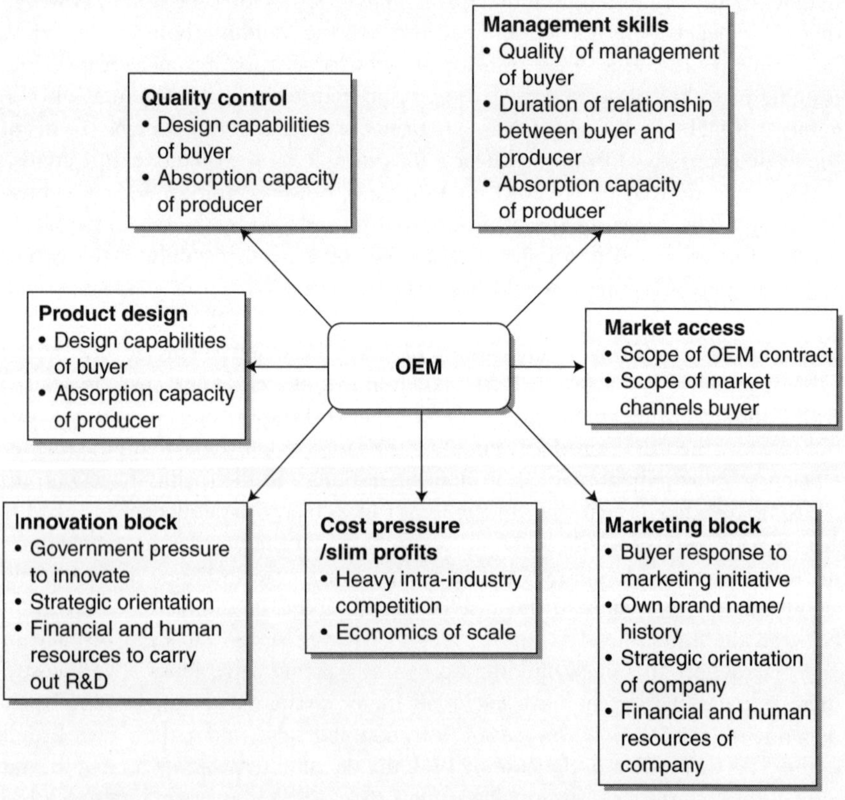

Figure 7.8 Determinants of benefits and costs of OEM contracts

less advanced company placing an OEM order can be expected to add little in this respect whereas OEM arrangements with a global leading company may bring about important new insights. In addition, the producing company has to possess the human capital to absorb this knowledge and use it to its own benefit in other projects. Similar preconditions have to be present to realise an adequate transfer of knowledge about quality control systems and more general managerial practices. With regard to the learning possibilities regarding less tangible managerial practices (ranging from management of production operations to perceptions about corporate culture) the length of the relationship (and the resulting mutual trust) is also an important determinant. The degree to which (overseas) market access can be gained from OEM contracts appears to primarily depend on the scope of the contract (does the buyer 'use' a single or various companies to produce a certain good) and the coverage of the overseas distribution channels available to the buyer(s).

The reluctance to put serious efforts into building up technological capabilities as observed in the case studies cannot, of course, be solely ascribed to the nature of OEM contracts. It does appear, however, that the fact that OEM buyers often provide the knowledge needed to produce a certain good has not stimulated producers to develop technological capabilities stretching beyond production activities. Another factor that certainly also contributed to the lack of eagerness to innovate, are limited financial and human resources available. In addition, the absence of strong government policies to pressure companies in Korea and Taiwan to step up their R&D efforts in the past also influenced this rather passive behaviour. The potential damage caused by a 'buyer response' (i.e. losing sales to an OEM buyer who feels threatened by the emergence of a new competitor) first of all, of course, depends on the extent to which the producer's sales depend on the buyer(s) who default(s) in case of a more aggressive own marketing strategy. In addition, there are some indications that the response of buyers is less fierce in the case of the producer who already has promoted its own brand – albeit in a less aggressive manner – in the past (i.e. has a record of 'brand history'). Furthermore, the strategic orientation of the company is of vital importance in catching up technologically and commercially. As that is a long-term process, a company which primarily focuses on realising profits in the short term probably will not make the long-term commitment that is also needed to internationalise further. In the case of FIC, overseas subsidiaries only get one year to become self-sustained; in cases where they are not able to realise the target within this period, they are closed down. The case studies clearly illustrated that building up international distribution channels and brand names is a very costly affair requiring substantial 'tacit' knowledge. The fact that typically, profit margins realised by OEM producers are relatively slim is partly inherent in the very nature of the production (assembly) activity business which adds less value than such activities as R&D and the marketing of goods. In addition, however, the intensity of the competition among potential producers to acquire OEM contracts as well as the capability of a company to realise economies of scale are important determinants of the profits realised.[89]

The case studies also showed that, in the course of time, the importance of OEM contracts as a source of technological upgrading has decreased. Many examples were provided which illustrated that the companies studied by now have accumulated substantial knowledge about product designs, quality controls systems as well as managerial practices. Moreover, all of them are in the process of (or have already established) building up their own global networks to sell their own output. For Samsung Electronics, by now, the disadvantages of OEM business outweigh the benefits gained. As a result, the company is attempting to minimise the importance of this mode of business in the coming years. SEC perceives any dependence on OEM as an obstacle to its aim to become a global leading company. For FIC, on the other hand, OEM sales are still considered of crucial importance. The same holds for Acer, although nowadays it primarily perceives OEM contracts as a means of realising greater economies of scale. The importance attached to OEM is also reflected in the increase of their share in Taiwan's total PC production from 37 per cent in 1995 to 53 per cent in 1996 (III/MIC, various issues). Although LG Electronics, in principle, would like to get rid of its OEM sales, the company has not yet taken the strategic decision to do so; it appears that, at present, OEM is still accepted as a 'necessary evil'.

The question arises whether these findings would also hold for other companies, for other industries, and perhaps for other countries. As Yin (1994: 10) rightly pointed out, case studies do not represent a 'sample' so that drawing statistical generalisations from the findings is impossible. Therefore, only tentative remarks can be made here. Interviews with other Korean electronics companies and Taiwanese PC firms suggest that for most of them, OEM has played a similar role as for the companies examined in this chapter. This observation is confirmed by Ernst (1994a) and Hobday (1995). However, especially in the case of Taiwan, not all companies turned out to be strong enough to cope with the disadvantages attached to OEM business. In other words, for these firms, OEM appears not to be an intermediate but also their final stage of development. Some of the companies that were visited during the field work have gone bankrupt whereas others have had to change their core business. The latter held especially for many small-sized companies which initially assembled final goods (PCs and/or notebooks), but later on were forced to re-focus to the production of such goods as keyboards or add-on cards. Also for many Korean companies there is still a long way to go. Daewoo Electronics, the third largest consumer electronics company in Korea, is known to be still much more heavily dependent on OEM sales than LG Electronics and certainly more so than Samsung Electronics. It therefore appears that the 'upgrading' into electronics companies which take care of their own marketing and have built up substantial innovational capabilities is reserved to a small number of firms.

In view of the fact that electronics in Korea and Taiwan are the most important export sectors, the focus on subsectors of this industry was a logical decision. To what extent is OEM of importance to other industries? Hobday (1995) concludes that industries in which OEM is (can be) an important mode of business are characterised by a fast export growth (needed to realise economies of scale) and the possibility of an extensive division of labour across national boundaries.[90] An additional requirement appears to be that the technology embedded in the production process has to be relatively easily transferable. It will mostly, therefore, concern rather more mature technology. It is important to stress that, for LIE companies as well, OEM is certainly not the only mode of accumulating O advantages or to be successful overseas. In fact, in such industries as petrochemicals and steel OEM has played no role at all.

Hobday (1995) showed that OEM also played a crucial role in the development of the electronics industry in Singapore and Hong Kong. Whether, and under which conditions, this mode of international business could also play a similar role in other parts of the world is as yet unclear. To get a more adequate and complete overview of the potential of these mechanisms, more research is needed.

The differences in internationalisation patterns

Next to the characteristics that the four companies have in common, the case studies also revealed important differences. As among others Yip (1994) pointed out, important variations exist in the internationalisation degree of industries. Although the consumer electronics and PC industries are both part of the electronics industry, the case studies showed that (sub-)sectoral characteristics have also influenced the way in which these companies have internationalised. A major difference relates to the fact that Korean CE companies were hit on a large scale, by protectionist measures in major target markets (namely, USA, EU) – forcing them to shift production to these countries – whereas Taiwan's PC products have not faced such trade hurdles. On the other hand, we saw that the very nature of the PC industry itself more or less forces companies to keep close contact with their (industrial and/or consumer) clients.[91] In other words, the basic motivations to invest in production activities in major target markets have in part, been quite different. This does not hold for FDI in production facilities in developing countries for which, in both cases, natural resource seeking (i.e. cheap labour seeking) arguments were decisive.

Even if industrial characteristics are taken into account, however, interesting differences still remain between the two Korean companies on the one hand and the two Taiwanese firms on the other. The case studies illustrated that the international presence through FDI of the two Korean companies is more widespread than that of the two Taiwanese companies – certainly more so than FIC. In part, this can be ascribed to the fact that Samsung Electronics and LG Electronics were established much earlier than Acer and FIC and therefore had more time to build up their overseas presence. Also, however, in their case, the actual internationalisation beyond export activities only commenced in the 1980s. In fact, large-scale foreign direct investment only gained momentum in the 1990s. It appears that the magnitude of their overseas engagements as well as the more aggressive strategies pursued to minimise their dependence on OEM is embedded in the way in which the economic system in Korea – as compared to Taiwan

– is organised. The characteristics of both systems were discussed at length in Chapter 4. Of course, the large size of the Korean conglomerates has been very supportive in providing vast resources and, closely related to this, the long-term profit orientation needed to internationalise on such a scale. In addition, a crucial element in the internationalisation of Korean companies has been the 'interwovenness' of the chaebol and the Korean financial system. Until recently, the authorities did not allow well-connected chaebol to go bankrupt, a fact that stimulated rather high-risk internationalisation strategies. It can be safely assumed that these conditions were also beneficial to the rather aggressive M&A strategies pursued (especially by SEC) to catch up with leading electronics companies in terms of technological and commercial capabilities. The collapse of the financial system at the end of 1997 has changed this situation completely. We have seen that the expansion strategies of the chaebol were immediately put on hold. In view of the pressure from the IMF (and the international community in general) on the Korean authorities to reform the financial system, it is expected that this major O advantage will lose its importance in the future.

The differences in internationalisation are not limited to diverging patterns between the Korean firms on the one hand and Taiwanese firms on the other. Discrepancies were also observed between companies originating in the same economy. At present, for instance, it looks as if Samsung Electronics is making a more serious bid to become a genuine competitor to conventional leading electronics companies than LG Electronics. The budgets available to build up a strong marketing position overseas (resulting, *inter alia*, from profits realised by SEC in the semiconductor business in the first half of the 1990s) certainly supported the company's efforts to realise this goal.[92] However, this does not appear to be the only explanation. Besides a rational appraisal of the feasibility of such a strategic move, the decision to attempt to grow into a globally competitive company appears to be partly a belief in one's own capabilities and the attitude of management towards risks. With regard to the latter, the policy of LGE confirms the more conservative and stable image the company has always had.

The differences between Acer and FIC are probably more substantial than between the two Korean companies. As opposed to the latter, no follower strategies were discerned with regard to the strategic choices made by Acer and FIC. There are indications that, at present, Acer has achieved a firmer position in terms of its technological capabilities. The case study showed that Acer – albeit in a specific field (namely, the upgradability of PCs) – has achieved some technological edge. Moreover, in recent years the Acer brand name has received a rapidly growing acceptance in many markets – in the industrialised world as well as elsewhere. In addition, fairly revolutionary organisational changes were pursued by Acer to prepare the company for a more global role. Although quite successful in terms of sales growth, FIC has not yet realised similar achievements. It is therefore no surprise that Acer has progressed much further in terms of the magnitude of its overseas operations than FIC.

Problems and challenges in industrialised economies

To gain a deeper insight into the problems that the companies have faced in operating in industrialised economies, their position in Europe was investigated in more detail. Notwithstanding the present debate on the need to globalise, the firms studied here are certainly not yet 'footloose'. Although mounting factor incongruity has caused a large-scale shift of labour-intensive, relatively simple production activities to cheaper, primarily neighbouring countries, all four cases still showed a strong embeddedness in their home region. In other words, the L advantages of the home region (such as cheap labour and government support) are still important determinants of the competitiveness of these firms. This is also reflected in the relatively low level of integration of their production activities into the European economy. This holds especially in the case of the two PC companies, whose activities in Europe for an important part are limited to assembly operations. But also in the case of the two CE firms, integration into the local economy is still rather modest.

The strong reliance on home (region) L advantages is closely related to the relative lack of *ownership-specific* advantages that these companies, at least until recently, possessed. The discussion on the technological, commercial and organisational capabilities of these companies illustrated that, in general, their O advantages were still not very obvious. We have seen that for Samsung Electronics especially, the commercial position is a major challenge. According to an interviewee, building up brand awareness and a positive image takes much more time than the accumulation of technological know-how.[93] The process is complicated because there were no real precedents from which one could learn about building up brand awareness in a highly competitive market in a very short period.[94] Moreover, in its efforts to become a global company, SEC also has to step up its localisation efforts. The latter are not, however, confined to the production process. The company will also have to strengthen local product planning, development and design capabilities. Similar challenges are faced by LG Electronics. In addition (although it does not follow from Table 7.22), it was indicated that production costs in Europe are too high. Given their focus on the lower end of the market, this was considered the biggest hurdle faced in the region.[95] Likewise, we noted that the biggest challenge for Acer is to improve its relative weak commercial position. The challenges faced by FIC in Europe appear to be somewhat less clear-cut. If the company does not wish to be squeezed between low cost manufacturers of motherboards (and other components) and those firms that aim at final users who have invested heavily in their corporate and product image, in the future FIC probably has to refocus its strategy. At present, as far as their final products are concerned, FIC in Europe still pursues a niche strategy targeting at customers who know exactly what technological standards their PCs have to comply with. If the company wants to broaden its target group in the future, its brand name and international distribution channels in Europe will have to be strengthened. This, however, among other things, requires substantial investments and a long-term commitment. A more general challenge confronts the management style which, given the aspirations of the company, may have to become more professionally oriented and less reliant on members of one single family in the future.

This brings us back to the observation, that in all four cases the decision to invest in industrialised economies did not follow from certain O advantages as such. On the contrary, the Korean CE firms were initially pushed into the EU as a result of protectionist measures; in the case of the Taiwanese PC companies the characteristics of their line of business more or less forced them to establish assembly lines in their important target markets. In addition, the need to gain access to new technological developments to upgrade their own capabilities also prompted the companies to invest in advanced economies ('strategic asset seeking FDI').

We conclude this chapter by drawing attention to the recent changes that were observed with regard to the O advantages possessed by the companies studied in the cases. Especially in the case of Samsung Electronics, LG Electronics and Acer, a high risk strategy of spending unprecedented resources to join leading companies from EIEs was observed. Interestingly, to reach that status, these companies have in many respects increasingly attempted to resemble (in terms of their 'new' O advantages) MNEs from EIEs. This certainly holds true for the two Korean CE companies. Also Acer at present, is in a stage of transformation trying to accumulate new O advantages that resemble the assets conventionally possessed by EIE MNEs (such as strong marketing and management capabilities, as well as innovatory strengths). FIC, on the other hand, has decided to expand somewhat further on its conventional strength in manufacturing capabilities before refocusing its emphasis. In the case of FIC, the accumulation of 'conventional' O advantages is still limited. It appears that, until now, the success of present-day overseas assembly operations has primarily been based on a combination of favourable L conditions in the Asian region and at the same time being close to the market. The most important O advantage is the capability to utilise the global network efficiently. Recently, components have increasingly been purchased in the European region instead of centrally by the Taiwanese headquarters suggesting some trend of regionalisation of their purchasing networks. However, an interesting remark was made by one of the FIC executives which also appears applicable to the other companies, namely that 'the company will never give up its manufacturing capability, since that is what we are much better at than western companies'.[96] It also shows that the late industrialisation roots of these four successful companies still play an important role in their strategic behaviour and are expected to do so in the future.

Appendices

Appendix 7.1 The Samsung Group – major companies

Business fields	*Principal companies*
Electronics	Samsung Electronics Co., Ltd. Samsung Display Devices Co., Ltd. Samsung Electro-Mechanics Co., Ltd. Samsung Corning Co., Ltd. Samsung Data Systems Co., Ltd.
Engineering	Samsung Heavy Industries Co., Ltd. Samsung Engineering & Construction Co., Ltd. Samsung Engineering Co., Ltd. Samsung Aerospace Industries, Ltd. Samsung Watch Co., Ltd.
Chemicals	Samsung Petrochemical Co., Ltd. Samsung General Chemicals Co., Ltd.
Financial & information services	Samsung Corporation Samsung Life Insurance Co., Ltd. Ankuk Fire & Marine Insurance Co., Ltd. Samsung Winners Card Co., Ltd. Samsung Securities Co., Ltd.
Consumer products	Cheil Foods & Chemicals Inc. Cheil Industries Inc. Cheil Synthetics Inc.
Amenity services	The Joong-Ang Daily News Hotel Shilla Co., Ltd. Joong-Ang Development Co., Ltd.

Source: Samsung

Appendix 7.2 The LG Group – major companies

Business fields	*Principal companies*
Electronics	LG Electronics Co., Ltd. LG Information & Communications Co., Ltd. LG Software, Ltd. LG Semiconductor Co., Ltd. Goldstar-Alps Electronics Co., Ltd. LG Precision Co., Ltd. LG Industrial Systems Co., Ltd. LG Electro-Components, Ltd. LG Cable & Machinery, Ltd. LG-Honeywell Co., Ltd.
Chemicals & energy	Global Petroleum Co., Ltd. Hoyu Sales Co., Ltd. Hoyu Tanker Co., Ltd. Hoyu Energy Co., Ltd. Honam Oil Refinery Co., Ltd.
Machinery & metals	LG Metals Co., Ltd. Kukje Electric Wire Co., Ltd. LG Metals Corp.
Trade & finance	LG International Corp. LG Securities Co., Ltd. LG Insurance Co., Ltd. LG Merchant Banking Corp. LG Credit Card Co., Ltd.
Construction & services	LG Construction Co., Ltd. LG Engineering Co., Ltd. LG Mart Co., Ltd. LG Leisure Co., Ltd. LG Ad Inc. Systems Technology Management Corp. LG Economic Research Institute
Public service & sports	The Yonam Foundation The Yonam Educational Institute LG Welfare Foundation, Inc. LG Sports, Ltd.

Source: LG

Appendix 7.3 The Acer Group

Business fields	*Strategic Business Unit*
Computer systems, components, CE products	Information Product Business Unit
Computer peripherals, communications products	Acer Peripherals Inc.
Semiconductors (DRAMS)	Texas Instruments-Acer Inc.
Hybrid and PCB microsystems	Ambit Microsystems Inc.
ASICs and VLSI semiconductors	Acer Laboratories Inc.
Periodical/book publishing, marketing and sales of international software, development of entertainment software	Acer Taiwan Corp.

Regions	*Regional Business Unit*
North America	Acer America Co.
Asia, Africa, M. East, Australia, N. Zealand, CIS	Acer Computer Intern. Ltd.
Europe	Acer Computer B.V. (Europe)
Latin America	Acer Computec Latino Amer.
Taiwan, PRC	Acer Sertek Inc.

Source: Acer

Appendix 7.4 FIC and the Wang Web

I FIC GROUP

Business fields	**Principal Companies**
LEO PC, workstations, motherboards, notebooks, add-on cards (Personal Computer Division)	FIC Inc.
Filenet, Facom, CV/Calma, Sun Micros,	
Concurrent, Stratus, Siemens (Product distribution)	
Display system, plant design, plant simulation, co-generation (Factory Automation)	
Unix, Mis system, Image management, Cim (Software development)	
SRAM, DRAM, Flash memory, I/C cards, chipsets (IC testing)	
Monitors, moniputers, portable systems, notebook development and OEM notebook	Leo Systems Inc.
Multimedia, network conferencing	Formosa Industrial computing

II Other members of the Wang Web

Business fields	**Principal companies**
Petrochemical products	Formosa Plastics
Rayon, nylon, other synthetics	Formosa Chemicals & Fiber
PVC products, polyester fibre, printer circuit boards	Nan Ya Plastics
Multi-user computers	Everex Systems

Source: FIC

Appendix 7.5 Antidumping measures taken by the EU against Korean CE exports (1986–1994)

	Period	*Companies involved*
Videotapes	1989–	Sunkyoung, Kolon Electronics, Saehan Media
CTVs 16″	1990–	Samsung Electronics, Goldstar, Daewoo Electronics
CD players	1990–93	Haitai Electronics, Inkel, Goldstar, Samsung Electronics
Audiocassette tapes	1991–92	Sunkyoung Magnetics, Kumsan Electronics, Sungnam Electronics, Saehan Media, Goldstar
Car radios	1992–	Samsung Electronics, Daewoo Electronics, Hyundai Electronics, Sunkyoung, Goldstar, Tongkook GE, Haitai Electronics, Namsung Electric Industry, etc.
Floppy disks	1992–	SKC, Goldstar
Under investigation in 1994		
CTVs > 16″	1992	Goldstar, Daewoo Electronics, Samsung Electronics
MWOs	1993–	Samsung Electronics, Daewoo Electronics, Hyundai Electronics

Sources: Eurocat database and KFTA (1994)

Note
Only confirmed antidumping measures are included

8 Summary and conclusions

Introduction

In recent years, the cross-border economic activities of Korea and Taiwan have clearly expanded beyond exports from their home bases. The rapid growth of outward FDI raised many interesting questions about the nature and theoretical implications of this relatively new phenomenon. The fact that Korea and Taiwan belong to the so-called 'late industrialisers' made a thorough investigation of their overseas investment activities even more worthwhile. In Chapter 1, we hypothesised that the late industrialisation background of Korea and Taiwan may influence the internationalisation pattern of their companies. Subsequently, a number of more specific questions were formulated that served as a guideline for this book:

- How does the industrialisation pattern in Korea and Taiwan look and which characteristics may influence their outward direct investment pattern?
- To what extent is received MNE theory applicable to outward FDI from LIEs?
- What is the actual size and nature of Korean and Taiwanese FDI flows?
- What are the characteristics of Korean and Taiwanese MNEs and what is the nature of their internationalisation strategies?
- What problems do companies face in operating in industrialised economies and to what extent are these hurdles related to their late industrialisation background?
- What differences in the outward investment pattern of Korea and Taiwan can be discerned and how can they be explained?

In this final chapter, the findings of the preceding chapters are interpreted. Subsequently, some theoretical suggestions based on these findings will be discussed. And finally, the restrictions of the present study are addressed and suggestions for future research will be made.

An interpretation of the findings

As was illustrated in Chapter 3, the economic successes of Korea and Taiwan are quite unique. Although the foundations were already laid in the colonial era and in the import substitution period of the 1950s, since the 1960s in particular, the industrialisation and economic transformation occurred at an exceptionally high pace. Although differences in emphasis existed,[1] in many respects the various stages of development of both economies show a remarkable resemblance. Moreover, the Korean and Taiwanese economies share some other important characteristics, such as the relatively small size of the home markets, a poor natural resource endowment and a strong outward-oriented industrialisation strategy, which are supposed to influence the idiosyncratic investment development paths of individual economies.

The way in which the economic activities in Korea and Taiwan are organised, a fourth factor presumed to influence the IDP, shows much fewer similarities (Chapter 4). Most striking is the different nature of the business actors which dominate the local economies and the business activities overseas. In Korea, the chaebol dictate both the national and the overseas business scenes. With regard to business activities abroad, another distinctive feature is the role played by the General Trading Companies almost all of which are associated with the chaebol. Although in Taiwan efforts were made to establish a similar system of trading companies, these attempts have failed. Moreover, we noted that in Taiwan – although tendencies of scaling up and industrial concentration are observed – overseas activities are still dominated by companies which, when compared with their Korean counterparts, are relatively small in size. These differences in size are not an isolated phenomenon but are a reflection of differences in historical, political and ethnic backgrounds of Korea and Taiwan. Whereas in Korea the success of the political establishment is strongly interwoven with the performance of the chaebol, in Taiwan the relationship between the private sector and political power is much more distant. In fact, an opposite relationship appears to exist between the increasing influence of the ethnic Taiwanese business community and the gradually eroding position of the ruling KMT authorities. The distinct business systems that have evolved in Korea and Taiwan no doubt have influenced the strategies of companies – also in the international (FDI) context.

A first step in analysing their overseas investment operations was made in Chapter 5 through the examination of aggregate FDI data. This exercise illustrated that Korea and Taiwan have in recent years emerged as two of the most important new home bases of MNEs. As compared with the early home countries of MNEs (the 'early industrialised economies'), however, their position is still relatively modest. According to the Late Industrialisation paradigm, even leading companies from these economies, at least until recently, lacked the O advantages needed to go abroad on a large scale through FDI – certainly in the industrialised world. In order to determine to what extent the late industrialised economies of Korea and Taiwan are also late in terms of outward investment, a 'Relative Investment Development' (RID) index was defined. The results suggest that, although the gap *vis-à-vis* early home economies of MNEs is clearly becoming smaller, outward FDI from Korea and Taiwan in most cases does indeed still lag behind their relative welfare positions. The RID values for Taiwan *vis-à-vis* some other early economies in recent years have rapidly approached 1, suggesting a closure of this gap. However, this development is observed especially in comparison with early industrialised economies characterised by very different home features, such as large home economies and/or a much stronger inward orientation which, according to the IDP, generally result in relatively modest outward FDI activities.

Chapter 5 also showed that in both economies, the electronics industry is the largest source of outward FDI. Because of this, the firm level analyses as described in Chapters 6 and 7 focused on (parts of) this industry. To map the characteristics of the companies that invest abroad, the firm-level determinants of outward investment by Korean and Taiwanese electronics companies were estimated using a tobit model in Chapter 6. The results showed that at least some features are shared with MNEs from early industrialised economies. The most obvious example is the importance of size which, *inter alia*, enables companies to realise economies of scale. It is a very important determinant for electronics companies from both economies in explaining their propensity to invest abroad and, consequently, the size of these investments.[2] The importance of overseas sales ('export orientation') except in one case (namely, Taiwanese FDI in developing countries) also turned out to be an important explanatory variable. Similarly, the foreign ownership of companies established in Korea and Taiwan had a significant (negative) impact on their inclination to invest abroad. Another interesting finding was that, at variance with EIE MNEs, marketing capabilities were not an important determinant in the case of Korean FDI in industrialised economies. Due to data restrictions, for the Taiwanese sample, only the importance of the intention to build up marketing capabilities could be estimated and turned out to be significant. This, in combination with the findings from the case studies (Chapter 7), suggests that, although importance is attached to this potential O advantage, at present in most cases it is not yet an important determinant of outward investment. Unlike Taiwanese investments, for Korean FDI in industrialised economies (and in general) a positive relationship between the quality of human capital embedded in the company and the propensity to invest abroad was found – again in accordance with EIE MNEs. Interesting also, is the negative relationship found between labour productivity and outward investment by Korean electronics companies. This suggests that the more (labour) efficient a company is, the more likely it is that it will keep its production operations at home. In the case of Taiwan, no significant influence could be determined of human capital and labour productivity on the decision to invest abroad. The same held for labour intensity, which it was hypothesised played a role in determining FDI in developing countries. In the case of Korea, surprisingly, a negative (instead of a positive) relationship was found. Moreover, it was shown that protectionist measures taken by host country governments against Korean electronics imports do have an impact on the decision to shift production overseas. No influence could be determined of the membership of a chaebol on outward investment.[3] A last determinant that, unfortunately, could only be tested in the case of the Taiwanese sample, concerned the importance of technological capabilities. The results confirmed a positive influence on the probability and size of overseas investment.

We emphasised that the determinants tested in Chapter 6 cannot be equated with O advantages since the latter only hold *vis-à-vis* other (foreign) companies in host countries. It was also underlined that, for various reasons, the econometric (tobit) study renders only a partial insight into the characteristics of MNEs from Korea and Taiwan. Therefore, the econometric analysis was complemented with case studies of two leading Korean and two leading Taiwanese companies that invest abroad. From the empirical chapters, a picture emerged of the characteristics of outward FDI, which, in a simplified manner, are summarised in Table 8.1. As was pointed out in Chapter 2, the 'first wave' refers to the FDI activities from (by that time) developing countries as described in publications in the end of the 1970s and the first half of the 1980s. In the table, a number of characteristics are listed that were typically ascribed to these 'Third World MNEs'. Subsequently, some of the features that came to the fore in this study on more recent ('second wave') Korean and Taiwanese FDI are shown in the table. Although substantial differences in some respects exist between the outward investment pattern of Korea and Taiwan, the present study, *by and large*, appears to confirm the hypothesis that we brought forward elsewhere (Dunning *et al.* 1997, 1998, van Hoesel 1992, 1993) and which was also suggested in Chapter 2, namely that the first and second wave of FDI from late industrialised economies are indeed quite different, bearing their own characteristics. Finally, in the last column of the table, some distinctive features of MNEs from early industrialised economies are mentioned. Before we discuss the table, it should be noted that during the first wave, Korea and Taiwan were not among the leading home nations of new MNEs, whereas in more recent years both economies have emerged as leading new home bases of outward FDI. Nevertheless, Korea and Taiwan did belong to the group of first-wave outward investors (Khan 1986a, Kumar 1982) so that the characteristics as described in the first-wave literature in general also applied to them.

A first observation concerns the geographical spread of overseas investment operations. During the first wave, the majority of (at that time still modest) outward investment flows from LIEs was aimed at neighbouring developing countries. In the case of Korea we noted that in recent years outward FDI flows have been less confined to the own region and gradually spread to the various continents. As was illustrated in the case studies, this pattern is indeed also observed at the micro level among the leading chaebol which account for the vast majority of outward FDI flows. In the mid-1990s, the first cases emerged of Korean suppliers following the chaebol in establishing production activities in the latter's target markets (van Hoesel 1998a). In the case of Taiwanese FDI, the picture is somewhat more heterogeneous. In Chapter 5 it was shown that the overseas investment activities are more concentrated in the own, Asian, region. Only the United States have also attracted a substantial share of Taiwanese companies. Other studies (cf. Kuo 1994) revealed that a very substantial share of these FDI projects in Asia reflect the overseas production operations of a large number of small-sized, often labour-intensive companies, whose existence is jeopardised as a result of mounting factor incongruity. For many entrepreneurs who were not able to upgrade their O advantages to match the rapidly changing L conditions (such as wage increases, a shortage of skilled labour and unfavourable exchange rates), the only way to survive was to shift their production activities abroad, especially to the People's Republic of China. In Table 8.1, characteristics of these investments appear under 'Taiwan-II'. These latter companies are very different from the firms that were analysed more in-depth in this study (Chapter 7) and indeed have invested in various (industrialised and developing) parts of the world ('Taiwan-I' in Table 8.1). Whereas in Korea, a dichotomy appears to exist between outward investors (primarily the chaebol) and other, domestically-oriented companies, in Taiwan, two distinct groups can be distinguished among the outward investors themselves. Given the unreliability of the data it is very hard to estimate the relative importance of each group of companies. It appears, however, that only a limited number of Taiwanese companies also invest 'beyond' the own region. The companies that have 'fled' to neighbouring countries ('Taiwan-II') certainly do not belong to the leading companies in their sector and cannot be expected to extend their investments to other regions in the foreseeable future. This different composition of the outward investor population in Korea and Taiwan again can be traced to the different organisation of their economic activities.

This brings us to the motivations for LIE firms investing abroad. As was pointed out in Chapter 2, although some 'natural (labour) resource seeking' FDI did take place, the first wave of manufacturing FDI from LIEs was primarily aimed at supplying goods to (developing) markets ('market seeking' investments) which by that time were often still more protectionist against the import of foreign goods than nowadays.[4] As Dunning (1993) noted, if companies become more experienced as foreign investors, the motivations partly change. In fact, we observed that the investment arguments have become more diverse for the second wave of outward FDI. With the rapid increase of domestic industrial wages natural resource-seeking FDI in LDCs meant to make use of cheap labour abroad have become much more important. Different from the first wave, the majority of this production is exported from these cheap production locations to major markets in the world. However, in recent years the growing purchasing power in these production locations has again resulted in the market seeking argument growing in importance. Investments in the industrialised world are, to a large extent, market seeking or retaining (like in the case of Korean consumer electronics companies) in nature. In addition, overseas 'R&D' centres enable Korean and Taiwanese companies to scan the latest technological developments ('internationalised learning') and local consumer demands. This 'reverse technology transfer', as well as the acquisition of high tech EIE companies or firms with strong brand names or distribution channels, can be labelled 'strategic asset seeking' FDI.[5] These new investment motives are relevant both in the case of Korean and Taiwanese (type I) companies. They also underline the observation already made in Chapter 2 that the Product Cycle model, which primarily explained market-seeking investments, is not applicable as an explanation to the second wave of FDI from LIEs. Especially in the case of the Korean chaebol, 'efficiency seeking' FDI has also become more important in recent years. These investments are typically associated with large and diversified MNEs aiming at making optimal use of different factor endowments, cultures, economic systems and policies, and market structures by concentrating production in a limited number of locations to supply multiple markets (Dunning 1993). This phenomenon is primarily observed within the regional trade blocks of the European Union and NAFTA.

The greater diversity of investment motives cannot be separated from a greater variety of the sectors in which companies at present invest abroad. During the first wave, investments were primarily made in labour-intensive ('Heckscher–Ohlin' or HO) industries.[6] During the second wave, investments in assembly based ('differentiated Smithian') industries (e.g. consumer electronics, PCs, automobiles) emerged as important targets meant to seek or maintain markets in the industrialised world. At the same time, labour-intensive (intermediate) components which are used to assemble these final goods are increasingly produced in countries with low wage levels (e.g. keyboards and motherboards in the case of PCs). These overseas production activities might be labelled 'backward linkage natural (labour) seeking' FDI. The investment projects in industrialised countries that are also motivated by strategic asset arguments may lead to an upgrading of the company's activities in that industry. *Ipso facto*, this may eventually lead to operations beyond the assembly based stage bearing characteristics of innovation-intensive ('Schumpeterian') industries.[7] The investments made by Taiwanese type II companies in neighbouring countries are limited to the HO-type of industries of which the output is partly sold in local markets and partly abroad – primarily in the own region (Kuo 1994).

Another important feature concerns the nature of the O advantages on which these overseas investments are based. As was pointed out in Chapter 2, during the first wave, the O advantages (such as small-scale, labour-intensive technology, low input costs and final goods appropriate for consumers in LDCs) were primarily accumulated in the home economies of the new MNEs. Although this also holds for most O advantages observed during the second wave of outward FDI, it appears that the *de facto* competitive edges have changed considerably. The high efficiency level of the shopfloor operations is seen as a key component of their success. It appears that the 'old' O advantages with regard to their operations in LDCs rapidly lose importance for Korean and Taiwanese (type I) investors. At the same time, the case studies illustrated that some leading companies from Korea and Taiwan heavily invest in developing new O advantages such as marketing capabilities (brand names, distribution channels) and leading edge technology. Whereas in some developing countries this has already resulted in strong or even dominant market positions, in the industrialised world most companies have not yet been so successful. As was noted earlier, by acquiring technologically or commercially advanced companies and by undertaking research and scanning activities in these markets, new O advantages are also increasingly accumulated outside the home country or region. In the case of Korea, the access which until recently chaebol had to large funds of financial capital enabled a rapid expansion abroad. As we saw in Chapter 4, whereas in the past government-controlled banks were the most important sources for (often cheap) credit, the chaebol increasingly invest in financial (bank and non-bank) institutions themselves. The conglomerate structure, which according to the Late Industrialisation paradigm is characteristic of many late industrialised economies, has also enabled Korean companies to expand rapidly through cross-lending and cross-guaranteeing of loans and through cross-holdings of equity among members. Moreover, the strong interwoveness of the performance of these investors with the position of the political establishment has also guaranteed a strong overall support by the latter – notwithstanding occasional power struggles between individual chaebol and the authorities. The international and domestic criticism that arose after the recent financial and economic crisis may lead to structural changes in the government–business relationships. In the short term, the crisis led to a drastic reduction of the pace of overseas investing. In the longer term, it may imply that the chaebol will lose an important determinant of FDI, namely the government support they have had for decades. For Taiwanese companies some other examples of O advantages are added in Table 8.1. With regard to FDI in (Asian) LDCs it is beyond doubt that the links with overseas Chinese partners makes setting up production operations in these countries much easier. In addition, it appears that the emphasis on the development of flexible and efficient logistics networks (a feature typically associated with Chinese family businesses; cf. Whitley 1994) and strong sourcing capabilities (Kuo 1994) in sectors such as the PC industry, have proved to be important O advantages. Korean companies have also built up strong links with cheap component suppliers, although they are less known for their flexibility.[8]

The study also pointed out that the L advantages that were relevant during the first wave of FDI from late industrialised economies are quite different for those that are relevant for present-day Korean and Taiwanese FDI. During the first wave, the protectionist economic policy in many LDCs *vis-à-vis* the import of foreign products prompted LIE firms to shift production to within the borders of the markets. Moreover, the limited purchasing power of most potential consumers in the home markets implied an important L disadvantage. Another L characteristic that influenced FDI in the developing world at that time was the buyer uncertainty that existed among consumers in LDCs regarding products offered by LIE firms, products which were much less well known than those of their counterparts from EIEs. Analogous to the O advantages, however, the L conditions have changed considerably since then. As noted earlier, a shortage of skilled labour, high wage levels at home and deteriorating exchange rates *vis-à-vis* major currencies in the world, for instance, made it more attractive to export output from overseas production locations. The same holds for the GSP status that many LDCs still have. Moreover, stricter environmental regulations in Korea and Taiwan have made the production of polluting operations more attractive in

Table 8.1 Some characteristics of outward manufacturing FDI

	'First wave' (*General characteristics*)	*'Second wave'*			
		Korea	*Taiwan-I*	*Taiwan-II*	*EIE MNEs*
Destination	Regional FDI: Neighbouring developing countries	Less regional, moving towards global	Primarily regional and USA	Regional	Global
Motivation	Market and natural (labour) resource seeking	LDCs: Natural (labour) resource and market seeking DCs: Market seeking/ retaining and strategic asset seeking	LDCs: Natural (labour) resource and market seeking DCs: Market seeking/ retaining and strategic asset seeking	LDCs: Natural (labour) resource and market seeking	Natural resource, market, efficiency and strategic asset seeking
Sectoral destination	Labour-intensive HO industries	LDCs: Labour-intensive HO industries DCs: Assembly type Smithian industries	LDCs: Labour-intensive HO industries DCs: Assembly type Smithian industries	LDCs: Labour-intensive HO industries	Heavy and chemical Smithian, assembly based Smithian and innovation-intensive Schumpterian industries
Origin O advantages	Primarily country-of-origin-specific	Primarily country- or region-of-origin-specific, but moving towards more global	Primarily country- or region-of-origin-specific, but moving towards more global	Primarily country-of-origin-specific	Global
Examples O advantages	• Small-scale, labour-intensive 'appropriate' technology • Low input costs (incl. managerial and technical personnel) • 'Appropriate' final goods • Specific market knowledge • Ethnic factor	LDCs: • Economies of scale • Production technology • Brand names • Access to financial capital • Sourcing capabilities • Home government support DCs: • Economies of scale • Access to financial capital • Production technology • Sourcing capabilities • Home government support	LDCs: • Production technology • Ethnic factor • Brand names • Flexible logistics networks and sourcing capabilities DCs: • Production technology • Flexible logistics networks and sourcing capabilities	LDCs: • Production technology • Ethnic factor • Flexible logistics networks and sourcing capabilities	• Economies of scale • Access to financial capital • Advanced technology • Product differentiation • Marketing know-how • Cross-country management skills • Vertical control over factor/product markets

Table 8.1 Some characteristics of outward manufacturing FDI (*cont.*)

	'First wave' (*General characteristics*)	*'Second wave'*			
		Korea	*Taiwan-I*	*Taiwan-II*	*EIE MNEs*
Examples L advantages	• Protectionist policy of host countries • Growth limits to home markets • Buyer uncertainty	LDCs: • Availability of cheap labour • Host country incentives • Home country FDI policy • Environmental restrictions of home country • GSP status of host country • Deteriorated exchange rate DCs: • Host country incentives • Home country FDI policy • Size of target market/limited size of home market • Deteriorated exchange rate • Protectionist measures	LDCs: • Availability of cheap labour • Host country incentives • Home country FDI policy • Environmental restrictions of home country • GSP status of host country • Deteriorated exchange rate DCs: • Host country incentives • Home country FDI policy • Size of target market/limited size of home market • Deteriorated exchange rate	LDCs: • Availability of cheap labour • Host country incentives • Home country FDI policy	LDCs: • Availability of cheap labour • Host country incentives • Home country FDI policy • Environmental restrictions of home country DCs: • Host country incentives • Home country FDI policy • Size of target market/limited size of home market

Source: Adapted and extended from Dunning *et al.* (1997)

LDCs where pollution control is less strict. Furthermore, as was discussed in Chapter 4, the government policy in Korea and Taiwan with regard to outward FDI has liberalised considerably since the second half of the 1980s and, to some extent, has even become supportive. Also in a large number of developing host countries, the attitude towards MNEs in recent years has turned more friendly. For Taiwanese type II companies, it will be no surprise that the most important L condition is the relatively low wage levels in the host countries. In the industrialised world, we noted that protectionist measures, for instance against Korean electronics products, have become an important L condition to reckon with. At the same time, individual governments – among others, in the European Union – increasingly offer various incentives to attract investments. Although in Korea and Taiwan the purchasing power of consumers nowadays is quite high, the limited number of inhabitants remains an important locational push factor. Furthermore, we noted that the available technological and commercial knowledge in advanced economies can be important L advantages for Korean and Taiwanese investors.

The preceding overview illustrated the considerable discrepancy that exists between the first and second waves of outward investment from late industrialised economies. Although the characteristics of the second wave of FDI have shifted in the direction of those from early industrialised economies, in accordance with our central hypothesis, important differences still exist that can be ascribed to their late industrialisation background. In addition, the case studies yielded more interesting insights related to outward FDI from Korea and Taiwan. The 'route' to overseas investments, for instance, in the case of Korean and Taiwanese companies was quite different from the patterns that are often associated with EIE MNEs and was depicted by, among others, Scandinavian scholars in the 1970s. The pace through the various stages of internationalisation was unprecedented and the geographical pattern of Korean and Taiwanese FDI also differed from what would be expected. Perhaps even more important was the crucial role played by Original Equipment Manufacturing (OEM) in this process.

Large-scale OEM contracts concluded with leading EIE companies implied a relatively easy access to major overseas markets for leading LIE firms. As a result, and perhaps different from what was expected, these firms often first sold their products in advanced economies and only later in the developing world. The impact of these OEM contracts was not confined to the access to international markets. The case studies revealed that important benefits (such as knowledge transfer related to product designs, quality control and managerial skills) were major contributions to the accumulation of potential O advantages. In the longer run, however, these gains are not without costs. We noted that continuous cost pressure resulted in slim profits. Moreover, innovational strengths and marketing capabilities are not stimulated by a heavy dependence on OEM sales.

A closer look at the operations of the four case study companies in Europe showed that the commercial position is considered the biggest challenge and not, perhaps contrary to expectations, their technological capabilities. The technology embedded in most goods produced in Europe is fairly mature. Since until recently, certainly in advanced economies, these products were primarily sold via OEM contracts, own brand names and images as well as distribution channels have been poorly developed. The company interviews underlined that developing these commercial capabilities is a very cumbersome process that not only requires considerable financial sacrifices but also important managerial and organisational changes. In view of the fact that the companies studied here are among the most successful in Korea and Taiwan in their fields, it can be safely assumed that this is predestined only for a limited number of companies.

Some theoretical suggestions

Based on the findings in this study, suggestions with regard to some of the theoretical concepts that were discussed in Chapter 2 are made. Whereas the questions that were formulated in Chapter 2 served as a guideline, the suggestions in particular relate to the following issues:

- The Late Industrialisation paradigm and outward FDI;
- Late Industrialisation and the IDP;
- 'Decomposition' of the IDP;
- The Eclectic Paradigm of International Production and FDI in advanced economies.

We would like to underline that most of the suggestions made here are tentative in nature and still need further rigorous empirical testing.[9]

The Late Industrialisation paradigm and outward FDI

The Late Industrialisation paradigm as enunciated in Chapter 2 basically provides a broad historical explanation of how countries have been able to industrialise in recent periods. Based on the present study, it appears that what is lacking in the original analysis, however, is a further distinction between R&D (or innovational),[10] production and marketing capabilities of leading companies (van Hoesel 1998b). This nuance is especially important for understanding the internationalisation pattern of the companies involved. In addition to superior technological (R&D) know-how, leading enterprises from *early* industrialised countries have for a long time also possessed – to a greater or less extent – a 'marketing monopoly' in many industries. Figure 8.1 illustrates – in a simplified manner – how the strategic (or 'competitive') focuses of leading companies from these early industrialised economies have evolved over time.

As is argued by the LI paradigm, in leading companies from early industrialised nations, inventions or innovations – resulting from research and development efforts – usually precede commercial production. After the product is introduced into a market, marketing usually becomes more important. The emergence of new competitors forced firms to develop marketing strengths (such as promotional skills and extensive distribution channels) to maintain or expand

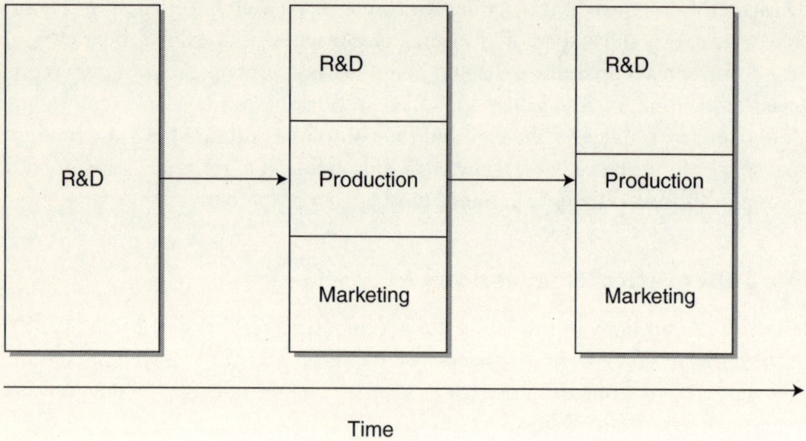

Figure 8.1 Strategic focuses of leading companies from early industrialised economies

their market share. In a continuous effort to keep up with other inventing or innovating companies, R&D remains a relatively important function. In time, as is shown in Figure 8.1, production itself has become less of a strategic focus. One of the reasons for this shift in strategic focuses is the international distribution of operations that has emerged since the 1960s in which leading companies from early industrialised countries have, increasingly, started to subcontract first part of the production process (namely, intermediate goods) and later on, the production of final goods ('Original Equipment Manufacturing') as well, to indigenous companies in LIEs. We saw that, according to the LI paradigm, in the latter economies the strategic focus has always been historically, on their shopfloor operations. This has led to the situation that, although substantial export volumes were realised through OEM sales, for these LIE firms no need and/or basis existed to internationalise further by means of direct investment abroad. Only recently, have marketing as well as research and development gained some importance. Figure 8.2 illustrates this shift in the relative importance of their strategic focuses. The figures show that not only is the origin of the strategic focus of leading LIE and EIE firms different, but so is the shift in these focuses over time.[11] In fact, at present, the strategic focus of most EIE companies (and the O advantages derived thereof) still appears to be on production ('shopfloor') activities. With regard to the problems LIE firms face in operating in industrialised economies, the present study suggests that only a limited number of companies have advanced to the second or the third stage depicted in Figure 8.2. This situation prevents them from building up strong market positions (especially in the industrialised world) using their own brand names and distribution channels. Strengthening commercial capabilities has to result from individual company efforts; the role of the government in this process is very limited. The possible government contribution in advancing towards innovational leadership

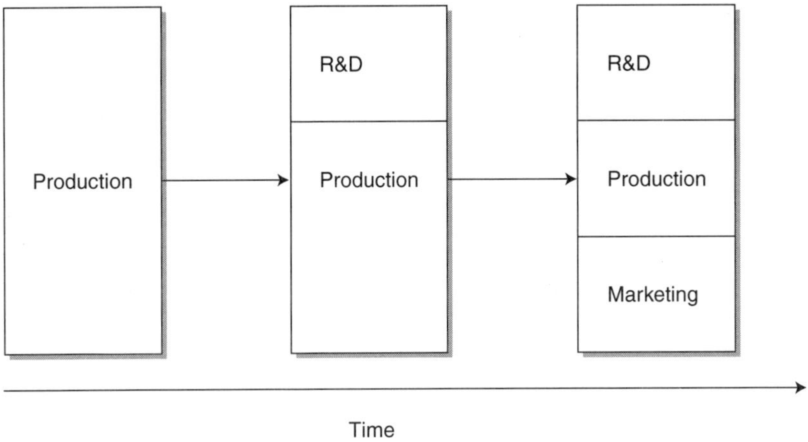

Figure 8.2 Strategic focuses of leading companies from late industrialised economies

is also restricted. As was shown in Chapter 7, in earlier stages of development, the home authorities in Korea and Taiwan, for instance, not only provided an adequate technology infrastructure, but in some cases even forwarded knowledge acquired abroad to individual domestic companies. Such a policy would be less effective in the case of acquiring leading edge technology. As we saw in Chapter 2, this knowledge is usually not readily available through arm's-length transactions and can only be acquired through substantial (time-consuming) internal company efforts and/or the acquisition of other companies.

Late industrialisation and the IDP

The Investment Development Path (IDP) as discussed in Chapter 2 theoretically connects the level of economic development with the inward and outward direct investment position of a country. In the present study, the emphasis was on the outward FDI pattern of Korea and Taiwan. The RID ratios *vis-à-vis* other countries calculated in Chapter 5 showed that these two economies had reached relatively high levels of prosperity before outward FDI assumed any significance. Moreover, we demonstrated that this 'lagged' pattern was related to their late industrialisation background. The most important explanation for this deviant FDI pattern is that overseas business activities were successfully undertaken at a relatively high welfare level at home without having developed the (commercial and/or technological) O advantages needed to shift production abroad on a large scale. Given the fact that in many cases the international marketing was undertaken by foreign buyers of their products, very little experience had been accumulated by indigenous companies in doing business in overseas markets themselves.

Although we only investigated FDI from Korea and Taiwan, it also appears that in general late industrialised economies, *ceteris paribus*, witness a relatively late outward FDI pattern in the earlier stages of development.[12] This fundamental distinction between early and late industrialised economies was not made in earlier versions of the IDP. We noted, however, that once wage levels (and other factor conditions) reach a level similar to those of countries in stage 4 (see Chapter 2), outward FDI increases rapidly. Although at present it is too early to make a definitive statement, it could be that the present trend reflects a catching up leading to outward FDI positions similar to that of early industrialised economies. It is therefore not certain that a fifth factor (next to a country's resource endowments; market size; industrialisation strategy; and government and organisation of economic activities) influencing the idiosyncratic IDP of individual economies, should be added to the original concept, namely whether an economy was early or late industrialised. The latter appears less justified if the impact of their late industrialisation background on outward FDI diminishes in the short-term future. If that turns out to be the case, we can conclude that late industrialised economies only temporarily witness a deviant IDP which would not justify the addition of a fifth factor.

'Decomposition' of the IDP

The Investment Development Path is no doubt a useful concept for analysing the interaction between economic development and inward and outward FDI, certainly if non-conventional FDI is also taken into account. The aggregate nature of the concept does, however, imply certain important restrictions. Especially when the transformation of individual industries is examined (as was done in Chapter 7), important additional insights can be acquired by focusing on the role played by (inward and) outward FDI at the meso level. An argument for constructing such 'meso-IDPs' brought forward by Ozawa (1996) is that it is very likely that, whereas some industries show a positive net outward investment position, others still witness a negative position. In addition to the meso-IDPs, we would like to add that, in order to understand the role of FDI in the transformation of an industry more precisely, a further distinction between FDI in industrialised and in developing countries is needed (see Figure 8.3).[13]

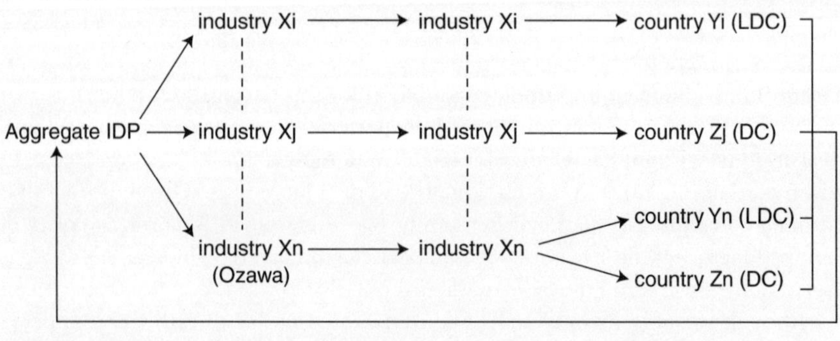

Figure 8.3 'Decomposition' of outward FDI in the Industrial Development Path

Although the IDP distinguishes between FDI in countries at earlier and later stages of their path, it does not explicitly deal with the *simultaneous* investment from one industry in both types of economies. However, in our discussion about the transformation of the Korean consumer electronics and Taiwanese PC industries we noted that, in order to stay competitive, labour-intensive parts are produced in locations with abundant, non- or semi-skilled labour whereas in high income markets, final goods are assembled for such reasons as protectionism or because of fast-changing consumer demands.[14] Given the importance of this mechanism for the upgrading of industries, it appears that this notion should be integrated into the present IDP concept. In Figure 8.3, we illustrate the alternatives that companies from home country X that are active in industries i, j and n respectively invest in a LDC ('Y'), DC ('Z') or in both simultaneously (Y and Z).

Related to this (although perhaps beyond the boundaries of the present study) is the following final remark about the IDP. It is generally accepted that the more technologically advanced an industry is, the less reasons there are to invest in production activities in economically less-developed locations. In Schumpeterian industries, for instance, the importance of a highly developed technological infrastructure (including physical and human capital) is a much more important consideration in choosing an overseas location than the availability of cheap labour.

This relationship – in a very simplified manner – is depicted in Figure 8.4, which is intended to connect the meso-IDP to the 'general' IDP. At any given point in time, the (inward and) outward investment position as reflected by the IDP is the aggregate of the investment positions of the individual industries of which an economy is composed. Depending on the level of sophistication of the various industries, the geographical destination of the FDI differs. Decomposing the geographical destination of FDI flows as depicted in the IDP therefore

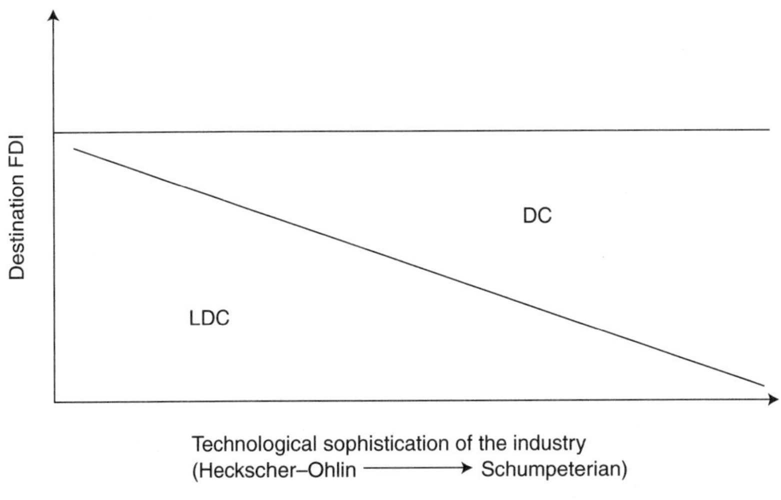

Figure 8.4 Geographical destination of FDI and tiers of industry

gives an indication of the level of sophistication of the various industries of an economy.

The Eclectic Paradigm of International Production and FDI in advanced economies

After having discussed the possible implications of our findings for theoretical concepts at the more aggregate level, a final suggestion on the applicability of received MNE theory concerns the Eclectic Paradigm of International Production (EPIP). As we saw in Chapter 2, this paradigm offers a powerful tool for analysing the factors that determine the route that individual companies choose in serving foreign markets. In the present study it became apparent that, in particular, Korean and Taiwanese investments in the industrialised world need extra theoretical attention. Until now, very few O advantages similar to those of EIE MNEs were observed among the investing companies. Illustrative is a remark made by Jun and Yoon (1995) who noted that the decision of Korean electronics companies to invest in the European Union was a result of protectionist measures taken against them and should be seen as a 'life or death decision'. Most studies of Korean or Taiwanese FDI are confined to their operations in the developing world which, from a theoretical point of view, are much easier to understand. However, this does not relieve us from the responsibility of addressing the question of whether the Eclectic Paradigm is capable of explaining FDI from such late industrialised economies as Korea and Taiwan in advanced economies.

Reviewing the strengths that were observed earlier in this chapter, it is striking that most of these are strongly rooted in the home country and region. As was discussed at length, this obviously holds for the strengths accumulated with regard to production operations. The economies of scale observed among the chaebol, for instance, were consciously made possible by the Korean authorities. Moreover, the favourable credit facilities that chaebol enjoyed in the past have been replaced by their growing involvement in the domestic financial sector which again enables them to have relatively easy access to financial capital.[15] An important strength for leading companies from both economies are the links with cheap component suppliers based in the (Asian) home region ('sourcing capabilities').

Such O advantages make it possible for LIE companies to produce in advanced economies as long as the local content is relatively low. If the local content has to increase (e.g. because a host country requires the aquisition of more intermediates locally or as a result of a strategic reorientation) the O advantages have to change accordingly. As we also argued in Chapters 6 and 7, a strong marketing position has to be built up in overseas markets. Moreover, we saw that the accumulation of R&D capabilities usually takes place at home as well as increasingly abroad in a technologically advanced environment. As was illustrated in Chapter 7, internal organisational changes are also unavoidable in a process of increased localisation. In addition, besides overseas backward vertical integration, in many cases external linkages with local or regional component suppliers will have to be developed.

The foregoing illustrates that the EPIP framework is also very useful in explaining FDI in advanced economies. It was also shown, however, that lower value-added activities require different O advantages than overseas production activities which are more deeply integrated into local overseas economies. We therefore argue that a conceptual distinction between these two modes of international production within the EPIP would imply a useful addition to the present framework. Does this imply that a similar additional distinction should be made for OEM contracts as an additional mode of international business? Unfortunately, this is a less clear-cut case. As was discussed in Chapter 7, Original Equipment Manufacturing is a heterogeneous concept that itself covers a wide range of business contracts between companies ranging from incidental arm's-length deals to long-term, strategic, alliances. Whereas the first type of deals can be grouped under export transactions, Dunning (1995) in a recent contribution illustrated that for strategic alliances, very different O, L and I advantages apply.

A research agenda

As has already been underlined in Chapter 1, as with all research, the scope of the present investigation was necessarily limited. To investigate the generalisability of our findings, more empirical work is needed at the various levels of analysis. Let us first turn to the country selection. Although the choice of Korea and Taiwan as two of the most active new outward investors was a logical one, we cannot a priori argue that FDI from other late industrialised economies shows similar characteristics. Since this study pointed out that in recent years other countries have also emerged as outward investors (such as Hong Kong, Malaysia, the People's Republic of China, Brazil and Mexico), more research into these new home countries of MNEs is needed. Although we saw that Korea and Taiwan are in some respects quite dissimilar, such investigations will also shed more light on the question as to what extent both economies are unique in terms of their investment development path.

A similar consideration as in the case of the country selection applies to the choice of (parts of) the electronics industry to be studied in more depth. Here too, the question arises to what extent the findings can be generalised. As was noted earlier, important industrial discrepancies between industries exist as to their propensity to invest abroad. Furthermore, there are industries from which FDI does take place but where mechanisms such as OEM played no or only a modest role in terms of building up O advantages or gaining access to international markets. To gain a more complete overview of which industries show internationalisation patterns similar to those of the ones discussed in this study, and to get a better understanding of the internationalisation patterns that are observed among other industries from late industrialised economies, more empirical research is needed.

Another implication of the restrictions made for the present study is that the question formulated in the beginning of the study about the potential of 'reverse technology transfer' was only partly addressed and needs further examination.

Since the PC industry (which was looked at more closely) is in Europe not considered a centre of excellence and most of the consumer electronics is a rather mature industry, we were not able to estimate the full potential of 'reverse technology transfer' through subsidiaries in advanced economies.

As was indicated in the discussion of the econometric analysis in Chapter 6, company data with regard to Taiwanese investment projects in developing countries are very scarce and in the case of the PRC even completely absent. As a result, a bias in the Taiwanese sample implied that the possibilities of measuring the firm-level determinants of investment in that part of the world were limited. Given the large number of companies involved (and the considerable share of total FDI represented by these firms' activities) the results would benefit greatly from an improved sample. Although it would be very time-consuming, this goal perhaps could be achieved by collecting primary data among Taiwanese companies involved and/or acquiring information in the host countries. Another restriction on the firm level, concerns the case studies. The companies investigated in this study are leaders in their domestic industries. In the Taiwanese PC industry most companies, however, focus completely on acquiring OEM contracts or developing subcontracting relationships with other, leading companies. In the case of the Korean consumer electronics industry, it is obvious that only the largest conglomerates are in a position to grow into full-fledged MNEs. To get a more comprehensive picture of the problems that less advanced companies experience in their further development (including their internationalisation), in-depth analyses into these companies are also needed.

Since outward investment from Korea and Taiwan is such a recent phenomenon, it is important to keep track of future developments. As came clearly to the fore in this study, most Korean and Taiwanese companies investing abroad are themselves in a process of rapid transformation. In particular, longitudinal research into the further development of the O advantages can sharpen our insights. It would be interesting, for instance, to look into the way in which, in the future, (own brand name) products are positioned in industrialised markets *vis-à-vis* the various competitors. Moreover, one wonders how the relationship with OEM buyers in major markets develops if own brand names are more strongly promoted (e.g. in the case of FIC). Given the limited experience that most of these investors have in running their own affiliates abroad it is also worthwhile investigating in more detail how successful the incorporation of local management into the Korean and Taiwanese organisational systems will be in the coming years. With regard to their technological capabilities, it is important to examine what their main sources will be in the future. What will the role of mergers and acquisitions and strategic alliances be and what are the prerequisites for the success of such ventures?

The research presented in this study attempted to draw a picture of the nature of Korean and Taiwanese investments abroad, especially in the industrialised world. However, as was pointed out in the foregoing, many questions still remain and more empirical research is needed. The theoretical suggestions made in the previous section are still tentative and leave a lot of scholarly work to be done.

Notes

1 Introduction

1 The economic problems with which Korea in particular, has been faced since the end of 1997 does not alter its status as one of the most important economic success stories of the post-war period. The crisis as well as its impact on the outward investment behaviour are nevertheless also addressed in this book.

2 Throughout this study, 'Taiwan' is used to denote the economy that for reasons of diplomacy is sometimes also called the 'Republic of China (ROC)' or 'Taiwan Province of China'. The term 'Korea' refers to the 'Republic of Korea (ROK)' or 'South Korea'.

3 The most important publications of that period are Khan (1986b), Kumar and McLeod (1981), Lall (1983), Wells (1983).

4 The terms and abbreviations foreign direct investment, outward direct investment, outward FDI and FDI are used interchangeably.

5 The latter is added to distinguish FDI from portfolio investments in which case the investor is primarily interested in the profits yielded by the firm and is not involved in managerial issues.

6 The terms 'transnational' and 'multinational' are used interchangeably.

2 Multinational enterprises and late industrialisation

1 This does not imply that no attempts have been made to incorporate MNEs into traditional trade theory. Examples are: Helpman (1984), Krugman (1983, 1990), Helpman and Krugman (1985) and Markusen and Venables (1996). For an interesting more general discussion on why economists have contributed so little to our understanding of MNEs, we refer to Vernon's recent article with the provoking title 'The economist's role in research in transnational corporations: or why the dogs have barked so softly' (Vernon 1994).

2 Theoretical concepts dealing with MNE behaviour which however are not relevant for the issues studied here, are left out of consideration. An example is Aliber's model (1970) that specifically deals with the impact of the failure of financial markets on MNE behaviour.

3 In general, however, the analysis of the various theories is not strictly confined to the micro level. Meso as well macro characteristics of the company's environment may very well be important determinants in the explanations.

4 Another important author who propagated the internalisation theory as a general FDI theory is Rugman (1980, 1981, 1982).

5 Buckley (1988) in later versions, however, did at least to some extent, acknowledge the importance of locational variables by stating that firms choose the least cost location for each activity they perform.

6 The proceedings of the Nobel Symposium on The International Allocation of Economic Activity, where this concept was introduced, were published in Ohlin *et al.* (1977).

7 In principle, two types of O advantages can be distinguished. Asset advantages (Oa) arise from the exclusive or privileged possession of assets *vis-à-vis* other firms. Assets refer here to tangible (e.g. manpower, capital) as well as intangible assets (such as technology, information, managerial skills). Transaction cost minimising advantages (Ot) result from the ability of MNEs *vis-à-vis* external markets to reduce transaction costs by the common governance of separate, but interrelated activities located in different countries. Since in practice it is not always easy to distinguish between these two types of O advantages, the distinction is not used in this study.

8 Oman (1984), for instance, distinguishes joint ventures, licensing agreements, franchising, management contracts, turnkey contracts, product-in-hand, production-sharing contracts, risk service contracts and international subcontracts.

9 Internationalisation is defined by these authors as the process of increasing involvement in international operations (Welch and Luastorinen 1986: 36).

10 In his empirical study, Luostarinen (1979) used a much more refined categorisation distinguishing between non-investment marketing operations (with four subcategories), non-investment production operations (four subcategories), direct investment marketing operations (four subcategories), and direct investment production operations (two subcategories).

11 Johanson and Vahlne (1977) already indicated that bigger steps can be taken when the firm: has very large resources; market conditions are stable and homogeneous; or the firm has considerable experience in markets with similar conditions.

12 Various labels are connected with approaches that all address issues related to technology and technological change. Cantwell and Tolentino, for instance, talk about the 'theory of technological accumulation' whereas Lall labelled his conceptualisation as the 'theory of localised change'. Since there is an important overlap between these approaches, the slightly more general term 'technology and technological change' (TTC) theories is used here.

13 In equilibrium models the dynamic aspect is limited to the process of change between the introduction of external disturbances and the settlement of a new equilibrium.

14 The economic system reflects the way in which resources and markets are organised.

15 In addition, the attitude of the host country government with regard to the presence of MNEs is crucial.

16 In addition, investment abroad to exploit natural resources may of course still take place.

17 It was recognised that not all countries can be in Stage 4, as the stock of world-wide outward investment must be equal to that of inward investment (Dunning 1986: 30).

18 Although many examples will underline this relationship (e.g. China, India), the particular case mentioned by Narula (1996a) – the United States – appears an unfortunate example. As Graham (1996) pointed out, inward FDI in the US lags behind outward FDI, which is consistent with the picture of a post-World War II lead by the US firms in technology (and other O advantages) over their non-US rivals.

19 Originally, Dunning and Narula (1994) labelled this factor as 'economic system'. Since the argument made by the authors is limited to the choice between an export orientated and import substitution industrialisation strategy, the latter term is used here instead.

20 Imposing import barriers can also lead to inward investment. In case the authorities aim at developing an indigenous industrial base, inward FDI – if allowed – is mostly seriously restricted, for instance by allowing only minority shares in joint ventures with local companies.

21 According to Singer (1989), the most important weakness of the IS industrialisation strategy is that it did not really substitute for imports. Due to the lack of vertical integration, it shifted imports from finished products to inputs of intermediate and capital goods, saving very little foreign exchange in the process.

22 In the exposition of the IDP (cf. Narula 1996a) a distinction in this regard is made between macro-economic policy and macro-organisational strategy. The macro-economic policy, Narula argues, is often associated with the economic system (or industrialisation strategy – see sub-section c) and is relatively well defined. Considerable variance among countries exists, however, in the role of the government in determining macro-organisational strategy that is of particular relevance here. According to Narula (1996a: 25) 'macro-organisational policy primarily influences the structure and organisation of economic activity, and the nature of the policy most appropriate at a particular stage should, in an 'ideal' situation, change as the economy evolves, reflecting the nature of market imperfections that the policy is designed to circumvent.'

23 The roots of Ozawa's Paradigm can be traced to the work of Kojima (1973, 1982). Kojima distinguished between 'trade-oriented' (Japanese) and 'anti-trade' (American) FDI. According to Kojima, the first category of FDI is preferred since it goes from a comparatively disadvantageous industry in the home country and will promote the upgrading of the industrial structure of the host country and stimulate trade between the two countries. The second type of FDI, on the other hand, is based on the exploitation of oligopolistic factor and product markets and results in balance-of-payments difficulties, the export of jobs, the prevention of structural adjustment, and trade protection. The normative pretension in particular, combined with the fact that in recent years Japanese FDI increasingly became similar to American FDI pushed Kojima's theory into the background. For a critical discussion of the concept, see for instance Gray (1982), Dunning (1993) and Tolentino (1993).

24 The city states of Hong Kong and Singapore are excluded from the analysis since they began neither from an agrarian nor from a raw material base typically seen as a starting-point for industrial transformation.

25 According to Kang (1995), the neo-classical explanation of the economic success in Taiwan and Korea basically comes down to two factors, namely that (a) state intervention was lower than was typical in developing countries and (b) a stable macro-economic environment and a commitment to exports promoted domestic and international competitiveness.

26 Recently, Hikino and Amsden (1994) made a further distinction between successful cases (e.g. Japan, Korea, and Taiwan), those that stumbled back (e.g. Argentina and the Philippines) and the ones that stayed behind (e.g. Pakistan, Bangladesh).

27 According to Amsden and Hikino (1993), a major difference in this respect between successful and less successful late industrialisers is that the former attach performance standards to their subsidy allocation that are subsequently strictly monitored to ensure increases in productivity.

28 The motivations to go abroad (such as market, resource or strategic asset seeking) mentioned by the concept do give an indication of relevant L advantages.

29 For an overview of the discussion on these issues, see van Hoesel (1993). In general, we can conclude that the prevailing opinion was that LIE MNEs rendered a more positive contribution to the development of developing nations than MNEs from EIEs.

30 The importance of this argument was seriously questioned by Lall (1983) and Riemens (1989), however.

31 Most ethnic investments take place in relation to Chinese and Indian communities overseas (Linge 1984).

32 In later publications Lall (cf. 1991a: 15) questions Wells' hypothesis that downscaling and more labour intensive adaptations are generic sources of advantages, since,

in his perception, MNEs would be able to realise such advantages much more easily.

33 This last internalisation argument is not restricted to LIE MNEs, but also seems valid for EIE MNEs operating in LDCs.

34 Evidence seems to suggest mixed reactions by host governments, however (Wells 1983).

3 High-speed economic development of South Korea and Taiwan

1 As opposed to many other authors, we did not include variables such as indicators for industrialisation, exports and savings in this table. The latter are considered underlying factors influencing economic development and are as such – if relevant – mentioned elsewhere.

2 Since information on per capita GNP in constant dollars was not available for all countries, data are depicted in current prices. Although this impedes a longitudinal comparison for each country (inflation is not taken into account), it does enable a comparison between welfare levels in the same year.

3 Similar treaties were signed with the United States, England and Germany in the 1880s.

4 During most of the colonial period, rice imports from Korea made up over half of Japanese rice imports.

5 This law gave the Governor General the authority to approve the establishment of firms. To prevent the establishment of new industries that might compete with industries in Japan, it also restricted private capital flows from Japan to Korea (Ho 1984).

6 This development did not imply that the Japanese neglected agriculture. On the contrary, even more efforts were put into increasing productivity in order to meet Japanese needs in terms of rice, as well as raw materials such as silk and cotton (Pilat 1993).

7 The mounting relative importance of manufacturing was at the expense of the share of agricultural products in the Net Commodity Product which went down from 85 per cent in 1910–12 to 50 per cent in 1939–41.

8 The same held for other important positions within the companies. For instance 81 per cent of all technical and engineering personnel had Japanese nationality.

9 In 1937, for instance, only about 17 per cent of all children between 9 and 14 years enrolled in school.

10 By 1940 there were 1.2 million Koreans in Japan and 1.5 million in Manchuria.

11 Next to the provision of food aid, the US also invested in setting up educational facilities and carried out limited land reforms.

12 Kuznets (1977) labels this policy as 'easy' (or first stage) import substitution because the goods involved (in this case foods, apparel and other consumer non-durables) can be produced on a small scale by labour-intensive means with relatively unskilled labour and without employing particularly advanced technology.

13 The EPB was advised by the US Agency for International Development, the IMF, the World Bank and American academics working as independent advisers (Das 1992).

14 Synthetic fibre yarns and threads, petroleum products and metal products are among the (intermediate) products which received strong government support in that period.

15 In the beginning the implementation of the new policy did not proceed without failures and part of the early reforms had to be cancelled more than once. An example of this is the introduction of a uniform exchange rate in 1961, a measure which had to be undone in 1963 because the devaluation of the won was fully eroded by inflation (Linnemann *et al.* 1987).

16 The won was devalued by almost 50 per cent (Sakong 1993).

17 The latter implied that exporters were entitled to duty-free imports of intermediary inputs up to certain limits, including a generous allowance to cover defective inputs and wastage in the production process. The allowances granted seem to have significantly exceeded actual wastage, which could be legally and profitably sold in the domestic market (Linnemann *et al.* 1987).

18 The HCI share in manufacturing rose from 38.0 per cent (in terms of value added) in 1973 to 54.4 per cent in 1980.

19 The unit labour costs, for instance, rose 20 per cent annually between 1976–78.

20 In 1979 president Park Chung Lee was assassinated.

21 In 1981 the Monopoly Regulation and Fair Trade Act was introduced for this purpose.

22 The number of industries in which inward investment was prohibited none the less remained the same between the 1987 and 1992 (OECD 1994).

23 The land under irrigation went from 227,000 hectares in 1910 to 545,000 hectares in 1942 (Fei *et al.* 1979).

24 According to Fei *et al.* (1979: 21–6), the colonial industrialisation was barely more than 'an extension of agriculture'.

25 As opposed to the Korean case, there are, unfortunately no sectoral shares in total national economic activity available for Taiwan.

26 Meagre mineral resources kept mining production from rising rapidly in Taiwan.

27 The Japanese had introduced compulsory primary education. About one-third of the school age (6–14) children were in school in 1930–31, and this share eventually climbed to 71 per cent in 1943–44. As a result, literacy rates went up from 1 per cent in 1905 to 27 per cent in 1940.

28 Death rates dropped from 33 to 19 deaths per 1,000 between 1906 and 1940.

29 Since the fall of the Kuomintang in mainland China is partly attributed to their failure to curb hyperinflation on the mainland, inflation is a particular sensitive subject in Taiwan's economic policy.

30 In 1952, 35.9 per cent of net domestic product was realised by agriculture whereas manufacturing represented only 10.9 per cent.

31 Kaohsiung and Nantze mainly accommodate electronics, garments and plastics industry, while the smaller Taichung zone is especially designed for precision instruments.

32 The 1960 Statute for Encouragement of Investments which offered incentives to stimulate private investments had been reviewed more than a dozen times by 1980.

33 On January 1, 1979, the United States and the PRC established formal diplomatic ties; as a direct consequence, the US embassy in Taiwan was closed.

34 Money supply rose by 30 per cent in 1971–72 and 50 per cent in 1973 (Clark 1989).

35 In January 1974 deposit rates jumped 33 per cent and the average rise in loan rates was 25 per cent.

36 An important exception was expenditure on basic education, which continued to rise.

37 Other sectors that received state support were: electrical machinery, electronics, precision machine tools, computer terminals and peripherals.

38 The current account surplus was not only the result of the successful export performance, but also because of extensive import, currency and financial market controls which restricted domestic purchasing power and consumption (Lim and Pang 1991).

39 Hsinchu mainly caters for firms in information technology, precision instruments, new materials and biotechnology, see also Chapter 7.

40 These industries were: information technology, consumer electronics, semiconductors, automation machinery, aerospace, new materials, speciality chemicals and pharmaceuticals, medical care and hygiene, and environmental clean-up equipment.

4 Governments and the organisation of economic activities

1 Although the chaebol are often compared with the Japanese *keiretsu*, important distinctions can be observed (Steers *et al.* 1989). In Japan, for instance, the percentage of family ownership is much more limited than in Korea. In addition, the power of the chief executive is much less substantial in Japan, where more professional managers are also employed. Whereas in Korea, finance is strongly government dependent, in Japan the group-connected banks play a substantial role. In addition, Korean chaebol depend much less on stable subcontracting relations with smaller and medium sized enterprises than the Japanese *keiretsu* do (Orrù *et al.* 1991).

2 The concept of a 'holding company' does not apply to the chaebol. Korean law explicitly bans holding companies which implies that, in a legal sense, business groups do not exist.

3 Only three of the present top 30 chaebol (namely, Samsung, Daelim and Kia) were established in the pre-war era; this was a result of Japanese colonial policy which made the growth of indigenous firms particularly difficult.

4 Although some debate exists about how the government allocated these resources (with corruption playing an important role or the government being confronted with only few qualified entrepreneurs), in either case it is beyond doubt that the transfer of such resources in a usually non-competitive environment was the primary source of chaebol growth (Kang 1996).

5 The steel industry, which was also selected as a strategic industry, was government owned and controlled.

6 These licenses were traded at a very high premium and many chaebol that initially did not own a construction company or an overseas construction license took over smaller companies owning such license. In addition, the Korean government provided the same type of subsidy for overseas construction as it did for other exports.

7 Although the companies are credited for their role in building up the country's economy, the ceaseless growth and continued dominance of the chaebol is seen as a result of an improper collaboration between the companies and the government (*Business Korea Yearbook* 1996). Moreover, the ownership of the chaebol to a large extent remains in the hands of the founding families who have accumulated enormous wealth.

8 Examples are the establishment of the Industrial Bank of Korea, the Korean Federation of Small Business and government supported funds for loans to SMEs.

9 Between 1965 and 1980 their share in gross fixed capital formation amounted to a considerable 23 per cent (Chen 1995). By 1995, this share had fallen to 14.8 per cent.

10 The most famous example is POSCO (Pohang Iron and Steel) which is the biggest steel producer in Korea and one of the largest in the world.

11 Fields (1995: 66) defines *guanxiqiye* as 'a cluster of enterprises owned and controlled by a group of persons tied by a network of various *guanxi*'; *guanxi* refer to 'particularistic' connections between persons that is based on some common or shared identification, such as family, friendship or same place of birth.

12 According to Fields (1995), this attitude can be traced to Confucian disdain for pursuing personal profit or material advantage and the injunction to avoid inequality.

13 This nevertheless implies a substantial decrease since 1982 when the share was still 33.8 per cent. The peak was in 1975 when no less than 40.3 per cent of gross fixed capital formation was realised by public enterprises.

14 The activities of most state-owned enterprises are confined to the domestic economy.

15 The Japanese trading companies, themselves modelled after the European trading houses that had emerged in the 17th century, initially were set up by the Meiji government in the nineteenth century primarily as importers to counter the dominance of European traders in Japan (Cho 1987). Since then, the '*sogo shoshas*' (as they were called for the first time in the 1950s) have developed into the most advanced trading companies in the world. According to Cho (1987), their strength is based on a

'multiplicative synergy effect' derived from the integration of operations like transportation, distribution, insurance, finance and information collection.

16 Koryo, by far the smallest of the eight survivors, was established by the government to stimulate exports by small and medium sized enterprises.

17 The average inflation rate of 30 per cent in Korea further stimulated such strategies.

18 These two companies are used here as illustrations because the internationalisation of their sister companies in the electronics industry is analysed in detail in Chapter 7.

19 The government's Office of Supply for the Republic of Korea handles the lion's share of Korean imports. Raw materials and agricultural products which represent 70 per cent of total imports are closed to the GTCs. Petroleum and defence products are also closed.

20 In cases where a company realised less than US$ 200,000 (on an annual basis), its export license was terminated.

21 Another example relates to the fact that the activities of the overseas branches were severely restricted.

22 The local banks have always been privately owned. Their major clients are regional entrepreneurs and public entities (Fields 1995).

23 The curb market is an unregulated semi-legal credit market in which loan suppliers and lenders can transact freely at uncontrolled interest rates (Fields 1995).

24 Their share went down from 36.2 per cent in the period 1966–71 to 3.1 per cent in 1987–91.

25 According to one estimate (Fields 1995), at the end of 1988, for instance, the top thirty chaebol had US$ 66 billion in cross-payment guarantees!

26 In 1991, 15 new private banks were allowed to be set up.

27 The conservatism of the bank sector in Taiwan was certainly stimulated by the lack of competition leading government-appointed management and staff to operate the banks as bureaucracies, seeking neither profitability nor development but safety and the avoidance of error (Fields 1995).

28 For the period 1971–86 these export loans averaged only 11 per cent of total loans of domestic banks (Fields 1995).

29 Public enterprises primarily borrow from the formal financial institutions.

30 Projects that were eligible for approval, for instance, were those aiming at the development and import of raw materials that could not be supplied domestically or projects to secure a fishery area.

31 In that case, FDI can be made if: (a) FDI in the leasing and selling in lots of real estate are made by overseas construction companies or GTCs; (b) the construction and operation of golf courses are carried out by overseas construction companies as fringe facilities of other housing construction projects and the costs of those golf course projects amount to less than one third of total expenses of the investment project.

32 Restricted business sectors will be fully liberalised; quantitative ceilings of FDI by individuals or individual entrepreneurs will be gradually lifted and FDI procedures are further liberalised (*Business Korea Yearbook* 1996).

33 The financing ratios amount to maximums of 80 per cent (Korea Mining Promotion Cooperation) and 70–80 per cent (Korea Petroleum Development Co-operation), respectively.

34 This initiative allows companies based in this area to export to the US against favourable terms. An additional (political) motive for Taiwan to support such investments was that quite a number of countries in the Latin American and Caribbean area have formal diplomatic ties with Taiwan. Investment relationships are considered to strengthen the mutual ties between the host and home country.

35 The minimum paid-in capital went down from NT$ 100 million in 1962 to NT$ 50 million in 1979 and NT$ 20 million in 1984.

36 In addition, if an investment project is provided with cash in the form of foreign

exchange, the paid-in capital has to exceed NT$ 20 million and the liability ratio (i.e. liability/net worth) has to be smaller than 300 per cent.

37 The maximum amount of insurance covered is 85 per cent of the capital remitted out by the firm plus the income therefrom.

38 Subsidies are available for travel expenses, preferential loans and preferential insurance.

5 Korea and Taiwan as outward investors

1 The acquisition of securities and real estate, the provision of technical engineering services abroad for fees and royalties for which the payment period exceeds one year, and the operation of private firms in foreign countries through investments in secondhand equipment also constitute outward investment.

2 The capital contribution can comprise: foreign exchange; machinery and equipment, parts and components; raw materials, semi-finished products, and finished products; technical know-how, patent rights, and exclusive rights of use of trademark or copyright; net profit or other income derived from outward investment; royalties or other income derived from technical co-operation; and securities in foreign countries (IDIC 1993).

3 With regard to *home* countries of MNEs, a distinction is made between early industrialised economies (EIEs) and late industrialising economies (LIEs). *Host* countries of MNEs are divided into developed countries (DCs) and less developed countries (LDCs). LIEs are usually (at least until recently) considered to be LDCs.

4 Since Japan's industrialisation had already taken off in the nineteenth century, it is considered here to be an 'early industrialised' economy although one might also argue that it was the first 'late industrialised' economy.

5 All LIEs with outward FDI stock exceeding US$ 100 million in 1980 or US$ 250 million in 1990 are listed. Since India – an often cited home country of 'Third World MNEs' – had, by the end of 1988 registered only US$ 76 million of outstanding FDI, it is not included in Table 5.2.

6 Due to differences in estimation methods, the exact size of FDI outflow from the PRC is unknown. By the end of 1994, over 900 Chinese companies had established 4,600 overseas affiliates in 130 countries representing estimated accumulated FDI outflows ranging between US$ 5.2 billion and US$16 billion (UNCTAD 1996). Notwithstanding the large number of host countries, it was estimated that a majority (61 per cent) of total investment was made in Hong Kong and Macao and therefore could be seen in anticipation of the integration of both economies into the PRC in 1997 and 1999, respectively.

7 Singapore's case is moreover rather exceptional since the city state used to be part of Malaysia. As a result, a major part of outward FDI has flown into the previous parent country (Aggarwal 1985, Lim and Teoh 1986).

8 Since for some countries GDP figures for 1995 were not yet available, 1994 is the most recent year for which these indicators could be calculated.

9 It is not argued here that exports and FDI are perfect substitutes. As Narula (1996a) shows, the relationship between exports and FDI is quite complex and is not always easily determined.

10 Miscellaneous products refer to a wide range of goods such as toys, sportswear, musical instruments and wigs.

11 As will be illustrated in Chapter 7, with the increased welfare levels in many Asian countries, market seeking (next to natural labour resource seeking) becomes an increasingly important motivation to invest there.

12 The main reasons for POSCO to buy the company were to escape an increase in quota restrictions on POSCO exports, to penetrate the local market and to obtain new technology (Hwang 1994).

13 According to Kuo and Chen (1996), for instance, PRC sources reveal that between 1986–96 no less than US$ 30.1 billion of Taiwanese investments had been approved.
14 Of the total amount invested in 'other areas', 83.1 per cent has gone to Latin America.
15 In view of the overriding dominance of the US in the North American region, a further distinction between the US and Canada is not made here.
16 The importance of the Netherlands as a host country of Korean FDI in fact is overestimated due to the establishment of a holding company in the tax haven of the Netherlands Antilles which accounted for about 80 per cent of the total investment amount. Therefore, the actual Korean investments in the Netherlands are considerably lower.
17 Since the investments by Daewoo in Eastern Europe took place in 1996 and 1997, they are not yet reflected in Table 5.8.
18 A first example of Taiwanese investment activities in the region is a 25 per cent stake that PC producer Aquarius took in Robotron, the former East German state monopoly supplier of computers and office equipment.
19 The large share of the transportation equipment business is fully accounted for by the (not yet materialised) project in the UK.
20 Firm level data show that the distinction between approved projects in manufacturing and trade is not always clear. Quite a number of companies have received approval for trading activities abroad but instead carry out assembly operations and vice versa.
21 Taiwanese banks (among others ICBC, First Commercial Bank, Chunghua Commercial Bank, Hua-Nan Commercial Bank) have established offices in the European Union to support the local business activities of other (manufacturing and trading) Taiwanese companies.
22 At the end of 1980, there were approximately 132,000 workers in the Middle East, equivalent to 4.5 per cent of all workers in Korea's manufacturing sector, or 15.7 per cent of all construction workers in 1980 (Sakong 1993).
23 Illustrative in this respect is that the total number of labour disputes went up from 265 in 1985 to 3,749 two years later (Cho 1994)! In Taiwan the number of labour disputes went from 359 cases in the period 1961–70 to 10,441 cases in 1981–88 (Hsiao 1992).
24 1971 is the first year for which investment stock data for Korea were available (published in Bong 1989).
25 For both countries, investments into Oceania have been excluded because of the mixed DC/LDC character of the region. Insufficient data were available to single out countries belonging to either category. In both cases, however, FDI into Oceania represents only a very small fraction of total outward investments (2.4 per cent and 0.4 per cent for Korea and Taiwan, respectively). In the case of Korea, also investments into Japan had to be excluded since no separate data are available for the pre-1991 period. As a result, a small amount of investments into Japan (namely US$ 16 million) is 'hidden' in the LDC category for the years before 1991. Since this represents only 0.3 per cent of total Korean investments in developing countries, it does not distort the overall picture.

6 Firm-level determinants of outward FDI from the electronics industry

1 Although the firm-level determinants of FDI do not necessarily coincide with their O advantages, according to Lall (1986: 30) this kind of analysis does shed some light on the nature of the 'perceived competitive advantage' of these firms in their overseas operations.
2 For an extensive summary of the principal empirical evidence regarding determinants of FDI, see UNCTC (1992b).

3 Already in 1980 Lall had emphasised the distinction between industrial and firm characteristics that influence (outward) FDI behaviour.

4 The dummy included in the LDC sample was not meant to distinguish between FDI prone and non-FDI prone industries but rather to test the importance of the 'invest-to-safeguard-raw-materials' motive for certain industries.

5 The only distinction made in this respect was a dummy included to measure the importance of dependence on raw materials.

6 Initially, I also considered the use of logit or probit models (in which case the dependent variable takes a value of 0 or 1) to estimate the equations. In that case, 1 represents an investment in a production location abroad, whereas 0 indicates that no investment has taken place. Although in principle the data allow for a distinction between investments by manufacturing companies in overseas production units from investments in sales subsidiaries, in practice there were strong indications that the distinction did not always match reality. Companies that had received approval to set up production sites abroad sometimes only operated a sales subsidiary and vice versa.

7 For an extensive discussion on the properties of tobit models, we refer to Maddala (1983) and Greene (1993).

8 Linear formulations were also tried out; however, they yielded similar but less significant results.

9 Since for six companies no information was available on marketing capabilities, the equation estimating FDI in industrialised economies (for which marketing capabilities were assumed to be a determinant) was based on 135 observations.

10 The census data tapes were made available at Chung-Hua Institution for Economic Research where the author carried out part of the research.

11 Some authors argue that 'size' itself can be perceived as a composite variable for other variables affecting propensity to invest abroad. We nonetheless argue here, that size *per se* is a determining factor.

12 In anticipation of the criticism that many surveys have failed to distinguish between motives for and determinants of FDI (cf. Dunning 1993: 139), we would like to point out that export orientation is a determinant of FDI whereas defending the (former) export market is the motive to invest.

13 The most notable exception concerns outward FDI from Hong Kong. As discussed earlier, Taiwanese companies that are not allowed to invest directly in the PRC set up shell companies in the city state from where investments in the PRC are made.

14 The maximum amount of such a duty is the 'margin of dumping' – that is, the difference between the price at which the good is sold on the home market of the exporter and the price of the good when exported.

15 In some cases, the threat of import barriers constituted by the start of an antidumping investigation was already sufficient to induce companies to start local production.

16 In Chapter 7, an overview of antidumping cases against Korean (consumer) electronics companies in the EU will be given.

17 In 1991, Daewoo, Lucky Goldstar, Hyundai, Samsung and Sunkyoung had indicated that electronics was a key business sector. Including a different dummy with a value of 1 for electronics companies belonging to a chaebol (and 0 otherwise), however, still renders a 'mixed' variable since this would also include the (substantial number of) electronics firms that are part of a chaebol but *in this line of business* mainly focus on domestic markets and therefore are not potential outward investors.

18 It should not be forgotten that the data on which the estimation is based refer to the situation in 1993. As will be illustrated in Chapter 7, there are indications that (at least some of the leading) firms in Korea in recent years have increased their marketing budgets substantially.

19 The relative high pair-wise correlation between HUMCAP and LABPROD2 and standard errors (the latter are not presented here) can probably be attributed to the presence of multicollinearity. Since deleting LABPROD2 (which showed the highest standard deviation) leaves the significance of other variables intact and slightly lowers the t-value of HUMCAP, we suspect that only the statistical significance of LABPROD2 is negatively influenced.

20 A high standard deviation and substantial pair-wise correlation coefficients with LABPROD2 and FOR suggests that this may be attributed to multicollinearity. Deleting LABPROD2 from the question leaves the other estimates and levels of significance intact which underlines their robustness. The result for HUMCAP, however, remains very low. Since FOR is an important significant explanatory variable that cannot be deleted, the equation is kept intact.

21 To counter the correlation among the variables, separate equations were estimated in which each time one 'suspect' variable was eliminated; however, this did not improve the results significantly.

22 The (mainly non-firm level) determinants that were included in previous studies but were not tested in the present investigation are not listed in the table. Since Jeon (1992) in particular, included a relatively large number of non-firm level explanatory variables in his model, the possibilities of comparing the results of both studies are rather limited.

23 Jeon (1992), who tested the importance of company size only for FDI in industrialised economies, found a similar result.

24 In the study by Chen *et al.* (1995) the relationship between export orientation and FDI was significant in LDCs but insignificant in DCs.

7 Four case studies of Korean and Taiwanese multinationals

1 In the case of Taiwan, only one non-industrial sector (namely banking and insurance) was a larger source of outward FDI.

2 Where necessary, the information given was updated during the course of the research project.

3 An extra check on the consistency of opinions was rendered by the possibility of responding to a draft version of the findings.

4 Since most companies operating in consumer electronics are also active in other subsectors of the electronics industry, no distinction is made here between consumer and other electronics companies. An additional argument is that such a distinction requires that separate (sales and employee) figures would have to be available for the various subsectors – however, this is not the case.

5 Specific items still had to receive a recommendation from the manufacturers association in order to allow them to be imported. Effective import liberalisation in Korea has therefore, only been a very recent phenomenon.

6 An EPZ can be defined as a relatively small, geographically separated area within a country, the purpose of which is to attract export-oriented industries, by offering them favourable investment and trade conditions as compared with the remainder of the host country. In particular, the EPZs provide for the importation of goods to be used in the production of exports on a bonded duty free basis (Dicken, 1992: 181).

7 The most important direct benefits for Korea were the provision of jobs (especially for young females) and the generation of foreign exchange.

8 Unfortunately no separate figures are available for the CE sector.

9 Unfortunately, 1995 data for technology exports were not yet available.

10 Following Kawakami (1996), the 'PC industry' is used as an abbreviation for the 'personal computer-related industry', covering a wide range of products such as personal computers, electronic notebooks as well as peripheral products such as monitors, scanners, motherboards, add-on cards and computer mice.

11 Already in 1981 IBM had chosen TECO, a Taiwanese television manufacturer, to mass produce its PC monitors (FT 08.10.93).

12 These figures do not include offshore production by Taiwanese owned companies.

13 There are indications that this situation – at least to some extent – is changing. In the 1990s, the fierce competition has driven many, especially smaller, companies out of business – a development that is expected to continue in the coming years (III, 1993).

14 The government provided grants and low interest loans to support investments in product design, technology innovation and upgrading, quality improvement, automation and personnel training. Some direct subsidies were also offered to induce firms to become more involved in R&D activities. The subsidies were supplemented by incentives such as tax exemptions, accelerated depreciation and other provisions.

15 Unfortunately, as in Korea, no separate figures are available for the computer industry. Based on other sources (such as *Industrial Panorama*), we can safely assume that a very substantial share of these investments are computer related.

16 In 1993, for instance, R&D expenditures as a share of GDP amounted to only 1.76 per cent as compared to 2.33 per cent in Korea and 2.68 per cent in Japan. Moreover, R&D efforts remained primarily a public sector affair.

17 SEC is not the only company of the Samsung Group which operates in the field of electronics. See Appendix 7.1 for an overview of all companies of the Samsung Group active in this line of business.

18 In 1994, LGE again was the largest *consumer electronics* manufacturer in Korea (FT 02.03.95). For electronics as a whole, however, SEC remained the biggest.

19 They were separated in the late 1960s; to strengthen the company's efforts in the multi-media market, it was considered important to merge both companies again (FT 09.09.94).

20 Appendix 7.2 is an overview of all LG members operating in electronics.

21 Starting from 1993, TI–Acer's revenues and staff count are included in the figures.

22 Acer has also acquired a global top position in peripherals. In monitors, the company has achieved a global top three position; in addition, Acer is in the top five of world-wide CD-ROM drive producers.

23 Perhaps contrary to what the name suggests, this division also includes components such as motherboards, add-on cards and workstations.

24 Next to FIC itself, there are two companies which are also considered part of the 'FIC Group' but have an independent status, namely LEO Systems Inc. and Formosa Industrial Computing (see also Appendix 7.4).

25 Strictly speaking, O advantages refer to assets that are unique to a firm and not available (at least at such favourable terms) to other companies in the country in which the company operates. (Dunning, 1993). Since the companies studied here are relative newcomers on the international markets and are still developing at a high pace, the word 'potential' is added to indicate that a priori it is unclear whether the characteristics discussed in this section comply with the requirements of genuine O advantages.

26 'Samsung-NEC'; later renamed 'Samsung Electronics Devices'.

27 Initially 'Samsung-Sanyo Electronics'; later named 'Samsung Electronic Parts'. Sanyo pulled out of this joint venture in 1983.

28 These activities are distinguished here from (equity or non-equity) strategic alliances specifically intended to generate technological know-how.

29 By the end of 1995, SEC announced an increase in its stake to almost 50 per cent. This acquisition was not only made to acquire technological know-how, but also to strengthen SEC's marketing position.

30 SAIT was established in 1987 to enable the group, which was engaged in a diverse range of manufacturing sectors, to secure basic and core technologies cross-cutting sectoral lines in an efficient way.

31 In the case of HDTV, joint research efforts were also conducted by a sister company Samsung Display Devices Co., and Goldstar Co., and Orion Electric Co. (a subsidiary of the Daewoo Group) to develop tubes for HDTV sets.
32 The transfer of 16 Mbyte synchronous DRAM technology to OKI Electric of Japan also illustrates the company's strength in this field. In addition, SEC became the first company in the world to receive an ISO 9001 rating for a semiconductor plant and an ISO 9002 rating for its finished semiconductor products.
33 NEC pulled out of Goldstar Electric in 1987. According to Bloom (1991), NEC indicated that its withdrawal from Goldstar was at the Korean government's insistence after certain of the CE factories of Goldstar Electric had been nominated as being for military use.
34 LG Electronics had already owned a minority share of 4.97 per cent in Zenith since 1991.
35 Some caution is due here. Samsung Electronics' image in recent years as the technology leader in Korean electronics industry is primarily shaped by its successes in the semiconductor industry. It should be realised that LG Electronics is not active in that line of business.
36 It should be realised that both Samsung's as well as LG's claim to be an innovator in (consumer) electronics primarily relates to their operations in Korea. In most or even all cases these 'innovatory' products were developed (and first launched elsewhere) by other non-Korean companies.
37 Interview in August 1993 with Mr H.W. Lee (LG Electronics headquarters – overseas planning and coordination department).
38 Notwithstanding the majority share owned by Acer, the company has no control over TI–Acer. The reason is that Texas Instruments Inc. provides the majority of technical know-how.
39 In addition, output is sold to other, external, clients. In the first quarter of 1995, however, the factory still only supplied about 20 per cent of the memory chips used by Acer. (Source: Interview with Mr. T. Lu, president Acer Europe, 24.03.95.)
40 In practice, a distinction between OEM and ODM is not always easily made. In the text, 'OEM' therefore refers to both OEM and ODM business.
41 Without trading revenues, this share would be 4.4 per cent.
42 According to Flamm (1993: 44–45), in the US computer companies on average spend 13.4 per cent of net sales on R&D.
43 After completion of the so-called 'Super IO Chip', no further interaction between the companies has taken place. Since then both companies have started separate families of the chip.
44 FIC is sometimes even asked to purchase components from other suppliers on behalf of their clients.
45 This impression is confirmed by EIAK (1996).
46 As was shown in Chapter 6, a significant positive relationship was determined between the technological capabilities of Taiwanese electronics firms and their likelihood of investing in industrialised economies.
47 Interview in June 1995 with Mr H.J. Lee (Samsung Electronics headquarters – regional strategy team (Europe), global operations division).
48 In the new logo, Samsung's name is inscribed 'in a dynamic ellipse, symbolising the world's movement through space and thus conveying a distinctive image of change and innovation' (Samsung Electronics, 1993).
49 The latter observation among others led the company to decide to distance itself from a Taiwanese campaign to improve the country's image, and for which purpose Acer's notebook computer was used in the USA (*Business Week* 04.07.94).
50 An important difference from Acer in this respect is that the latter company promoted its own brand name from the beginning in overseas markets as well.
51 This can be partly ascribed to the closed nature of the company. In addition, as it is

primarily active on the industrial market, FIC attaches less importance to public relations.

52 Attempts to establish the Vanda brand name for (by that time) new products such as notebooks, failed.

53 The enormous losses incurred by the major computer companies in the world – primarily as a result of overstock losses – have driven them to companies such as FIC to contract out major parts of their production activities.

54 The company had gone bankrupt as a result of its high R&D expenses and huge overheads (*Business Week* 15.08.94).

55 The company had more than US$ 2 billion in contracts to supply the US government and army.

56 We saw that for the Taiwanese sample used in the econometric analysis in Chapter 6 the *intention* to build up marketing capabilities was in fact a significant determinant of FDI.

57 John Koo, who is a member of the founder's family, was installed as president, however.

58 For many Taiwanese these two characteristics 'proves' that personal success is not limited to those who have emigrated from mainland China and those who obtained a university degree abroad. Next to being the top man of Acer, Shih has taken up many other public functions, such as chairman of the board of the Taipei Computer Association; member of the board of directors of ITRI; member of the board of the Asian Institute of Management; director of the ROC–USA Economic Council; president of the South East Asian Information Technology Organisation; and founding chairman of the Brand international promotion association (Bipa).

59 Another example of the importance attached to the value of the company's 'human capital' is the decision to move the headquarters back to Taipei after it turned out that the establishment of a new headquarters in Taoyuan county in 1990 had made many qualified employees leave the company because of the travelling distances involved. Acer had not only invested substantially in office buildings at the new site, but also in a huge cafeteria and recreational facilities.

60 An additional advantage of this set-up is that, for instance, in the American market, Acer saves 4 per cent on tariff charges by shipping machines without the microprocessors and buying them in the USA from Intel.

61 This shift in focus can, at least partly, be ascribed to substantial problems which FPG faced in the USA. In Texas, Formosa Plastics Co. was fined US$ 4.3 million by the Environmental Protection Agency in 1991 for mishandling hazardous waste. In addition, the company has been cited for dozens of workers'-safety violations (*Business Week* 15.08.94).

62 In 1996, Nan Ya Plastics – in an alliance with Japanese OKI Electric Industry – opened a big DRAM plant in Taiwan and announced plans to construct a plant manufacturing liquid crystal displays (FT 24.07.95).

63 Y.C. Wang, for instance, is known for pressuring managers to squeeze costs through financial incentives and sheer intimidation. The chairman is assisted in his attempts by no less than 200 auditors who monitor possible unnecessary expenditure by departments (*Business Week* 15.08.94, FT 24.07.95).

64 Other parts, such as the computer disk, motherboard and floppy disk drive, are considered less important and are therefore already inserted.

65 For a much more detailed description of the internationalisation process, see Luostarinen (1979).

66 Formally, this company is a subsidiary of Acer Peripherals, Inc.

67 The expectation is that, now most Taiwanese computer companies re-focus on the production of high value-added products, the mass production of low value-added keyboards in a low-wage country will again become a rewarding business (interview with Mr. Teddy Lu, president Acer Europe, 24.03.95).

68 New sites will be established in South Africa (2), Argentina, Colombia, India (2), Korea, Thailand, the Philippines, Israel, the PRC and New Zealand.
69 Interview in October 1995 with Mr V. Lee (FIC Europe).
70 In addition, the new plant will produce barebone systems primarily meant for OEM clients in overseas markets. The systems will be shipped to a target market (for instance Europe or the USA) where final assembly takes place.
71 The importance of Asia is clearly growing, going up from 20 per cent in 1994 to 30 per cent in 1995. In 1996 for the first time, FIC became active in new markets such as Latin America, South Africa and Russia.
72 The European Union had withdrawn GSP benefits in 1988, but restored them on 1 February 1992.
73 Although Acer did not mention the limited size of the home market as a motive to invest abroad, the orientation on foreign markets from the onset suggests otherwise.
74 The exceptions relate to host government incentives in the EU (LGE and FIC) and the possibility of scanning for new technological developments in the USA (Acer).
75 In the same period, Samsung made public a similar plan in Mexico.
76 The latter plant first started as a joint venture of which LGE orginally owned 30 per cent, until buying the local partner's shares.
77 Acer indicated that small orders by companies that merely request changing the label on the computer are also handled from the European sites. About 90 per cent, however, sold from the European offices is under the own brand name.
78 Notwithstanding the unification programme of the European Union, SEC basically still perceives Europe as consisting of separate markets.
79 In Spring 1996, four local nationals were appointed managing directors to SEC's European operations.
80 Based on data provided by the factory in Worms.
81 In practice, however, the localisation of production is hard to realise. In 1995, in Germany for instance, the share of locally acquired inputs amounted to only 20 per cent for VCRs and 50 per cent for TV sets.
82 Interview in April 1995 with Mr W.D. Lee (LG Electronics – Central European headquarters).
83 Ernst (1994b) categorises OEM and ODM contracts (together with subcontracting) under the 'supplier networks'.
84 This observation is confirmed by Hobday (1995) and Kawakami (1996).
85 This applies to (sub-)sectors in which brand names are not considered important (e.g. intermediates or components industries), although the term 'international subcontracting' will usually be used instead of 'OEM', it appears that some of the learning mechanisms described in this chapter are also observed (Dicken, 1992).
86 Wong's study was confined to subcontracting relations between MNE subsidiaries based in Singapore and indigenous, Singaporean firms.
87 This holds for buyers who themselves are producers. In case of a wholesales/retailer, R&D is of course not usually carried out.
88 Ernst (1994a) showed that, especially in the beginning of these OEM contracts, restrictions were sometimes imposed on the export of goods that were made with technology acquired for OEM purposes. An example concerns VCRs produced by SEC for JVC (Japan) which SEC was not allowed to export until 1985.
89 The Taiwanese PC industry has witnessed the bankruptcy of a large number of smaller-sized companies that were unable to realise sufficient economies of scale to cut costs further.
90 Examples of industries that comply with these requirements are bicycles, clothing, athletic footwear and sewing machines.
91 As was illustrated earlier, the importance attached to an efficient logistics system by Acer and FIC is also reflected in the choice of the Netherlands as their regional headquarters and central assembly location.

92 However, the collapse of the memory chips market in 1996 underlines the highly volatile nature of this capital intensive industry.
93 Interview in March 1995 with Mr H.J. de Jong (Samsung Electronics – European headquarters).
94 These attempts were not without costs. The fact that profits can be realised much more rapidly in Eastern Europe than in Western Europe in 1995 led the parent company to decide to shift the *relative* emphasis of its efforts to the former East bloc. This again suggests that – with regard to the 'conventional' CE products currently being sold – SEC does not expect to play a leading role in Europe.
95 The shift of (CTV) production from Germany to the new plant in the UK is expected to lower production costs.
96 Interview in October 1996 with Mr V. Lee (FIC Europe).

8 Summary and conclusion

1 An obvious difference concerns the strong emphasis on the development of the heavy and chemical industry in Korea in the 1970s, something that was not observed to the same extent in Taiwan.
2 Although the Taiwanese outward investors are on average considerably smaller than their Korean counterparts, *within their respective samples*, size is a very significant variable in both economies.
3 As was discussed in Chapter 6, the results may be influenced by the absence of an adequate proxy variable.
4 FDI in export platforms ('natural labour resource seeking') were acknowledged but, in view of the prevailing relative low wages in the home economies at the time, was not considered an important argument.
5 Many investment projects are motivated by more than one argument at the same time and therefore should not be perceived in isolation.
6 Examples of important target industries for Korean companies in the first wave are textiles and shoes (Jo 1981, Kumar and Lim 1984) and for Taiwanese firms, food and beverage processing and textiles (van Hoesel 1996).
7 Of course, this involves additional activities besides investing abroad, such as large-scale R&D and possibly concluding collaborative agreements.
8 Whereas in the case of Taiwanese companies, external suppliers will, in most cases, be dominant, the Korean chaebol to a larger extent source from within their own group.
9 Although the suggestions made in this section all concern the three 'core theories' used in this study (namely, The Eclectic Theory of International Production, the Investment Development Path and the Late Industrialisation paradigm), we would like to stress that very useful additional insights for the examination of Korean and Taiwanese FDI were gained from the other theoretical concepts discussed in Chapter 2. The distinction of the various 'categories' of industries made in Ozawa's Stages Theory, for instance, in the present section is directly integrated into the discussion of the IDP (see under 'Decomposition' of the IDP).
10 In his analysis of the development of industrial technology in developing countries, Kim (1980) distinguished three stages. During the initial stage, the technology task is to implement transferred foreign technology in order to produce products whose technology and markets has been tested and proved elsewhere. For this purpose, limited engineering efforts are required. In the second stage, foreign technologies are assimilated to produce differentiated products. In this stage, the emphasis is on engineering and limited development. In the third stage, foreign technology is improved through local efforts in research development and engineering. The term 'research and development (R&D)' as denoted in the figures starts with activities as described by Kim in the third stage and refers to efforts aimed at realising innovations (or inventions) of any kind.

11 It goes without saying that the time frames of Figures 8.1 and 8.2 are not the same.
12 This is not to ignore the possibility of modest FDI flows from relatively poor ('stage 1 or 2') countries such as India. However, given the lack of growth of these investment flows in recent years, it appears that these overseas projects should be considered the exceptions to the rule.
13 We acknowledge that these figures are not always available. This does not, however, make the conceptual distinction less useful.
14 Interestingly, Dunning (1993: 64) noted a similar trend among Japanese electronics MNEs in the 1970s.
15 The recent establishment of non-government controlled private banks suggests that in Taiwan, private companies are also becoming increasingly active in the financial sector, which may overcome their difficulties to access formal financial capital.

Bibliography

Aberbach, J. D., D. Colar, K. L. Sokoloff (eds) (1994), *The Role of the State in Taiwan's Development*, New York: M. E. Sharpe.

Acer (various years), *Annual Report*.

Adriaansen, W. L. H., S. Storm and J. G. Waardenburg (1989), 'Forty years of experience in development and practice. An introduction', in W. L. M. Adriaansen and J. G. Waardenburg (eds) *A Dual World Economy*, Groningen: Wolters-Noordhof.

Agarwal, J. P. (1980), 'Determinants of foreign direct investment: a survey'. *Weltwirtschaftliches*, vol. 116, pp. 739–73.

—— (1985a), *Pros and Cons of Third World Multinationals: Case Study of India*, Tübingen: Paul Siebeck.

—— (1985b), 'Intra-LDCs foreign direct investment: a comparative analysis of Third World multinationals', *Developing Economies*, September, XXIII-3, pp. 236–53.

—— (1986), 'Third World multinationals and balance-of-payments effect on home countries: a case study of India', in K. M. Khan (ed.), *Multinationals from the South: New Actors in the International Economy*, London: Pinter Publishers.

Aggarwal, R. (1985), 'Emerging Third World multinationals: a case study of the foreign operations of Singapore firms', *Contemporary South East Asia*, vol. 7, no. 3, pp. 193–208.

Aggarwal, R. and J. K. Weekly (1982), 'Foreign operations of Third World multinationals: a literature review and analysis of Indian companies', *Journal of Developing Areas*, vol. 17, October, pp. 12–30.

Agmon, T. and C. P. Kindleberger (eds) (1977), *Multinationals from Small Countries*, Cambridge, MA: MIT Press.

Aharoni, Y. (1966), *The Foreign Investment Decision Process*, Boston, MA: Graduate School of Business Administration, Harvard University.

Aliber, R. Z. (1970), 'A theory of direct foreign investment', in C. P. Kindleberger (ed.), *The International Corporation*, Cambridge, MA: MIT Press.

Amsden, A. H. (1985), 'The state and Taiwan's economic development', in P. E. Evans *et al.* (eds), *Bringing the State Back In*, Cambridge: Cambridge University Press.

—— (1989), *Asia's Next Giant: South Korea and Late Industrialization*, New York: Oxford University Press.

—— (1991), 'Diffusion of development: the Late Industrialization model and Greater Asia', *American Economic Review*, May, pp. 282–6.

—— (1992), 'Taiwan in international perspective', in N. T. Wang (ed.), *Taiwan's Enterprises in Global Perspective*, Armonk: M. E. Sharpe.

Amsden, A. H. and Y.-D. Euh (1993), 'South Korea's 1980s financial reforms: good-bye financial repression (maybe), hello new institutional restraints', *World Development*, vol. 21, no. 3, pp. 379–90.

Amsden, A. H. and T. Hikino (1993), 'Borrowing technology or innovating: an exploration of the two paths to industrial development', in R. Thomson (ed.), *Learning and Technological Change*, New York: St. Martin's Press.

Balassa, B. (1988), 'The lessons of East Asian development: an overview', *Economic Development and Cultural Change*, vol. 36, no. 3, pp. S273–S290.

Bank of Korea (1993), 'Korean Direct Investment Abroad', July, mimeographed.

Bark, T. (1991), *Anti-dumping Restrictions Against Korean Exports: Major Focus on Consumer Electronics Products*, Seoul: KIEP Working Paper No. 91-02.

—— (1993), 'The Korean consumer electronics industry: reaction to antidumping actions', in J. M. Finger (ed.), *Antidumping: How it Works and Who Gets Hurt*, Ann Arbor: University of Michigan Press.

Baumol, W. J., R. R. Nelson and E. N. Wolff (eds) (1994), *Convergence of Productivity: Cross-Country Studies and Historical Evidence*, New York: Oxford University Press.

Belderbos, R. (1994), *Strategic Trade Policy and Multinational Enterprises; Essays on Trade and Investment by Japanese Electronics Firms*, Amsterdam: Thesis Publishers.

Bloom, M. D. H. (1991), *Globalisation and the Korean Electronics Industry*, Fontainebleau: EAMSA Conference Paper.

—— (1992), *Technological Change in the Korean Electronics Industry*, Paris: OECD Development Centre.

Boddewyn, J. J., M. B. Halbrich and A. C. Perry (1986), 'Service multinationals: conceptualization, measurement and theory', *Journal of International Business Studies*, vol. 17, no. 3, pp. 41–58.

BOK (various years), *Overseas Direct Investment Statistics Yearbook.*

Bong, C. S. (1989), 'The multinationalization of Korean business and the new direction of FDI from Korea', in A. R. Negandhi (ed.), *Multinationals and Newly Industrialized Countries: Strategies, Environmental Imperatives, and Practices*, London: Jai Press.

Buckley, P. J. (1988), 'The limits of explanation: testing the internalization theory of the multinational enterprise', *Journal of International Business Studies*, vol. 19, no. 2, pp. 181–93.

—— (1989), *The Multinational Enterprise: Theory and Applications*, Basingstoke: Macmillan Press.

Buckley, P. J. and M. C. Casson (1976), *The Future of the Multinational Enterprise*, London: Macmillan Press.

Business Korea: various issues.

Business Korea Yearbook (various years), Seoul: Business Korea Co. Ltd.

Business Times: various issues.

Business Week: various issues.

Busjeet, V. (1980), *Foreign Investors from Less-Developed Countries: A Strategic Profile*, dissertation, Boston: Harvard Business School.

Cantwell, J. A. (1991), 'The theory of technological competence and its application to international production', in D. McFetridge (ed.), *Foreign Investment, Technology and Economic Growth*, Calgary: University of Calgary Press.

—— (1992), 'The methodological problems raised by the collection of foreign direct investment data', *Scandinavian International Business Review*, vol. 1, no. 1, pp. 86–102.

Cantwell, J. A. and P. E. E. Tolentino (1990), *Technological accumulation and Third World*

Multinationals, Discussion Papers in International Investment and Business Studies, No. 139, University of Reading.

Casson, M. C. (1984), *General Theories of the Multinational Enterprise: A Critical Examination*, Discussion Paper Department of Economics, no. 77, University of Reading.

Caves, R. E. (1971a), 'International corporations: the industrial economics of foreign investment', *Economica*, no. 38, pp. 1–27.

—— (1971b), 'Causes of direct investment: foreign firms' shares in Canadian and United Kingdom manufacturing industries', *Review of Economics and Statistics*, no. 56, pp. 279–93.

—— (1993), 'Japanese investment in the United States: lessons for the economic analysis of foreign investment', *World Economy*, vol. 16, no. 3, pp. 279–300.

CBC (various years), *Balance of Payments – Taiwan District the Republic of China.*

CCIS (various years). *The Largest Corporations in Taiwan – Top 500.*

CEPD (various issues), *Taiwan Statistical Data Book – ROC.*

Chang, C.-C. (1992), 'The development of Taiwan's personal computer industry', in N. T. Wang (ed.), *Taiwan's Enterprises in Global Perspective*, Armonk: M. E. Sharpe.

Chen, E. K. Y. and P. Drysdale (eds) (1995), *Corporate Links and Foreign Direct Investment in Asia and the Pacific*, Pymble: Harper Educational.

Chen Chiu, L.-I. (1995), 'The pattern and impact of Taiwan's investment in Mainland China', in S. J. La Croix *et al.* (eds), *Emerging Patterns of East Asian Investment in China; from Korea, Taiwan, and Hong Kong*, Armonk: M. E Sharpe.

Chen, M. (1995), *Asian Management Systems: Chinese, Japanese and Korean Styles of Business*, London: Routledge.

Chen, T.-J. (1992), 'Determinants of Taiwan's direct foreign investment; the case of a newly industrializing country', *Journal of Development Economics*, no. 39, pp. 397–407.

Chen, T.-J. and Y.-P. Chen (1995), 'Foreign direct investment and deindustrialisation: the case of Taiwan', *Journal of Industry Studies*, vol. 2, no. 1, pp. 57–68.

Chen, T.-J., Y.-H. Ku and M.-C. Liu (1995), 'Direct foreign investment in low-wage and high-wage countries: the case of Taiwan', in E. K. Y. Chen and P. Drysdale (eds), *Corporate Links and Foreign Direct Investment in Asia and the Pacific*, Pymble, New South Wales: Harper Educational.

Chen, T. P. (1989), 'The internationalization of the Taiwanese economy', in C. F. Lee and S.-C. Hu (eds), *Taiwan's Foreign Investment, Exports, and Financial Analysis*, Greenwich, CT: Jai Press.

China Productivity Center (1996), *Taiwan's Buyers' Guide*, Taipei: China Productivity, Center.

Cho, D.-S. (1987), *The General Trading Company: Concept and Strategy*, Lexington, MA: Lexington Books.

Cho, D.-S., J. Choi and Y. Yi (1994), 'International advertising strategies by NIC multinationals: the case of a Korean firm', *International Journal of Advertising*, no. 13, pp. 77–92.

Cho, S. (1994), *The Dynamics of Korean Economic Development*, Washington: Institute for International Economics.

Choi, K. and Y. S. Lee (1990), 'The role of the Korean government in industrializaton', in Ch. L. Lee and I. Yamazawa (eds), *The Economic Development of Japan and Korea*, New York: Praeger.

Chu, Y.-H. (1994), 'The state and development of the automobile industry in South Korea and Taiwan', in J. D. Aberbach *et al.* (eds), *The Role of the State in Taiwan's Development*, New York: M. E. Sharpe.

Chung, K. H. and H. C. Lee (eds) (1989), *Korean Managerial Dynamics*, New York: Praeger.

Chung, K. M. (1991), 'Korean direct investment in European Community: strategies for Korean firms', in K. M. Chung (ed.), *The Economic Cooperation between EC and Korea: Problems and Prospects*, Seoul: Joint KIEP/FES Conference.

—— (ed.) (1991), *The Economic Cooperaton between EC and Korea: Problems and Prospects*, Seoul: Joint KIEP/FES Conference.

—— (1992), 'Growth of Korean Foreign Direct Investment: Problems and Prospective', KIEP, mimeographed.

Clark, C. (1989), *Taiwan's Development: Implications for Contending Political Economy Paradigms*, New York: Greenwood Press.

Coase, R. (1937), 'The nature of the firm', *Economica*, no. 4, pp. 386–405.

Cuyvers, L. and D. van den Bulcke (1989), 'Some reflections on the "outward-oriented" development strategy of the Far Eastern Newly Industrializing Countries', in W. L. M. Adriaansen and J. G. Waardenburg (eds), *A Dual World Economy*, Groningen: Wolters-Noordhof.

Das, D. K. (1992), *Korean Economic Dynamism*, Macmillan.

Datamation (01.12.92): *The PC Big 50*, vol. 38, no. 24, pp. 36–9.

DGBAS (Directorate-General of Budget, Accountancy and Statistics) (1993), *The Report in 1991 Industrial and Commercial Census: Taiwan-Fukien Area, the Republic of China*, Taipei: DGBAS.

Dicken, P. (1992), *Global Shift; the Internationalisation of Economic Activity*, London: Paul Chapman.

Dunning, J. H. (ed.) (1974), *Economic Analysis and the Multinational Enterprise*, London: Allen & Unwin.

—— (1977), 'Trade, location of economic activity and the multinational enterprise. A search for an eclectic approach', in B. Ohlin *et al.* (eds), *The International Allocation of Economic Activity*, New York: Holmes & Meier.

—— (1981a), 'Explaining outward direct investment of developing countries: in support of the eclectic theory of international production', in K. Kumar and M. G. McLeod (eds), *Multinationals from Developing Countries*, Toronto: Lexington.

—— (1981b), 'Explaining the international direct investment position of countries: towards a dynamic or development approach', *Weltwirtschaftliches Archiv*, vol. 117, no. 1, pp. 30–64.

—— (1986), 'The investment development cycle and Third World multinationals', in K. M. Khan (ed.), *Multinationals from the South: New Actors in the International Economy*, London: Pinter Publishers.

—— (1988), *Explaining International Production*, London: Unwin Hyman.

—— (1993), *Multinational Enterprises and the Global Economy*, Wokingham: Addison-Wesley.

—— (1995), 'Reappraising the eclectic paradigm in an age of alliance capitalism', *Journal of International Business Studies*, vol. 26, no. 3, pp. 461–91.

—— (1997), 'What's wrong – and right – with trade theory?', in J. H. Dunning (ed.), *Alliance Capitalism and Global Business*, London: Routledge.

Dunning, J. H. and A. M. Rugman (1985), 'The influence of Hymer's dissertation on the theory of FDI', *American Economic Review*, vol. 75, no. 2, pp. 228–32.

Dunning, J. H., J. A. Cantwell and T. A. B. Corley (1986), 'The theory of international production: some historical antecedents', in P. Hertner and G. Jones (eds), *Multinationals: Theory and History*, Aldershot: Gower.

Dunning, J. H. and J. A. Cantwell (1987), *IRM Directory of Statistics of International Investment and Production*, Basingstoke: Macmillan.

Dunning, J. H. and R. Narula (1994), *TransPacific Foreign Direct Investment and the Investment Development Path: the Record Assessed*, Columbia, SC: Essays in International Business, no. 10.

Dunning, J. H. and R. Narula (eds) (1996), *Foreign Direct Investment and Governments: Catalysts for Economic Restructuring*, London: Routledge.

Dunning, J. H., R. van Hoesel and R. Narula (1997), *Third World Multinationals Revisited: New Evidence and Theoretical Implications*, Reading: Discussion Papers in International Investment and Business Studies, no. 227, University of Reading.

—— (1998), 'Explaining the "new" wave of outward FDI from developing countries', *International Business Review*, (forthcoming).

EIAK (Electronics Industry Association of Korea) (1994), *Annual Directory of Electronic & Electrical Manufacturers in Korea*, Seoul: EIAK.

—— (1996), *Annual Directory of Electronic & Electrical Manufacturers in Korea*, Seoul (in Korean).

Ernst, D. (1994a), *What are the Limits to the Korean Model? The Korean Electronics Industry Under Pressure*, Berkeley, CA: A BRIE Research Paper.

—— (1994b), 'Network transactions, market structure and technology diffusion – implications for South–South co-operation', in L. K. Mytelka (ed.), *South–South Co-operation in a Global Perspective*, Paris: OECD Development Centre.

ESCAP/UNCTC (1985), *Transnational Corporations from Developing Asian Economies*, Bangkok: United Nations.

—— (1988), *Transnational Corporations from Developing Asian Economies – Host Country Perspectives*, Bangkok: United Nations.

Euh, Y.-D. and S. H. Min (1986), 'Foreign direct investment from developing countries: the case of Korean firms', *Developing Economies*, XXIV-2, June, pp. 149–68.

Evans, P. E., D. Rueschemeyer and T. Skocpol (eds) (1985), *Bringing the State Back In*, Cambridge: Cambridge University Press.

Fei, J. C. H., G. Ranis and S. W. Y. Kuo (1979), *Growth with Equity; the Taiwan Case*, New York: Oxford University Press.

FIC (First International Computer) (various years), *Annual Report*, Taipei: FIC

Fields, G. S. (1992), 'Living standards, labor markets and human resources in Taiwan', in G. Ranis (ed.), *Taiwan, from Developing to Mature Economy*, Boulder, CO: Westview Press.

Fields, K. F. (1995), *Enterprise and the State in Korea and Taiwan*, New York: Cornell University Press.

Finger, J. M. (ed.) (1993), *Antidumping: How it Works and Who Gets Hurt*, Ann Arbor: University of Michigan Press.

Flamm, K. (1993), 'The computer industry', in Wellenius *et al.* (eds), *Developing the Electronics Industry*, Washington: World Bank.

Flowers, E. B. (1989), 'Taiwan's potential for foreign direct investment in the United States', in C. F. Lee and S.-C. Hu (eds.), *Taiwan's Foreign Investment, Exports, and Financial Analysis*, Greenwich: Jai Press.

Fortune: various issues.

Froot, K. A. (ed.) (1993), *Foreign Direct Investment*, Chicago: University of Chicago Press.

FT (*Financial Times*): various issues.

Fujita, M. (1990), 'Non-conventional TNCs; small- and medium-size TNCs and TNCs from developing countries', *CTC Reporter*, no. 30, pp. 37–45.

Gerschenkron, A. (1962), *Economic Backwardness in Historical Perspective. A Book of Essays*, Cambridge, MA: Harvard University Press.

Ghoshal, S. (1988), 'Environmental scanning in Korean firms: organisational isomorphism in action', *Journal of International Business Studies*, vol. 18, Spring, pp. 69–86.

Giddy, I. A. (1978), 'The demise of the product cycle model in international business theory', *Columbia Journal of World Business*, Spring, pp. 90–7.

Gold, Th.B. (1986), *State and Society in the Taiwan Miracle*, Armonk: M. E. Sharpe.

Graham, E. M. (1996), 'The United States: some musings on its investment development path', in J. H. Dunning and R. Narula (eds), *Foreign Direct Investment and Governments: Catalysts for Economic Restructuring*, London: Routledge.

Gray, H. P. (1982), 'Macroeconomic theories of foreign direct investment: an assessment', in A. M. Rugman (ed.), *New Theories of the Multinational Enterprise*, London: Croom Helm.

Green, R. H. (1986), 'Operational relevance of Third World multinationals to collective self-reliance: some problems, provocations and possibilities', in K. M. Khan (ed.), *Multinationals from the South: New Actors in the International Economy*, London: Pinter Publishers.

Greene, W. H. (1993), *Econometric Analysis*, New York: Macmillan.

Guillén, M. (1995): *Breaking Through the Glass Ceiling*, Seoul: AIB Conference Paper.

Hagedoorn, J. (1993), 'Understanding the rationale of strategic technology partnering: inter-organizational modes of cooperation and sectoral differences', *Strategic Management Journal*, vol. 14, pp. 17–30.

Haggard, S. (1988), 'The politics of industrialization in the Republic of Korea and Taiwan', in H. Hughes (ed.), *Achieving Industrialization in East Asia*, Cambridge: Cambridge University Press.

Hamilton, G. H. and M. Orrù (1989), 'Organizational structures of East Asian companies', in K. H. Chung and H. C. Lee (eds), *Korean Managerial Dynamics*, New York: Praeger.

Han, C. M. and Th. L. Brewer (1987), 'Foreign direct investments by Korean firms: an analysis with FDI theories', *Asia Pacific Journal of Management*, vol. 4, no. 2, pp. 90–102.

Helpman, E. (1984), 'A simple theory of international trade with multinational corporations', *Journal of Political Economy*, no. 92, pp. 451–71.

Helpman, E. and P. Krugman (1985), *Market Structure and Foreign Trade*, Cambridge, MA: MIT Press.

Hennart, J. F. (1982), *A Theory of Multinational Enterprise*, Ann Arbor: University of Michigan Press.

—— (1989), 'Can the "new forms of investment" substitute for the "old forms?" A transaction costs perspective', *Journal of International Business Studies*, vol. 20, no. 2, pp. 211–33.

Hertner, P. and G. Jones (eds) (1986), *Multinationals: Theory and History*, Aldershot: Gower.

Hikino, T. and A. H. Amsden (1994), 'Staying behind, stumbling back, sneaking up, soaring ahead: late industrialization in historical perspective', in W. J. Baumol *et al.* (eds), *Convergence of Productivity: Cross-Country Studies and Historical Evidence*, New York: Oxford University Press.

Hindley, B. (1988), 'Dumping and the Far East trade of the European Community', *World Economy*, December, no. 4, pp. 445–64.

Ho, S. P. S. (1978), *Economic Development of Taiwan, 1860–1970*, New Haven, CT: Yale University Press.

Ho, S. P. S. (1984), 'Colonialism and development: Korea, Taiwan, and Kwantung', in

R. H. Myers and M. R. Peattie (eds), *The Japanese Colonial Empire, 1985–1945*, Princeton, NJ: Princeton University Press.

Hobday, M. (1995), 'East Asian latecomer firms: learning the technology of electronics', *World Development*, vol. 23, no. 7, pp. 1171–93.

Hoesel, R. van (1992), *Multinational Enterprises from Developing Countries with Investments in Developed Economies: Some Theoretical Considerations*, Antwerp: CIMDA Discussion Paper no. 1992/E/6.

—— (1993), *Third World Multinationals: Retrospect and New Issues*, Rotterdam: Discussion Paper No. 90, Centre for Development Planning.

—— (1996), 'Taiwan: foreign direct investment and the transformation of the economy', in J. H. Dunning and R. Narula (eds), *Foreign Direct Investment and Governments: Catalysts for Economic Restructuring*, London: Routledge.

—— (1997), *Beyond Export-led Growth; the Emergence of New Multinational Enterprises from Korea and Taiwan*, Amsterdam: Thesis Publishers.

—— (1998a), 'New challenges: Korean investment in Europe', In K. W. Radtke and J. Stam (eds), *Dynamics in Pacific Asia: Conflict, Competition and Cooperation*, London: Kegan Paul.

—— (1998b), 'The emergence of Korean and Taiwanese multinationals in Europe: prospects and limitations', *Asia Pacific Business Review*, vol. 4, no. 2, pp. 109–29.

Hong, Y. S. (1993), *Leveraging Technology for Strategic Advantage in the Global Market: Case of the Korean Electronics Industry*, Seoul: KIEP Working Paper no. 93–07.

Horst, Th. (1972a), 'Firm and industry determinants and the decision to invest abroad: an empirical study', *Review of Economics and Statistics*, no. 54, pp. 258–66.

—— (1972b), 'The industrial composition of US exports and subsidiary sales to the Canadian market', *American Economic Review*, no. 62, pp. 37–45.

Hsiao, H.-H. M. (1992), 'The labor movement in Taiwan: a retrospective and prospective look', in D. F. Simon and M. Y. M. Kan (eds), *Taiwan Beyond the Economic Miracle*, Armonk: M. E. Sharpe.

Hughes, H. (ed.) (1988), *Achieving Industrialization in East Asia*, Cambridge: Cambridge University Press.

Hwang, J. S. (1994), '*Economic Development and Internationalisation as Viewed Through the Investment Development Path: With Special Reference to the Korean Economy*', dissertation: University of Reading.

Hymer, S. (1976), *The International Operations of National Firms: A Study of Direct Foreign Investment*, Cambridge, MA: MIT Press. (Original dissertation with same title: MIT, 1960.)

IDIC (Industrial Development and Investment Center) (1993), *Outward Investment from Taiwan ROC; A New Source of International Capital*, Taipai: IDIC.

—— (various issues), *Taiwan Industrial Panorama*, Taipei: IDIC.

III (1993), *Information Industry Yearbook Taiwan, R.O.C. – 1993 Edition*, Taipei: III.

III/MIC (Institute for Information Industry/Market Intelligence Center) (various issues), *Asia IT Report*, Taipei: III/MIC.

IMF (International Monetary Fund) (various years), *Financial Statistics*, Washington: IMF.

Itaki, M. (1991), 'A critical assessment of the eclectic theory of the multinational enterprise', *Journal of International Business Studies*, vol. 23, no. 3, pp. 445–60.

Jeon, Y.-D. (1992), 'The determinants of Korean foreign direct investment in manufacturing industries', *Weltwirtschaftliches Archiv*, vol. 128, no. 3, pp. 527–41.

Jepma, C. J. and A. P. Rhoen (eds) (1996), *International Trade: A Business Perspective*, Harlow: Longman.

Jo, S-H. (1981), 'Overseas direct investment by South Korean firms: direction and pattern', in K. Kumar and M. G. McLeod (eds), *Multinationals from Developing Countries*, Toronto: Lexington.

Johanson, J. and J. E. Vahlne (1977), 'The internationalisation process of the firm – a model of knowledge development and increasing foreign market commitments', *Journal of International Business Studies*, vol. 8, no. 1, pp. 23–32.

Jones, L. P. and Il Sakong (1980), *Government, Business, and Enterpreneurship in Economic Development: the Korean Case*, Cambridge, MA: Harvard University Press.

Jun, Y. (1987), 'The reverse direct investment: the case of the Korean consumer electronics industry', *International Economic Journal*, vol. 1, no. 3, pp. 91–104.

—— (1993), *Strategic Responses of Korean Firms to Glabalization and Regionalization Forces – the Case of the Korean Electronics Industry*, Sapporo: Conference Paper.

Jun, Y. and D. F. Simon (1991), 'The Pattern of Korea's Foreign Direct Investment: Implications for the Internalization of China's Economy', Beijing: Conference Paper.

Jun, Y. and D. Yoon (1995), *An Exploratory Explanation of the Reverse Direct Investment: the Case of Korean Electronics Industry*, Seoul: AIB Conference Paper.

Jung, K. H. (1989), 'Business-government relations in Korea', in K. H. Chung and H. C. Lee (eds), *Korean Managerial Dynamics*, New York: Praeger.

Kang, G.-W. (1993), 'Marketingstrategic des Koreanischen Chaebol (Konglomerat) in den Weltmärkten, dissertation: Universität Zürich.

Kang, T. W. (1989), *Is Korea the Next Japan? Understanding the Structure, Strategy, and Tactics of America's Next Competitor*, New York: Free Press.

Kang, D. C. (1995), 'South Korean and Taiwanese development and the new institutional economics', *International Organization*, vol. 49, no. 3, pp. 555–87.

Kang, M. H. (1996), *The Korean Business Conglomerate: Chaebol Then and Now*, Berkeley, CA: Institute of East Asian Studies, Korea Research Monograph.

Kawakami, M. (1996), *Development of the Small- and Medium-Sized Manufacturers in Taiwan's PC Industry*, Tapei: CIER Discussion Paper No. 9606.

KFTA (1994), *Import Restrictions Against Korean Products by Major Industrialised Countries* (in Korean), Seoul.

Khan, K. M. (1986a), "Multinationals from the South: emergence, patterns and issues', in K. M. Khan (ed.), *Multinationals from the South: New Actors in the International Economy*, London: Pinter Publishers.

—— (ed.) (1986b), *Multinationals from the South: New Actors in the International Economy*, London: Pinter Publishers.

Ki, M. C. (1992), 'Growth of Korean Foreign Direct Investment: Problems and Prospective', KIEP, mimeographed.

Kim, L. (1980), 'Stages of development of industrial technology in a developing country: a model', *Research Policy*, no. 9, pp. 254–77.

Kim, S. J. (1995), 'Korean direct investment in China: perspectives of Korean investors', in S. J. La Croix *et al.* (eds), *Emerging Patterns of East Asian Investment in China; from Korea, Taiwan, and Hong Kong*, Armonk: M. E. Sharpe.

Kim, J.-D. and Y. J. Wang (1996), *Toward Liberalization of International Direct Investment in Korea: Retrospects and Prospects*, Seoul: KIEP Working Paper no. 96–02.

Kindleberger, C. P. (ed.) (1970), *The International Corporation*, Cambridge, MA: MIT Press.

Kindleberger, C. P. and Audretsch, D. B. (eds), (1983), *The Multinational Corporation in the 1980's*, Cambridge, MA: MIT Press.

KITA (Korea Industrial Technology Association) (1995), *Major Indicators of Industrial Technology 1994*, Seoul.

Knickerbocker, F. T. (1973), *Oligopolistic Reaction and Multinational Enterprise*, Boston, MA: Harvard University.

Kojima, K. (1973), 'A macroeconomic approach to foreign direct investment', *Hitotsubashi Journal of Economics*, vol. 14, no. 1, pp. 1–21.

—— (1982), 'Macroeconomic versus international business approach to direct foreign investment', *Hitotsubashi Journal of Economics*, vol. 23, no. 1, pp. 1–19.

Koo, S. M. (1994), 'Korea's big business groups and international competitiveness', in Y. K. Sung (ed.), *The Korean Economy at a Crossroad*, Westport, CT: Praeger.

Korea Development Bank (1994), 'Korea's foreign direct investment (FDI) outflow', *KDB Economic and Industrial Focus*, September, pp. 35–45.

Krugman, P. (1983), 'The "new theories" of international trade and the multinational enterprise', in C. P. Kindleberger and D. B. Audretsch (eds), *The Multinational corporation in the 1980's*, Cambridge, MA: MIT Press.

—— (1990), *Rethinking International Trade*, Cambridge, MA: MIT Press.

Kumar, K. (1982), 'Third World multinationals; a growing force in international relations', *International Studies Quarterly*, vol. 26, pp. 397–424.

Kumar, K. and K. Y. Lim (1984), 'The Korean manufacturing multinationals', *Journal of International Business Studies*, vol. 15, no. 2, pp. 45–61.

Kumar, K. and M. G. McLeod (eds) (1981), *Multinationals from Developing Countries*, Toronto: Lexington.

Kumar, N. (1992), *Single European Market, Multinationals and Industrial Reorganisation; Implications for Developing Countries*, Delhi: RIS Occasional Paper no. 37.

—— (1994), *Multinational Enterprises and Industrial Organization; the Case of India*, New Delhi: Sage.

—— (1995), *Changing Character of Foreign Direct Investment from Developing Countries: Case Studies from Asia*, Maastricht: UNU INTECH Discussion Paper no. 9516.

Kuo, W.-J. (1993), *Internationalisation of Taiwan Industry – Globalisaton of Taiwan's Small- and Medium-Sized Enterprises*, Taipei: CIER Paper.

—— (1994), *The Characteristics of Taiwan's Direct Investment Toward Asian Developing Countries: A Study on Firm's Behaviour*, Nagoya, Japan: International Economic Conflict Discussion Paper no. 76.

Kuo, W.-J. and S. H. Chen (1996), 'Policy Competition and Foreign Direct Investment: Views from Taiwan's Perspectives', Taipei: Report prepared for OECD (draft).

Kuznets, P. W. (1977), *Economic Growth and Structure in the Republic of Korea*, New Haven, CT: Yale University Press.

Kwok, R. Y.-W. (1995), 'Hong Kong investment in South China', in S. J. La Croix *et al.* (eds), *Emerging Patterns of East Asian Investment in China; from Korea, Taiwan, and Hong Kong*, Armonk: M. E. Sharpe.

La Croix, S. J., M. G. Plummer and K. Lee (eds) (1995), *Emerging Patterns of East Asian Investment in China; from Korea, Taiwan, and Hong Kong*, Armonk: M. E. Sharpe.

Lall, R. B. (1986), *Multinationals from the Third World; Indian Firms Investing Abroad*, Delhi: Oxford University Press.

Lall, S. (1980), 'Monopolistic advantages and foreign involvement by U.S. manufacturing industry', *Oxford Economic Papers*, vol. 32, pp. 102–22.

—— (1983), *The New Multinationals – The Spread of Third World Enterprises*, Chichester: Wiley & Sons.

—— (1990), *Building Industrial Competitiveness in Developing Countries*, Paris: OECD.

—— (1991a), *Developing country TNCs and Their Impact on Home Economies*, New York: Technical Paper for the UNCTC.

—— (1991b), 'Direct investment in South-East Asia by the NIEs: trends and prospects', *BNL Quarterly Review*, no. 179, pp. 463–80.

—— (1992), 'Technological capabilities and industrialization', *World Development*, vol. 20, no. 2, pp. 165–86.

—— (1996), 'The investment development path: some conclusions', in J. H. Dunning and R. Narula (eds.), *Foreign Direct Investment and Governments: Catalysts for Economic Restructuring*, London: Routledge.

Lecraw, D. J. (1977), 'Direct investment by firms from less developed countries', *Oxford Economic Papers*, November, pp. 443–57.

—— (1981), 'Internationalization of firms from LDCs: evidence from the ASEAN region', in K. Kumar and M. G. McLeod (eds.), *Multinationals from Developing Countries*, Toronto: Lexington.

—— (1991), *Third World MNEs Once Again: the Case of Indonesia*, Jakarta: IPM Paper.

Lee, C. H. (1992), 'The government, financial system, and large private enterprises in the economic development of South Korea', *World Development*, vol. 20, no. 2, pp. 187–97.

Lee, H.-K. (1996), *The Korean Economy: Perspectives for the Twenty-First Century*, Albany: State University of New York Press.

Lee, J. (1992), 'Capital and labor mobility in Taiwan', in Ranis, G. (ed.) *Taiwan, from Developing to Mature Economy*, Boulder, CO: Westview Press.

Lee, C. F. and S.-H. Hu (eds) (1989), *Taiwan's Foreign Investment, Exports, and Financial Analysis*, Greenwich: Jai Press.

Lee K. (1994), 'Structural adjustment and outward direct foreign investment in Korea', *Seoul Journal of Economics*, vol. 7, no. 2, pp. 179–211.

Lee, C. H. and K. Lee (1992), 'A transition economy and outward direct foreign investment: the case of South Korea', *Seoul Journal of Economics*, vol. 5, no. 1, pp. 89–111.

Lee, K. U. and J. H. Lee (1996), *Business Groups (Chaebols) in Korea: Characteristics and Government Policy*, Seoul: KIET Occasional Working Paper no. 23.

Lee, Ch. L. and I. Yamazawa (eds) (1990), *The Economic Development of Japan and Korea*, New York: Praeger.

Levy, B. and W.-J. Kuo (1991), 'The strategic orientations of firms and the performance of Korea and Taiwan in frontier industries: lessons from comparative case studies of keyboard and personal computer assembly', *World Development*, vol. 19, no. 4, pp. 363–74.

LG Electronics (various years), *Annual Report*.

Li, K.-T. (1988), *Economic Transformation of Taiwan, ROC*, London: Shepheard-Walwyn.

—— (1995), *The Evolution of Policy Behind Taiwan's Development Success*, Singapore: World Scientific.

Lim, L. Y. C. and E. F. Pang (1991), *Foreign Direct Investment and Industrialisation in Malaysia, Singapore, Taiwan and Thailand*, Paris: OECD.

Lim, M. H. and K. F. Teoh (1986), 'Singapore corporations go multinational', *Journal of Southeast Asian Studies*, vol. XVIII, no. 2, pp. 336–65.

Lin, A.-L. (1996), Trade effects of foreign direct investment: evidence for Taiwan with four ASEAN countries', *Weltwirtschaftliches Archiv*, vol. 131, no. 4, pp. 737–47.

Linge, G. R. J. (1984), 'Developing-country multinationals: a review of the literature', *Pacific Viewpoint*, no. 25, pp. 173–95.

Linnemann, H. (ed.), P. van Dijck and H. Verbruggen (1987), *Export-oriented Industrialization in Developing Countries*, Manila: Singapore University Press.

Liu, C. Y. (1992), 'Money and financial markets: the international perspective', in G. Ranis (ed.), *Taiwan, from Developing to Mature Economy*, Boulder, CO: Westview Press.

Lo, D. (1995), 'Techno-economic paradigm versus the market: on recent theories of late industrialization', *Economy and Society*, vol. 24, no. 3, August, pp. 443–70.

Luostarinen, R. (1979), *Internationalization of the Firm*, Helsinki: Acta Academie Oeconomicae, Helsinki School of Economics.

Maddala, G. S. (1983), *Limited-Dependent and Qualitative Variables in Econometrics*, Cambridge: Cambridge University Press.

Mae-il Kyung Jae Shin Moon Sa (1994), *Maekyung: Annual Corporation Reports* (in Korean), Seoul.

Markusen, J. R. and A. J. Venables (1996), *The Theory of Endowment, Intra-Industry and Multinational Trade*, London: CEPR Discussion Paper no. 1341.

McDermott, M. C. (1991), *Taiwan's Industry in World Markets: Target Europe*, London: Economist Intelligence Unit Special Report no. 2111.

McDermott, M. and S. Young (1989), *South Korea's Industry: New Directions in World Markets*, London: Economist Intelligence Unit Special Report no. 2005.

MFE (various issues), *Trends in Foreign Investment and Technology Inducement.*

Michell, T. (1988), *From a Developing to a Newly Industrialised Country: the Republic of Korea, 1961–92*, Geneva: ILO.

MOEA/IC (Ministry of Economic Affairs/Investment Commission) (various issues), *Statistics on Overseas Chinese & Foreign Investment, Technical Cooperation, Outward Investment, Outward Technical Cooperation, Indirect Mainland Investment, Taipei: MOEA/IC.*

Monkiewicz, J. (1986), 'Multinational enterprises of developing countries: some emerging characteristics' *Management International Review*, vol. 26, no. 3, pp. 67–79.

Myers, R. H. and M. R. Peattie (eds) (1984), *The Japanese Colonial Empire, 1885–1945*, Princeton, NJ: Princeton University Press.

Mytelka, L. K. (ed.) (1994), *South–South Co-operation in a Global Perspective*, Paris: OECD Development Centre.

Narula, R. (1993), 'An Examination of the Evolution and Interdependence of Foreign Direct Investment and Economic Structure: the Case of the Industrialised Countries', dissertation, State University of New Jersey.

—— (1996a), *Multinational Investment and Economic Structure; Globalisation and Competitiveness*, London: Routledge.

—— (1996b), 'Forms of international cooperation between corporations', in C. J. Jepma and A. P. Rhoen (eds), *International Trade: A Business Perspective*, Harlow: Longman.

Negandhi, A. R. (ed.) (1989), *Multinationals and Newly Industrialized Countries: Strategies, Environmental Imperatives, and Practices*, London: Jai Press.

NSO (National Statistical Office) (various issues), *Major Statistics of Korean Economy*, Seoul: NSO.

—— (1993), *Report On Mining and Manufacturing survey 1991 – Republic of Korea.*

O'Brien, P. (1980), 'The new multinationals; developing-country firms in international markets', *Futures*, August, pp. 303–16.

OECD (1994), *Economic Surveys: Korea 1994*, OECD, Paris.

—— (1996), *Economic Surveys: Korea 1996*, OECD, Paris.

Oh, W. W. (1993), *Export Processing Zones in the Republic of Korea: Economic Impact and Social Issues*, Geneva: International Labour Office Working Paper no. 75.

Ohlin, B., P. O. Hesselborn and P. M. Wiskan (eds) (1977), *The International Allocation of Economic Activity: Proceedings of a Nobel Symposium held at Stockholm (8–11 June 1976)*, New York: Holmes & Meier.

Oman, C. (1984), *New Forms of International Investment in Developing Countries*, Paris: OECD Development Centre.

Orrù, M., N. W. Biggart and G. G. Hamilton (1991), 'Organizational isomorphism in East Asia', in W. W. Powell and P. J. DiMaggio (eds), *The New Institutionalism in Organizational Analysis*, Chicago: University of Chicago Press.

Ozawa, T. (1992), 'Foreign direct investment and economic development', *Transnational Corporations*, vol. 1, no. 1, pp. 27–54.

—— (1995), 'Structural upgrading and concatenated integration – the vicissitudes of the Pax Americana in tandem industrialization of the Pacific Rim', in D. F. Simon (ed.), *Corporate Strategies in the Pacific Rim: Global Versus Regional Trends*, London: Routledge.

—— (1996), 'Japan: the macro-IDP, meso-IDPs and the technology development path (TDP)', in J. H. Dunning and R. Narula (eds), *Foreign Direct Investment and Governments: Catalysts for Economic Restructuring*, London: Routledge.

Pilat, D. (1993), *The Economics of Catch Up: the Experience of Japan and Korea*, Groningen: Groningen Growth and Development Centre, Monograph Series, no. 2.

Plummer, M. G. and M. F. Montes (1995), 'Direct foreign investment in China: an introduction', in S. J. La Croix *et al.* (eds), *Emerging Patterns of East Asian Investment in China; from Korea, Taiwan, and Hong Kong*, Armonk: M. E. Sharpe.

Porter, M. E. (1990), *The Competitive Advantage of Nations*, New York: Free Press.

Powel, W. W. and P. J. DiMaggio (eds) (1991), *The New Institutionalism in Organizational Analysis*, Chicago: University of Chicago Press.

Probert, J. (1991), *Asian Direct Investment in Europe*, Fontainebleau: Euro-Asia Centre.

Pugel, Th. A., E. S. Kragas and Y. Kimura (1996), Further evidence on Japanese direct investment in U.S. manufacturing', *Review of Economics and Statistics*, vol. LXXVIII, May, no. 2, pp. 208–13.

Radtke, K. W. and J. Stam (eds) (1998), *Dynamics in Pacific Asia: Conflict, Competition and Cooperation*, London: Kegan Paul.

Ranis, G. (1992), 'From developing to mature economy: and overview', in G. Ranis (ed.), *Taiwan, from Developing to Mature Economy*, Boulder, CO: Westview Press.

—— (ed.) (1992), *Taiwan, from Developing to Mature Economy*, Boulder, CO: Westview Press.

Redding, S. G. (1990), *The Spirit of Chinese Capitalism*, Berlin: Walter de Gruyter.

Riemens, P. H. J. (1989), *On the Foreign Operations of Third World Firms*, Amsterdam: Nederlandse Geografische Studies no. 100.

Rugman, A. M. (1980), 'Internalisation as a general theory of foreign direct investment: a reappraisal of the literature', *Weltwirtschaftliches Archiv*, vol. 116, no. 2, pp. 365–79.

—— (1981), *Inside the Multinationals*, London: Croom Helm.

—— (ed.) (1982), *New Theories of the Multinational Enterprise*, London: Croom Helm.

Sakong, Il (1993), *Korea in the World Economy*, Washington: Institute for International Economics.

Samsung Electronics (various years): *Annual Report*, Seoul: Samsung.

Samsung Newsletter (various issues), Seoul: Samsung.

San, G. (1992), *Taiwanese Corporations in Globalisation and Regionalisation*, Paris: OECD, Technical Papers no. 61.

Shea, J.-D. and Y.-H. Yang (1994), 'Taiwan's financial system and the allocation of investment funds', in J. D. Aberbach *et al.* (eds), *The Role of the State in Taiwan's Development*, New York: M. E. Sharpe.

Simon, D. F. (1992), 'Taiwan's emerging technological trajectory: creating new forms of competitive advantage', in D. F. Simon and M. Y. M. Kan (eds), *Taiwan Beyond the Economic Miracle*, Armonk: M. E. Sharpe.

—— (ed.) (1995), *Corporate Strategies in the Pacific Rim: Global Versus Regional Trends*, London: Routledge.

Simon, D. F. and M. Y. M. Kan (eds) (1992), *Taiwan Beyond the Economic Miracle*, Armonk: M. E. Sharpe.

Singer, H. W. (1989), 'Lessons of post-war development experience, 1945–1988', in W. L. M. Adriaansen and J. G. Waardenburg (eds), *A Dual World Economy*, Groningen: Wolters–Noordhof.

Singh, A. (1995), *How Did East Asia Grow So Fast? Slow Progress Towards Analytical Consensus*, Geneva: UNCTAD Discussion Papers, no. 97.

Smeets, M. (1991), 'Het succes van de Aziatische draken', *Economisch–Statistische Berichten*, 31 July, pp. 764–72.

Song, B.-N. (1994), *The Rise of the Korean Economy*, Hong Kong: Oxford University Press.

Steers, R. M., Y. K. Shin and G. R. Ungson (1989), *The Chaebol: Korea's New Industrial Might*, New York: Harper & Row.

Stewart, J. (1991), *Econometrics*, New York: Philip Allan.

Sullivan, D. and A. Bauerschmidt (1990), 'Incremental internationalization: a test of Johanson and Vahlne's thesis', *Management International Review*, vol. 30, no. 1, pp. 19–30.

Sung, Y. K. (ed.) (1994), *The Korean Economy at a Crossroad*, Westport, CT: Praeger.

Thomson, R. (eds.) (1993), *Learning and Technological Change*, New York: St Martin's Press.

Tolentino, P. E. E. (1993), *Technological Innovation and Third World Multinationals*, London: Routledge.

Tung, P. (1991). *The Taiwanese Presence in Europe*, Insead: Euro-Asia Centre.

UN (various issues), *Monthly Bulletin of Statistics*, New York: UN.

UNCTAD (1995), *World Investment Report – Transnational Corporations and Competitiveness*, Geneva: United Nations.

—— (1996), *World Investment Report 1996: Investment, Trade and International Policy Arrangement*, Geneva: United Nations.

UNCTC (1978), *Transnational Corporations in World Development: a Re-examination*, New York: United Nations.

—— (1992a), *World Investment Directory 1992; Foreign Direct Investment, Legal Framework and Corporate Data – Volume 1 Asia and the Pacific*, New York: United Nations.

—— (1992b), *The Determinants of Foreign Direct Investment; a Survey of the Evidence*, New York: United Nations.

UNDP (various issues), *Human Development Report*, New York: Oxford University Press.

Vernon, R. (1966), 'International investment and international trade in the Product Cycle', *Quarterly Journal of Economics*, vol. 80, no. 2, pp. 190–207.

—— (1974), 'The location of economic activity', in J. H. Dunning (ed.), *Economic Analysis and the Multinational Enterprise*, London: Allen & Unwin.

—— (1979), 'The Product Cycle hypothesis in a new international environment', *Oxford Bulletin of Economics and Statistics*, vol. 41, no. 4, November, pp. 255–67.

—— (1985), *Government Control Over Its Multinational Enterprises: the U.S. Case*, Cambridge, MA: John F. Kennedy School of Government.

—— (1994), 'The economist's role in research on transnational corporations: or why the dogs have barked so softly', *Transnational Corporations*, vol. 3, no. 3, pp. 81–9.

Vernon-Wortzel, H. and L. H. Wortzel (1988), 'Globalizing strategies for multinationals from developing countries', *Columbia Journal of World Business*, vol. XXIII, Spring, pp. 27–35.

Wade, R. (1990), *Government the Market – Economic Theory and the Role of the Government in East Asian Industrialisation*, Princeton, NJ: Princeton University Press.

Wang, C.-N. (1994), *The Development and Strategy of Taiwan's Information Technology Industry*, Taipei: CIER Occasional Paper no. 9405.

Wang, N. T. (ed.) (1992), *Taiwan's Enterprises in Global Perspective*, Armonk: M. E. Sharpe.

Wang, C.-N. and J.-L. Hsu (1992), 'The impacts of European integration on Taiwan and Taiwan's responses', in *1992 Sin-European conference on Economic Development: Impacts of the European Integration and Responses of non-European Countries*, Taipei: Chung-Hua Institution for Economic Research, pp. 169–222.

Wee, C.-H. (1994), 'National branding strategies and economic development: implications for NIEs and LDCs', *International Executive*, vol. 36. no. 2, pp. 119–45.

Welch, L. S. and R. Luostarinen (1986), 'Internationalization: evolution of a concept', *Journal of General Management*, vol. 14, no. 2, pp. 34–55.

Wellenius, B., A. Miller and C. J. Dahlman (eds) (1993), *Developing the Electronics Industry*, Washington: World Bank.

Wells, L. T., Jr. (1977), 'The internationalization of firms from developing countries', in T. Agmon and C. P. Kindleberger (eds), *Multinationals from Small Countries*, Cambridge, MA: MIT Press.

—— (1981), 'Foreign investors from the Third World', in K. Kumar and G. M. McLeod (eds), *Multinationals from Developing Countries*, Toronto: Lexington.

—— (1982), 'The Third World grows its own multinationals', *International Management*, January, pp. 39–40.

—— (1983), *Third World Multinationals – the Rise of Foreign Investment from Developing Countries*, Cambridge, MA: MIT Press.

—— (1986), 'New and old multinationals: competitors or partners', in K. M. Khan (ed.), *Multinationals from the South: New Actors in the International Economy*, London: Pinter Publishers.

—— (1993), 'Mobile Exporters: New Foreign Investors in East Asia', in K. A. Froot (ed.), *Foreign Direct Investment*, Chicago: University of Chicago Press.

Wells, C. (1988), 'Brazilian multinationals', *Columbia Journal of World Business*, Winter, pp. 13–23.

Whitley, R. (1994), *Business Systems in East Asia; Firms, Markets and Societies*, London: Sage.

Williamson, O. E. (1975), *Markets and Hierarchies: Analysis and Antitrust Implications: a Study in the Economics of Internal Organisation*, New York: Free Press.

—— (1979), 'Transaction cost economics: the governance of contractual relations', *Journal of Law and Economics*, vol. 22, pp. 223–62.

Wohlmuth, K. (1986), 'Practices and policies of host countries towards Third World multinationals: a competitive edge against old multinationals?', in K. M. Khan (ed.), *Multinationals from the South: New Actors in the International Economy*, London: Pinter Publishers.

Wong, P.-K. (1992), 'Technological development through subcontracting linkages: evidence from Singapore', *Scandinavian International Business Review*, vol. 1, no. 3, pp. 28–40.

World Bank (1993), *The East Asian Miracle – Economic Growth and Public Sector*, New York: Oxford University Press.

—— (various issues), *World Development Report*, OUP Washington.

Wu Qi (1988), *South Korean and Taiwan: a Comparative Analysis of Economic Development*, Brighton: Institute of Development Studies, discussion Paper no. 252.

Wu, R.-I. (1989). 'Economic development strategies and the role of direct foreign investment in Taiwan', in C. F. Lee and S.-C. Hu (eds), *Taiwan's Foreign Investment, Exports, and Financial Analysis*, Greenwich: Jai Press.

Yang, Y.-H. (1994), *A Study of the Small- and Medium-Sized Enterprises Financial System in Taiwan*, Taipei: CIER Occasional Paper no. 9401.

Yin, R. K. (1994), *Case Study Research – Designs and Methods*, Thousand Oaks, CA: Sage.

Yip, G. (1994), 'Industry drivers of global strategy and organization', *International Executive*, vol. 36, no. 5, pp. 529–56.

Yu, T.-S. (1993), *Taiwan's Economic Development and Its Economic Relationship with Mainland China*, Taipei: CIER Occasional Paper no. 9307.

Zander, I. (1994), *The Tortoise Evolution of the Multinational Corporation – Foreign Technological Activity in Swedish Multinational Firms 1890–1990*, Stockholm: Institute of International Busines, Stockholm School of Economics.

Zhang, H.-Y. and D. van den Bulcke (1996), 'China: rapid changes in the investment development path', in J. H. Dunning and R. Narula (eds), *Foreign Direct Investment and Governments: Catalysts for Economic Restructuring*, London: Routledge.

Index